Dynamic Modeling, Empirical Macroeconomics, and Finance

Pencil Portrait of Willi Semmler by the Anglo-French artist Alexia Carr, who is also a sculptor, an opera singer, and an actress.

Lucas Bernard • Unurjargal Nyambuu
Editors

Dynamic Modeling, Empirical Macroeconomics, and Finance

Essays in Honor of Willi Semmler

Springer

Editors
Lucas Bernard
Chairman
Department of Business
The New York City College of Technology
The City University of New York
Brooklyn, NY
USA

Unurjargal Nyambuu
Professor of Economics
Department of Social Science
The New York City College of Technology
The City University of New York
Brooklyn, NY
USA

ISBN 978-3-319-39885-3 ISBN 978-3-319-39887-7 (eBook)
DOI 10.1007/978-3-319-39887-7

Library of Congress Control Number: 2016954240

Printed on acid-free paper

This Springer imprint is published by Springer Nature
The registered company is Springer International Publishing AG Switzerland

Foreword

The 25 years from 1948 to 1973 were pretty good for middle-class and even poor households in the advanced world. Employment was high, inflation was low, productivity grew faster than ever, and real wages grew along with it. Innovation was strong and new sectors opened up everywhere, and the public sector was more fully funded than ever before. This was the Keynesian era. Governments—even so-called conservative ones, as in the UK—explicitly adopted Keynesian policies and supported the development of econometric models that could help to design their interventionist policies. (Roy Harrod was an advisor to Harold Macmillan.)

Since then, there has been a move away from interventionist policy and from econometric modeling. The models could not deal with the oil shocks, nor could they handle or explain "stagflation." And the political climate changed, first following the oil shocks and then with the election of Reagan and Thatcher. In the resulting confusion, the simplicities of Friedmanesque monetarism took flight and lit up the skies. By the time Keynesians had got their act together to shoot down monetarism, it had been replaced, or supplemented, by new classical thinking, together with real business cycles, all draped in rational expectations.

What followed was an era that might best be described as ideological argument over how an idealized (and therefore unrealistic) economy should work. New classicals, new Keynesians, and real business cycle theorists—all taking rational expectations seriously—offered accounts of how an economy based on unrealistic, sometimes impossible, assumptions would move toward an equilibrium or end up cycling in an equilibrium path in spite of the all-too-evident fact that actual economies like the USA or the UK were regularly exhibiting features wholly inconsistent with equilibrium and behaving in ways contrary to the models. Post-Keynesians and the various schools of Marxian and other heterodox economists, however, were simply ignored.

Keynes, however, was not ignored; he was vilified. He was not a real economist, according to Lucas. No real economist under the age of 40 takes Keynes seriously anymore, it was said. As for "involuntary unemployment," Lucas intoned, speaking ex cathedra, "it was not a fact to be explained, it was a hypothesis, offered to explain a real phenomenon, the variation in actual levels of employment." There wasn't even a Keynesian problem anymore.[1]

So from about 1980 to the crash of 2007–2008, there was a general atmosphere unfriendly to new research in macroeconomics—unless it was designed to show that macro-policies were ineffective or damaging. Showing that interventionist policies were "ineffective" was a particularly common argument, and it was often based on a demonstration that the multiplier was miniscule or nonexistent or offset by monetary developments. And of course, there was the crowding-out story, and the effect of an expansionist intervention on interest rates, undermining the expansion.

These issues go back to the beginning of Keynesian analysis, but work on them had largely stagnated or become repetitive. Willi Semmler and his associates have now developed new methods and new techniques to carry forward the work that had begun early in the Keynesian era but had become stagnant and unimaginative. For example, they show not only that multipliers are different when the economy is expanding and when it is contracting, they have worked out how to calculate different but related multipliers for different states of the economy generally, and they examine how the multiplier interacts in these different states with monetary and financial variables.[2]

This is pretty advanced, but it deals with issues that go back a long way right to the beginning. The story of the Keynesian multiplier can be dated to May 1929; Keynes gave two talks, one in London on May 28 and another in Leicester the next day. Both were talks supporting Liberal candidates in the ongoing general election, and in both cases Keynes was supporting the program of Lloyd George (a program he was largely responsible for writing). This program called for the creation of large-scale public works to offer employment, putting the unemployed to work and getting them off the dole. But the government with the support of the treasury objected. First, the only way to pay for such a program would be to take away money now financing other projects, so there would be essentially no net gain (crowding out). And secondly, any such effort to expand would certainly drive up interest rates. All too familiar, for the liberals to win, these arguments had to be overcome (and essentially the same arguments are out there today).

Keynes's notes for the talks those two days cover the different ways money can be found or created, so that nothing has to be diverted from financing ongoing projects, and he seeks to show that interest rates do not have to come under pressure to rise.

[1]Lucas, at the time of his Nobel Prize, was the only economist to have the word "macroeconomics" appear in his citation. The great macroeconomists, Hicks, Samuelson, Solow, Tobin, and many others, were all cited for other accomplishments, not for their work in macro. For the mainstream establishment, real economics is about scarcity, and how can there be scarcity if there is involuntary unemployment? Macro is the wrong way to go.

[2]See Stefan Mittnik and Willi Semmler (2012): "Regime Dependence of the Multiplier," *Journal of Economic Behavior and Organisation*, vol. 83, no 3: 502–522. Concerning regime-dependent effects of monetary policy, see Carl Chiarella, Peter Flaschel, and Willi Semmler, "Reconstructing Keynesian Macroeconomics." Macroeconomic Activity, Banking, and Financial Markets, vol. 3, Routledge, 2014.

We would certainly rephrase or refine his arguments today, but there can be no doubt that he opened the way. Willi and his colleagues offer us not only models but also empirical methods to demonstrate these points.

Even more remarkable is the fact that Keynes sought to calculate the precise, numerical extent of the expansion in employment as the result of "secondary" spending: this is the multiplier and it is the first time that we have it as the sum of an infinite series. In his notebook, he wrote out an arithmetical formula for an infinite series that summed to a finite amount. This was the secondary employment that would result from removing an unemployed worker from the dole and putting him to work—thus doubling his spending and setting up the series of respending. Moreover he then calculated the savings from the dole, as secondary workers were taken off it and put to work—this savings could be part of the financing of the public works.

Keynes considers several different possibilities of leakages, essentially different states of the economy. In other writings and in a letter to Harrod, he suggests that these conditions might be stable for long enough to study but that over time they would most likely change, though slowly. He recommended empirical studies of the multiplier in marked contrast to his reaction to Jan Tinbergen's early econometric modeling, which he criticized very strongly at about the same time.

Keynes's views on the likelihood that the multiplier might vary in value with the state of the economy seem to be an early version of the argument that James Duesenberry presented in the 1950s and 1960s. He proposed that the multiplier/accelerator mechanism in the upswing of the cycle would be different from that in the downswing, and every cycle would very likely differ in some ways from every other. Duesenberry asserts this but doesn't demonstrate it nor does he present much evidence. But he makes a plausible case for looking into it.

This is what Willi and his coauthors have done. They have not only studied and answered these questions, but they have developed empirical methods of analysis to make their studies precise and workable. They have moved the Keynesian project forward after decades of stagnation.

Edward J. Nell

Contents

Acknowledgments

The authors would like to thank several people, without whom this book would never have been possible. First, we need to thank Prof. Rachel Raskin of our own New York City College of Technology. Without her input, editorial expertise, and general suggestions, several of the manuscripts in this collection could not have been published. Second, Dr. Martina Bihn and Yuliya Zeh, of Springer International, have provided invaluable support throughout the project. Third, we would especially like to thank Dr. Aleksandr V. Gevorkyan who, back in the summer of 2009, came up with the original idea of a collected studies edition dedicated to Dr. Willi Semmler. Aleksandr's encouragement and assistance with the preparation of this book were instrumental in bringing the project to its successful completion.

Theory and Practice in a Sensible Economic Policy Mix: A Glance at Willi Semmler's Contributions to Empirical Macroeconomics

Aleksandr V. Gevorkyan

The Global Financial Crisis, which peaked in late 2008, accentuated the urgent need for reconciliation between the theoretical and practical components of macroeconomic theory. Worldwide, apart from a small minority, economists faced the profession's dire challenge to develop an ability to think outside conventional macroeconomic equilibrium models. Such ability, of course, pre-supposes an open approach to dynamic macroeconomic modeling, one founded in a profound understanding of the diverse currents within the many traditions of economic thought.

The challenge has been to popularize a critical alternative, which is intellectually open, diverse, and provides a methodologically rich approach to macroeconomic modeling. This view finds its real-time effective application across a range of related and tangential fields: from macroeconomic disequilibrium analysis, business cycles, role of the financial markets for macroeconomics, private and public debt analysis, prolonged periods of unemployment open economy dynamics, international risk drivers of a macroeconomy, such as capital flows, exchange rates, and resource boom- bust cycles, giving rise to resources' supply and price volatility, and macroeconomic policies. Furthermore, new challenges, such as the economics of climate change are rising as well.

This is where I am proud to endorse my teacher, mentor, co-author, and, indeed, true friend Professor Willi Semmler, in whose honor this book has been compiled, celebrating his life and grand scientific accomplishments. An avid opera connoisseur, Professor Semmler is first and foremost a world-renowned macroeconomist, a true scholar of the highest caliber, who possesses a unique and astounding ability to intellectually process the plethora of information on current economic trends. His work encompasses research on problems of dynamic modeling, computational methods in economics and finance, financial economics, exchange rate, credit

A.V. Gevorkyan (✉)
The Peter J. Tobin College of Business, St. John's University, New York, NY, USA
e-mail: gevorkya@stjohns.edu

L. Bernard, U. Nyambuu (eds.), *Dynamic Modeling, Empirical Macroeconomics, and Finance*, DOI 10.1007/978-3-319-39887-7_1

risk, and ultimately climate change and the linkages between financial and "real" economic activity.

He is particularly known internationally for addressing those issues in the context of non-linear dynamic mathematical models, modeling of instability, regime changes, and economic cycles. In this context, recently, his research has also included comparative empirical macroeconomics. A particular area of research covers both the US and Europe, where Professor Semmler is currently exploring the deficiencies in the design of the Euro-area, the Euro crisis, and financial stress as the ill-advised austerity policies for the macroeconomics of Europe.

Moreover, Willi Semmler's macroeconomic analysis penetrates the most deeply hidden and convoluted aspects of the complex modern global economy. His work on economic growth, economic crises, and the economics of climate change is pioneering and unprecedented. In fact, Professor Semmler is one of those few economists, who have been influencing economic methodological debate both before and, now, after the Global Financial Crisis.

1 Teaching

It is natural to begin with Professor Semmler's teaching activity. As the Henry Arnhold Professor of Economics at the New School for Social Research in New York City, Professor Semmler has dedicated his life to educating generations of new economists. Aside from his teaching responsibilities, Professor Semmler has chaired the NSSR's economics department encouraging the faculty and student body in their progressive and policy relevant work, shaping the modern understanding of applied macroeconomics and social policy.

In the latter context he is active as a Research Associate at The Schwartz Center for Economic Policy Analysis and the Tishman Environment and Design Center of the New School. A rare example of an effective educator, respected and appreciated by his very diverse students, Professor Semmler is always caring for every student in his class and respectful of each question raised. As his former student, myself, I can certainly attest to that.

One of the difficulties of teaching in New York City, albeit a benefit as well, is a very diverse and intellectually demanding student mix. Connecting with many on an often-dry topic of macroeconomics is not an easy task, even for experienced public speakers. Yet, from the moment Professor Semmler walks into the classroom he sets a professionally focused and intellectually informative and engaging dynamic that captivates his audience for the entire class period. Perhaps it is a specific topic he brings up in his remarks or a new method that he introduces to students. More often than not, his ability to capture the audience's attention rests on the basic premise of relating theoretical to practical by boldly posing a question from often most unexpected angle.

Over the years students from all over the world have either been taught by or somehow communicated with Professor Semmler. And equally diverse is his professorship experience outside of his home department at the New School. He has been a visiting researcher at Columbia University, Stanford University and the *Centre pour la recherche économique et ses applications* (CEPREMAP) in Paris. He has taught as a Fortis-Bank Visiting Professor at the University of Antwerp, Visiting Professor at the University of Marseilles/Aix-en-Provence as well as teaching at the European Quantitative Economics Doctorate (QED) Program at universities in Italy, Spain, and Portugal. He has taught at American University, DC, University of Bielefeld, and as Fulbright Professor at University of Economics, Vienna, as well as at numerous universities around the globe, for example at the National Autonomous University of Mexico in Mexico City, Mexico; the University of Orléans in Orléans, France and Chuo University in Tokyo, Japan.

And these are just few of Dr. Willi Semmler's appointments.

Aside from broad topics of empirical macroeconomics, public finance, industrial organization, and international finance, Professor Semmler's more focused teaching interests cover problems of dynamic modeling, financial economics, time series and computational methods in economics and finance, and the estimation of macroeconomic models. Yet, most distinct is Professor Semmler's immeasurable contribution to the teaching of topics such as economic growth and the economics of climate change.

Speaking on these latter topics, as an educator and empirical macroeconomist, Professor Semmler is always on top of current events and new trends in related fields. He is always alert to students' views allowing unparalleled flexibility for one to explore their original idea. For now, since we have mentioned few of the themes taught by Professor Semmler, it is worthwhile to talk about some of his accomplishments in research.

2 Research

Willi Semmler has authored or co-authored more than 160 articles in refereed journals and chapters in edited books, more than 75 of which have been published in the last decade. He has also served as guest editor for numerous special issues of well-ranked economic journals. His work is represented in more than 20 authored, co-authored or edited books and is featured by such established publishing houses Columbia University Press, MIT-Press, Springer Publishing House, Princeton University Press, Cambridge University Press, Oxford University Press, and others.

Professor Semmler is co-editor, with Stefan Mittnik, of the series on "Dynamic Modeling and Econometrics in Economics and Finance," published by Springer Publishing House (Heidelberg and New York), and is associate editor of the *Journal Econometrics and Statistics* and more policy oriented journals. As mentioned, Professor Semmler's research encompasses work in dynamic modeling, computational methods in economics and finance, financial economics, macroeconomics and employment, and economics of climate change. He has also worked on *knowledge-based economies* as well as modeling and estimating quite complex linkages in economics such as the interrelation between financial and real activity. In the latter context recent work is also focusing on issues of financial market regulation.

It is perhaps premature to catalogue Professor Semmler's work under a convenient label. There are at least two reasons for that. First, a brief view on the immense body of intellectual work produced by Professor Semmler would unveil the unprecedented level of intellectual comfort that he has working across the neoclassical, new-Keynesian, post-Keynesian, Schumpeterian, Marxist, institutional, behavioral, and other schools of economic thought. Applying methodology or critiquing findings from one or the other strand of very diverse economics literature, Professor Semmler is able to synthesize his own path in modern economic analysis. It is such new derived methodology that absorbs the best aspects of critical thinking from each school of economics that will be making the difference in the theoretical and policy debates in the years to come.

Furthermore, Professor Semmler's approach may even be traced back to the work of another famous German, a pillar in German Classical Philosophy, Georg Hegel whom Willi Semmler admires. Indeed, a careful reading of Semmler's work on any topic carries through one common element with Hegelian philosophy, i.e. the dynamic evolution of a complex system, be it a society or an applied problem within macroeconomics, financial economics, and other areas in economics. For Semmler, the evolutionary aspect is categorically fundamental. Seeking this element out theoretically, developing in a new methodology, leads to, perhaps, most valuable advice to policy makers: be prepared for things to change. Professor Semmler's most recent work on the topics of climate change (e.g. *The Oxford Handbook of the Macroeconomics of Global Warming* co-edited with Lucas Bernard, and other more recent work on resource booms and bust such as the oil and Shale Energy Sector—Game Changers in the Oil Industry co-authored with Arkady Gevorkyan) is the tangible scientific demonstration of the evolutionary vision in economics.

And so the second reason, why standard JEL classification fails to categorize Semmler, is because there is yet more to come. It is this continuous search for a new method, new angle to the age-old problems of economics, which Professor Semmler invokes in his dynamic modeling oriented cross-disciplinary and cross-theoretical work.

A regular participant in high-profile international conferences, seminars, and workshops Willi Semmler has served on the Board of Directors for the Center for Macroeconomics at Bielefeld University in Germany, serves on the advisory board of the Society for Computational Economics, and is Research Associate at the Schwartz Center for Economic Policy Analysis in New York City and Center for European Economic Research in Mannheim, Germany. He has been the Project Evaluator for National Science Foundations of Austria, Germany, Belgium, France, UK, and EU-Commission in different periods. He is a visiting scholar at the European Central Bank in Frankfurt, Germany, visitor of the World Bank and IMF and external evaluator of the IIASA in Vienna, Austria.

On a personal note Professor Semmler and I have collaborated on topics of emerging markets post-crisis development. In a paper, titled "Sailing Out of Crisis Emerging Markets Style: Blending Fiscal-Monetary Rules, Nominal Targets, and

Debt Dynamics in Some Transition Economies," back in 2011, we develop a Taylor-like central bank's model with international reserves as a target. Today, as emerging markets are experiencing renewed pressures to their financial systems, the research gains new urgency. We continue to discuss and collaborate on a range of other topics. Those topics range from problems of the modern welfare state, income and wealth inequality, to the discussions on the economy's evolutionary changes and socially oriented entrepreneurship.

More recently, our team, comprised of Will Semmler, Lucas Bernard and Thomas Palley, tackled a more advanced problem of recurring cycles in a paper titled "Time Scales and Mechanisms of Economic Cycles: A Review of Theories of Long Waves". Specifically, the problems of Kondratieff waves, Keynesian cycle theory, and Goodwin cycles were explored, in fact, in several related publications. A fantastic team of scholars it was and I was fortunate to be a part of the effort.

Professor Semmler has also been a consultant and team leader on research on "The Employment Effects of Climate Policies," published as a World of Work Report by the International Labor Organization in Geneva. He has advised on the World Bank project on Growth and Fiscal Policy; to the Deutsche Bundesbank in a discussion group on European unemployment; European Central Bank; and other world's leading financial and economic policy making institutions. Willi Semmler was also a member of the Columbia University Task Force on Global Warming, directed by Joseph Stiglitz.

As an active researcher in his field, Professor Semmler has gained prominent recognition across various prestigious grant-making institutions. For example, he is a recipient of the research grant from the German Science Foundation working on the topic of Global Warming. The foundation is also assisting in his work with other colleagues from the New School and in German universities on macroeconomics and inequality. The Walker Foundation provided financial assistance on "The Economics of Global Warming" and the Thyssen Foundation helped with the "The Economics of Global Warming" project. Other notable awards include the EU Commission sponsored project on the "European Knowledge Based Economies;" joint work with Malte Sieveking on "Optimal Resource Exploitation", financed by the German Science Foundation; research grant from American Council of Learned Societies on "Market Structure and Competition: Theories and Empirical Evidence," as well as research award from the New School for Social Research on "Finance and Macroeconomics" project undertaken at Stanford University.

It is also worth mentioning his journalistic work as a long time commentator for the German Der Spiegel, and Die Zeit, as well as interviews for radio stations and journals in Germany, Italy, Hungary, China, and elsewhere.

Again, above describes only a fraction of Willi Semmler's truly multifaceted research scope and career. Indeed, one requires an open and flexible mindset to intellectually digest and think about such diverse phenomena; apply new technical and econometric tools and develop new procedures tackling more unusual methods; pioneering and then leading research in cross-discipline topics.

As a result, Professor Semmler's work as a researcher establishes a direct link in our understanding of the functioning of the financial and real sectors of the economy with the dynamic evolution of our modern society.

3 Mentorship

But to complete the story on Professor Willi Semmler we must also comment on his long-lasting friendship and warm relations with his current and former students as well as colleagues. It is the ultimate manifestation of one's intellectual authority to not just be focused on one's work but to be able to effectively guide others to successful completion and realization of a given research project.

It has been mentioned above, how attentive to students' concerns Willi Semmler is as an educator. To add to this some personal flavor, Professor Semmler supported my research idea from the very start even though it was somewhat odd to be studying problems of economic and social transition of Eastern Europe and Former Soviet Union economies. Yet, that was exactly what we collaborated on: me as a full-time working graduate student and Professor Semmler as my dissertation advisor. In the end, the dissertation was completed in time and successfully defended. Following the defense and leveraging that foundation, I continued my research initiative, which eventually led to a book publication of my monograph *Innovative Fiscal Policy and Economic Development in Transition Economies* (2013 in paperback and 2011 in hard cover).

My story is not unique. Willi Semmler has served as advisor on a large number of dissertation committees at the New School in New York City and elsewhere (e.g. University Marseille/Aix, Bielefeld University, University of Darmstadt, Technical University of Vienna, etc.). For all of his former students the impression is the same: Professor Semmler went along with the originally proposed idea and we developed a practically sound macroeconomic model testing with real data and deriving policy implications. Moreover, assuming such individual approach, Willi Semmler strives to offer the best economist training out there preparing the student for a career in academia, corporate world, or public policy setting.

It is a rare gift of an experienced educator and a natural skill of a successful mentor to be able to work with sometimes before unknown concepts, guiding and helping his mentees in their research. The feeling of accomplishment and gratitude overwhelms successful student.

4 How This Book Is Organized

The rest of this book is structured around contributions authored by a solid group of prominent experts in their respective fields. Some of the authors in this volume are Professor Semmler's colleagues and friends. Many are co-authors. And some, just like myself, are his former students with many co-authored research papers.

As we saw, the book opened with a foreword from Professor Semmler's longtime colleague and friend Professor Edward J. Nell (and on a personal note, one of my esteemed professors who helped polish my dissertation to its final level). In his "Willi Semmler's Macroeconomics: An Appreciation" Nell offers us a concise and critical overview of challenges to macroeconomic theory since late 1940s up until present time. The focus is on Keynes and initial interventionist approach of post-war recovery being replaced by the separation from Keynesian views between 1980 to the 2007–2008 global financial crisis. And while the macroeconomic policy issues have been debated for a long time, Nell notes, that Professor Semmler and his co-authors have not only studied and answered many of the related questions, but "they have developed empirical methods of analysis to make their studies precise and workable."

Franz Wirl and Yuri Yegorov, in their chapter entitled "Iran's Nuclear Program and the West's Response: A Game Theoretic Approach" open this compilation. The authors tackle a complex problem of Iran's nuclear power program. This chapter develops a game theoretic model between Iran (to side step the non-proliferation treaty) and the West (imposing sanctions). The resulting analysis suggests that a substantial benefit (or as authors refer to it, a 'carrot') must be offered by the West for Iran to dismantle its program under the West's strategy of economic sanctions. The authors then attempt to extend the game to gain understanding as to why Iran chose this kind of a strategy and whether and if what this choice reveals about Iran's strength or weakness.

Next is a chapter by Askar Akaev, a well-known mathematician at the Russian Academy of Sciences and former President of Kyrgyzstan. In his work, entitled "From the Great Divergence to the Great Convergence," Dr. Akaev offers a comprehensive demonstration of the evolutionary process of transition from the Great Divergence between the West and the rest of the world up until the Great Convergence. These periods overlap; the former taking place between 1820 and 1970, while the latter originates in the mid-twentieth century, with most intense and rapid manifestation in the first decade of the twenty-first century. In this work, built on careful analysis historical economic data, the author discusses the evolution of the global model of divergence between developed countries of the West and developing countries of the rest of the world. The forecast, calculated in this chapter reveals that Great Convergence will occur by 2050 and the world center of production of goods and services will shift again to China, India and other Asian countries, like it was in pre-industrial era.

In the chapter entitled "Borrowing Constraints and Monetary Policy: The Networth Channel at Work," Tiziana Assenza and Domenico Delli Gatti review the impact of monetary policy on economic activity as a fundamental question in macroeconomics. In this chapter the authors model a monetary economy with financing constraints adopting the money-in-the-utility function. The occurrence of multiple equilibria is a likely outcome of the dynamics generated by the model. A change in the growth rate of money supply can affect real output through its impact on net worth. Such approach allows the authors to explore the net worth channel of the monetary transmission mechanism.

The next chapter, "Modeling Climate Change Effects on the Renewable and Non-Renewable Resources", is co-authored by my brother Arkady Gevorkyan and me—Aleksandr V. Gevorkyan—(both former students of Professor Semmler). The chapter addresses the problem of economics of climate change. Specifically, we are analyzing the link between primary commodities differentiated based on their exhaustibility factor and a pass-through to futures prices. We work in the best of Professor Semmler's traditions: developing a dynamic theoretical model simulating dynamic consumption paths of renewable and non-renewable resources. Dynamic paths are calculated applying the Non-linear Model Predictive Control (NMPC) methodology. The chapter relates the connection between intensifying impacts of climate change, commodity futures price volatility, rising urban-dependency pressures and the renewable and non-renewable resources production and consumption balance among advanced and emerging economies. We conclude with a call for attention to the accumulated evidence that suggests probable higher volatility in commodity prices in the near term, directly affecting small and large economies across all regions.

In their chapter entitled "A North-South Model with Technological Spillovers, Environmental Degradation and Structural Change" by Anton Bondarev and Alfred Greiner, the authors argue that boosting structural change in the world economy is a perspective way to mitigate climate changes. The chapter analyzes the dynamic R&D spillovers and cooperation between two sufficiently close economies under endogenous turnover of sectors. It turns out that the costless spillover of technology from one country to another accelerates structural change in both economies while increasing productivity in the lagging country only. The authors find that if new technologies are progressively cleaner then such an increase in the speed of adoption of newer technologies leads to a slowdown of environmental degradation as measured by surface temperature. As such, technological spillovers and cooperation are argued to be beneficial both for the economy and environment provided participating parties are close enough in their technologies.

The chapter "Foreign Exchange Volatility and Its Implication for Macroeconomic Stability: An Empirical Study of Developing Economies" is written by one of Professor Semmler's former graduate students, Unurjargal Nyambuu. In this chapter, Dr. Nyambuu analyzes foreign exchange fluctuations and how they impact both a sustainable macroeconomic environment and external debt. The chapter relies on a dynamic growth model in an open economy to calculate the excess debt, with added FX rates. Empirical findings show that sovereign credit default risk increases together with greater fluctuations in FX rates, especially in fragile emerging countries.

The chapter "Keynes' Microeconomics of Output and Labor" comes as a contribution from Professor Semmler's colleague at the Economics Department at the New School for Social Research, Duncan K. Foley. This paper outlines a market-crowding theory of industry output and labor markets. The chapter emphasizes the fact that Keynes's theory was derived from microeconomic methods, and based on conceptions of market crowding equilibrium, exchange and production mediated by money, increasing returns, and unresolved social coordination problems. However, the chapter argues that specific ideas of Keynes's macroeconomic analysis have had limited appreciation of their centrality to broader macroeconomic debates. This chapter explores a social coordination approach to the theory of output and employment, providing a unifying theoretical context.

In their "'Wavelet-Based' Early Warning Signals of Financial Stress: An Application to IMF's AE-FSI," Marco Gallegati, Mauro Gallegati, James B. Ramsey, and Willi Semmler, himself, construct a "wavelet-based" early warning indicator for the IMF financial stress index for advanced economies. Selecting across individual indicators those that display the best leading performance on a "scale-by-scale" basis does that. The leading properties of each country's "wavelet-based" early warning indicator for its corresponding financial stress index are evaluated using univariate statistical criteria and a pseudo out-of-sample forecasting exercise. The authors find that the "wavelet-based" early warning composite indicator largely outperforms at every horizon any individual financial variable taken in isolation in early detecting financial stress. The gains tend to increase as the time horizon increases.

The chapter "Corporate Liquidity Under Financial Constraints and Macroeconomic Uncertainty" comes from another former student, Daniel Samaan and his co-author Immo Schott. In their chapter, the authors, using a dataset of about 60,000 firms, test a hypothesis of the transaction motive for liquidity demand has declined over the past two decades given advances in information technology and financial instruments. In their chapter, the authors find that corporate cash holdings have sharply increased since the 1990s and peaked in most advanced economies during the financial and economic crisis in 2008. Their analysis also shows that rather than the transaction motive it have been risk considerations and financing constraints that defined demand for corporate liquidity. The results are firm heterogeneity dependent and vary across the range of various risk variables, including macroeconomic uncertainty. Here, smaller, younger, and financially more vulnerable firms were more subject to idiosyncratic and aggregate risk across most OECD economies.

Ekkehard Ernst in his "Might Tobin Be Right? The Impact of Portfolio Shifts on Growth and Inflation in the Presence of Market Frictions," starts off with endogenous money demand in the presence of search frictions are present on financial markets. Here, monetary and fiscal policies affect households' portfolio decisions through their impact on the value of real balances. With exogenous growth, money continues to be neutral. However, when growth is endogenous, money-induced portfolio shifts allow for the existence of an optimal inflation rate. Similarly, portfolio shifts induced by governments' fiscal interventions allow enhancing growth as public debt helps to overcome search externalities by providing additional assets on the financial market. Finally, when monetary and fiscal policies interact, growth reaches maximum in comparison to each of the two policies taken individually. The results of the paper suggest that balance-sheet recessions can be overcome by targeted monetary and fiscal interventions that provide additional assets and induce portfolio shifts whereby these additional assets are invested in productive capacity.

The chapter "Business Confidence and Macroeconomic Dynamics in a Nonlinear Two-Country Framework with Aggregate Opinion Dynamics" showcases the contribution of Matthieu Charpe, Carl Chiarella, Peter Flaschel, and Christian R. Proaño. In this analytical paper, the objective is to investigate the role of the state of confidence in the macroeconomic dynamics of two interacting. The authors follow the assumption that the overall state of confidence in the world (two-country) economy plays a role not only in the dynamics of the nominal exchange rate, but also in the dynamics of the real economy through the determination of aggregate investment, as common in some literature. Such approach, allows the authors to consider far richer international macroeconomic interactions than most standard models. Of particular interest are the wage-price dynamics interacting with output and employment fluctuations, as well as debt dynamics due to a credit-financed investment behavior. The resulting framework is both advanced as well as flexible enough to generate various types of persistent fluctuations, and also complex dynamics.

Finally, Rajiv Sethi in his "Self-Falsifying Prophecies," explores the hypothesis that general perceptions regarding the stability of the macroeconomic environment affect realized levels of stability: regime perceptions vs. regime realizations. Sethi uses a simple nonlinear business cycle model to argue that self-fulfilling regime perception may not exist. As such all regime perceptions are self-falsifying. The reason for that is that volatility need not be continuous. The economy transitions from a stable to a cyclic regime as a parameter passes through a bifurcation value, in such a manner as to result in a discontinuous change in volatility.

5 Final Remarks

I believe the rollup of high-caliber authors and diverse topics covered in this volume attests to this publication's goal of celebrating the life, work, research, accomplishments, and friendship of Professor Willi Semmler. In the true spirit of openness and plurality in debates on empirical dynamic macroeconomics and social policy, these collected studies are written in an accessible enough style for specialists and non-specialists alike. Such universality and intellectual affordability, without the loss of the main point, is what makes a successful work stand out in the wide range of original intellectual contributions.

Indeed, the symbolic significance of this volume does not in any way overshadow the practical value of the collected studies. Knowing Professor Semmler personally, many of us have labored on not just developing sound theoretical presentations: every chapter in this collection also carries a critical practical component, so common to Willi Semmler's work.

And I think I will speak for all of Professor Willi Semmler's students—past, present, and future—if I say how grateful we are for his insights, guidance, and wisdom. In fact, pushing us to "look into new data" at the early stage of research or, perhaps, "try a new model" in a paper or maybe add "just one more chapter" to the thesis is what an economist must be made of and be prepared to carry out if the final product is to be taken seriously.

Actually, the persistent push for improvement and intellectually grasping new horizons is a characteristic of a scientist. In that and many other ways, Professor Willi Semmler is a true guiding example for his colleagues, friends, and students. It is our hope that you, the reader, will enjoy working through the studies collected in this book, would feel challenged to learn, explore, and apply new macroeconomic ideas in policy decisions that are yet to come.

Iran's Nuclear Program and the West's Response: A Game Theoretic Approach

Franz Wirl and Yuri Yegorov

JEL Classifications C7, Q48

1 Introduction

This paper considers the long standing conflict between Iran's nuclear power program and the sanctions imposed by the West. Iran's brinkmanship strategy of going nuclear has led to the most serious political crisis of the recent years, turning the previous brinkmanship strategies of North Korea pale in comparison. This difference is due to a number of facts: the size and power of Iran, its role in the Islamic world strengthened by the Islamic awakening following the Arab spring, its role as an important energy supplier (oil and gas), and last but not least the still unsettled situation in the Middle East coupled with Iran's open threats towards Israel including the support of radical Israel enemies like Hamas in Gaza and Hezbollah in Lebanon. The counter threat of Israel to attack Iran when it finds it necessary makes this situation very explosive.

This paper was written prior to the recently achieved breakthrough, which somehow confirms our ex ante prediction that appear to some now to be ex post. Although the paper addresses a political issue, it attempts to apply rigorous modelling, which is very much in the spirit of Willi's work; the analysis is even dynamic albeit of a much simpler nature than the one Willi prefers.

F. Wirl (✉) • Y. Yegorov
Faculty of Business, Economics and Statistics, Department of Industry, Energy and Environment, University of Vienna, Oskar Morgenstern Pl., 1, 1090 Vienna, Austria
e-mail: franz.wirl@univie.ac.at; yury.egorov@univie.ac.at

L. Bernard, U. Nyambuu (eds.), *Dynamic Modeling, Empirical Macroeconomics, and Finance*, DOI 10.1007/978-3-319-39887-7_2

Given this dramatic situation, the West, i.e. the US and the EU, responded with strongly tightening their economic sanctions since July 2012 that followed earlier sanctions as will be documented below. It is well known and confirmed in this particular case that sanctions create negative economic externalities for all parties. While any sanction distorts an existing economic equilibrium, after all this is the purpose, those distortions are large and significant at the scale of the world economy as Iran is the second largest reserve holder of conventional natural gas and is one of the largest reserve holders for conventional oil. Experiencing those costs drove the recent negotiations with Iran with the possibility of a final agreement and an end to the sanctions. The goal of this article is to suggest a simple game theoretical model that can explain Iran—West game and to argue about necessity and effectiveness of sanctions for such case.

Several game theoretic proposals try to explain the imposition of economic sanctions and their effectiveness. Tsebelis (1990) considers six different scenarios accounting for a wide range of circumstances. The results show that any attempt to increase the severity of the sanctions leads to counterintuitive outcomes, and the size of the sanction has no impact on the behavior of the target country. Shidiqi and Pradiptyo (2011) modify the sanction game proposed in Tsebelis (1990). In contrast to Tsebelis's (1990) findings, their results show that any attempt to impose more severe economic sanctions may reduce the probability of the target country to violate international agreements. A similar result holds if the sender country (sanctioning for target country) offers incentives ('carrots') for complying, such as aid and assistance to the target country. However, the effect of economic sanctions is meager at best in particular if the sanctions are not applied universally. Indeed, the restricted set of countries imposing sanctions on Iran suggested at the time of launching that the consequences will not be severe. For example, banning oil exports from Iran (and even here some loopholes were left for Greece and others relying on Iranian supplies) should not severely harm Iran since other big consumers like China and India did not join and should be happy to exchange their contracted oil supplies for slightly discounted Iranian supplies originally targeted for the West. Nevertheless, and to our surprise, the sanctions seem to be more effective than expected as the devaluation of the Iranian currency documents.[1]

Dixit and Skeath (2004) provide a game theoretical description of the Cuban Missile Crisis 1962 emphasizing the role of "brinkmanship",[2] a strategy also employed by the Greek finance minister Yannis Varoufakis during Greece's exposure to bankruptcy during 2015. The set up in Dixit and Skeath (2004) is used as our starting point in spite of an entirely different focus and no role for brinkmanship. Brinkmanship can be added by considering tough, war-like strategies, e.g., air strikes against strategic targets (with Israeli support). Basuchoudhary and Meredith

[1]During 70 days before Oct 2012 rial dropped its value from 20,000 per to 35,000 per USD. Source: http://www.businessinsider.com/chart-of-iranian-rial-2012-10.

[2]In order to render credibility to strategies that move a situation to the brink it requires "*a player to relinquish control over the outcome of the game without completely losing control*".

(2009) investigate the effectiveness of such threats within the framework of Dixit and Skeath (2004) and try to answer the following question; "can the threat of war deter Iran from building a nuclear weapon?" Basuchoudhary and Meredith (2009) argue that a brinkmanship strategy would only lead to a war in the unlikely event where the international community dramatically misperceives the induced incentives. Mousavi and Norouzi (2010) investigate the current conflict between Iran and the USA. They argue that the main problem is the absence of mutual trust, so each player chooses the strategy that imposes the highest cost on other side. Therefore, the establishment of mutual trust is a critical step to achieve a permanent solution. Han (2012) analyzes triadic sanctions where the receiver of the sanction is not the actual target. Such a threat is only credible in a repeated game setting but not in stage games.

The effectiveness of sanctions has been debated and investigated empirically. Tsebelis (1990) reports that only 33 out of 83 cases of economic sanctions have been effective. Hufbauer et al. (2007) show that only about 34 % of economic sanctions introduced between 1914 and 2006 had been effective. Lacy and Niou (2004) argue that the threat phase is critical and important for understanding the outcome of sanction games. Fuhrmann and Kreps (2010) is an empirical investigation about intentions and motivations of some countries to attack other countries that develop nuclear weapons. They identify 18 dyads where such attacks have been considered during the twentieth century. Some countries, being target states at some stage, have finally developed the nuclear weapon (China, Israel, India, Pakistan) while others have not done this. Others have joint the nuclear club without facing the threat of sanctions: USA, USSR, UK, and France. However some attacks took place (e.g., against Iraq and Syria) in order to deter those states from entering the nuclear club, and others have stopped their nuclear programs when facing threats (such as Libya). The paper tests three hypotheses related to the link between an intention to attack nuclear developers and: (a) previous experience of military conflict with that state, (b) level of democracy in target states, (c) states with dissimilar interests in foreign policy. Their statistical analysis shows that preceding violent conflict increases the likelihood of considering an attack by 606 % and of an actual attack by 1521 %. Foreign policy similarity has a statistically significant negative effect, while dictatorship is an insignificant variable. Among the used control variables, strong states are actually more likely to be attacked than weak states (and this is not trivial, since states have to worry about the risk of retaliation) but international norms do not play an important role. Since Iran's military conflicts date back at least to the 1980-ies, the results of Fuhrmann and Kreps (2010) suggest that the risk of Iran being attacked is not very high, which is confirmed by the recent experience (so far). However, the political importance of the present situation motivates us to consider this particular game in more detail, combining both the history of sanctions and economic incentives.

The paper is organized as follows. Section 2 reviews the history of the Iranian nuclear program and sanctions against Iran. Section 3 presents the game. The game is solved in Sect. 4 using backward induction for two cases—small and big 'carrots' for a complying Iran. Section 5 extends the game beyond the current crisis by investigating Iran's incentive to go nuclear in the first place.

2 Review of Iran's Nuclear Program and Sanctions Against Iran

2.1 Iran's Nuclear Program

Since 1959 Iran is a member of International Atomic Energy Agency (IAEA). Iran signed the Nuclear Non-Proliferation Treaty (NPT) in 1970. However, concerns about Iran's nuclear program started in 2002, when Russian technicians started the construction of the first nuclear reactor in Busher. This is the start of the confrontation between Iran and IAEA. Iran claims that it is complying with the international agreements in its nuclear program. This is supported in some IAEA reports (the Iranian nuclear program is safe) but rejected in others. Figure 1 briefly outlines the history of Iran's nuclear program and its confrontation by the IAEA (for more details see Tables in Appendix).

2.2 The History of Sanctions Against Iran

Since the Islamic revolution of 1979 in Iran and the following hostage crisis within the American embassy in Tehran, sanctions have been imposed against Iran (the majority by the USA). These sanctions can be classified into four major categories: (a) UN sanctions, (b) EU sanctions, (c) U.S. sanctions (including sanctions imposed by U.S. states), and (d) sanctions imposed by other nations (e.g., Australia, Canada, Switzerland, Japan and South Korea introduced domestic sanctions packages on their own). These sanctions had the following targets: weapon development, trade and investment, nuclear materials, finance (transactions and assets), and gasoline. Table 1 summarizes the sanctions that were imposed on Iran since 1979. Experience suggests that these sanctions have not been very successful in changing Iran's goals, policies and behavior and thus have caused only harm on both sides (Hufbauer et al. 2007).

As we can see from Table 1, Iran has faced a long history of sanctions that started several decades before the recent controversy about its nuclear weapon program. Thus, we can suspect that the history of sanctions could in the end have motivated Iran to go nuclear. In other words, without this history of sanction it would have behaved differently and possibly better. This observation will be used in our model as the level of external pressure as a starting point for the nuclear program.

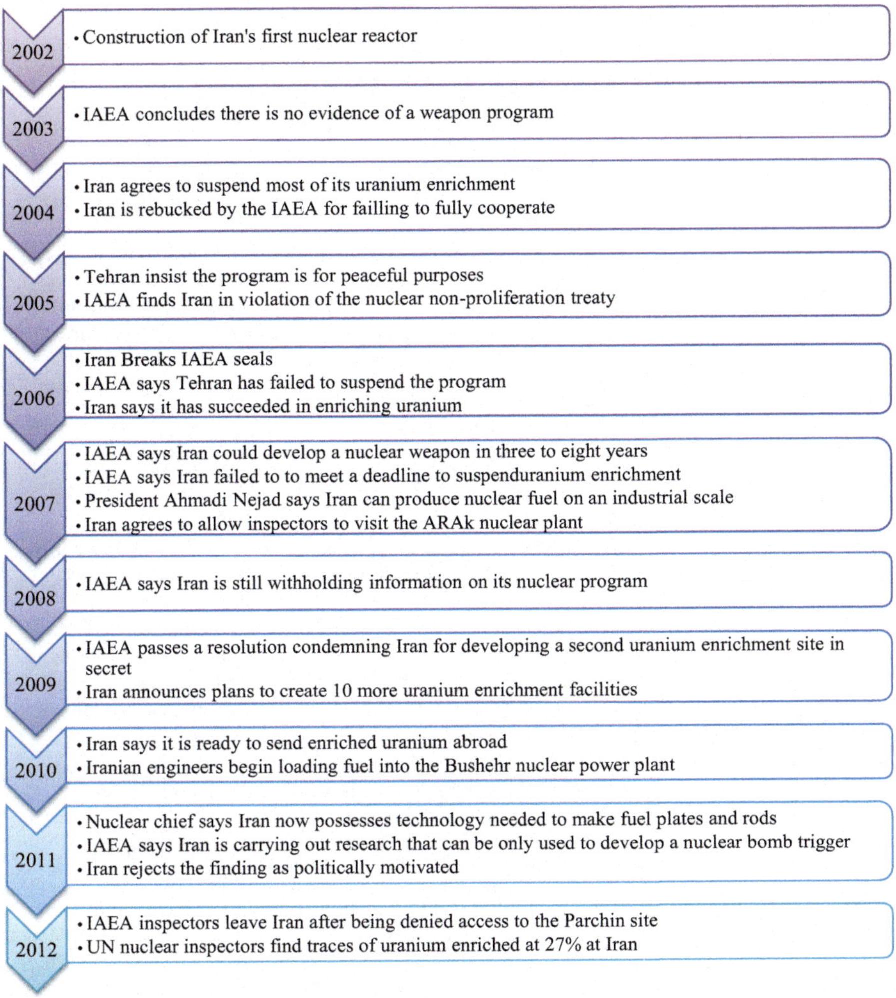

Fig. 1 The history of Iranian nuclear program

3 A Game Theoretic Approach

The following model of Iran continuing its nuclear program and the West's reaction by imposing sanctions departs from a framework in Dixit and Skeath (2004), although that addresses the much different Cuban crisis 1962 and emphasizes the role of brinkmanship strategies. At first glance the problem between Iran and the US and EU (short the West in the following) looks simple in terms of objectives and strategies. The West wants Iran to stop its nuclear program and to achieve this, it may impose sanctions. This leaves two options for Iran: either to stop its nuclear

Table 1 Sanctions against Iran since 1979

Year	Source of sanction	Reason of sanction stated by sanction senders	Sanction
1979	U.S.	Seizing the American Embassy	Freezing Iranian Assets
1983–1987 (During the Iran-Iraq War)		Supporting International Terrorism	1. Prohibit weapons sales and all U.S. assistance to Iran 2. Opposed all loans to Iran from international financial institutions 3. Blocking certain property or interest in property of the Iranian government 4. A ban on all commerce and travel between Iran and the United States
1993		The Iran-Iraq Arms Nonproliferation	Banning any transfer that aids Iranian or Iraqi attempts to acquire chemical, biological, nuclear, or destabilizing numbers and types of advanced conventional weapons
March 1995		Supporting Terrorism, destabilizing Iraq, works on weapons program	Prohibiting U.S. trade in Iran's oil industry
May 1995			Prohibiting any U.S. trade with Iran
1996		Iran-Libya Sanctions Act (ILSA)	All foreign companies that provide investments over $20 million for the development of petroleum resources in Iran will have imposed penalties against them
2004		Institute of Electrical and Electronics Engineers (IEEE)	Temporarily stopped editing manuscripts from Iranian researchers
2005		President Ahmadinejad lifted the suspension of uranium enrichment that had been agreed with the EU3	Freezing the assets of individuals connected with Iran's nuclear program

Table 1 (continued)

Year	Source of sanction	Reason of sanction stated by sanction senders	Sanction
2006	United Nations Security Council (UNSC)	Iranian nuclear program	Resolution 1696 and Resolution 1737
2007	States of USA		State of Florida enacted a boycott on companies trading with Iran and Sudan, while New Jersey's state legislature was considering similar action
2007	United Nations Security Council (UNSC)		Resolution 1747
2008			Resolution 1803 and 1835
2010			Resolution 1929
2010	United States Senate and House of Representatives		Comprehensive Iran Sanctions, Accountability, and Divestment Act of 2010 (CISADA) enhance restrictions in Iran
2012	European Union		A European Union embargo on purchases of Iranian crude oil that took full effect on July 1, 2012
2012	EU/US		The impending U.S. sanctions on the central bank of Iran

Sources: en.wikipedia.org/wiki/Sanctions_against_Iran, Fayazmanes (2003), Katzman (2012)

program or not. This last choice is influenced by the reward that the West offers to Iran if quitting its nuclear ambitions and by Iran's strength (or the stubbornness of its leaders, or some commitment including an element of brinkmanship). When the Western countries impose their sanctions they do not know Iran's strength (binary for simplicity, strong = s or weak = w, an alternative interpretation is tough or soft). Let p denote the West's prior probability assessment that Iran is strong and 1-p that it is weak. The game starts with an external determination (by 'nature') of Iran's type $\{s \text{ or } w\}$, which leads to two different subgames depending on whether Iran is strong or not.

Summarizing, we consider a game in extensive form shown in Fig. 2 and make the following assumptions:

1. Players are rational; there is perfect observation of the strategies and common knowledge of all payoffs.

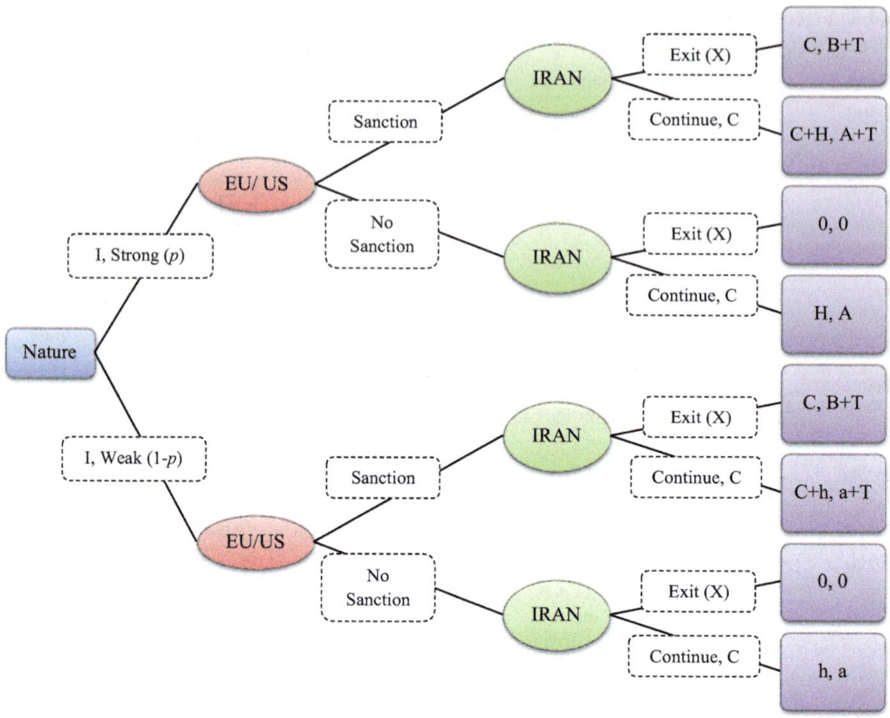

Fig. 2 Game tree Iran vs West

2. All players face binary choices—default and action, continuation of the nuclear program for Iran and imposing sanctions for the West. Therefore, Iran's strategies are $\{X, C\}$ for stopping (X) or continuing (C) with its nuclear program and $\{N, S\}$ for the West (here S means sanctions and N means no sanctions). Clearly, the restriction to binary strategies (in particular, for the West) is a simplification, since the degree of punishment is scalable, at least in principle. The consideration of an additional scenario of tougher, e.g., almost world-wide sanctions, compensates for this shortcoming by broadening the options of the West. Alternatively, the boundary solution (of either no or maximum action) results for declining average sanction costs, e.g., due to fixed costs (economical and/or political).
3. A dynamic game with sequential moves by the players. Figure 2 shows the game tree. The sequence is as follows. First nature moves and determines Iran's strength ($w = weak$ or $s = strong$). Then the West decides whether to impose sanctions or not unaware of Iran's strength. Finally, Iran continues (C) or exits (X) its nuclear program in full knowledge of its strength and its payoffs. If it stops the nuclear program, the West lifts the sanctions and may offer a reward.

3.1 Notation and Parameter Values

Table 2 introduces the notation and the chosen reference values and the values in the different scenarios are shown in Fig. 2. Payoffs are simply the sums of the benefit cost components; of course, one may consider sub- or super-additive versions. Considering the default of no action on both sides (i.e., Iran stops its program, the West imposes no sanctions, lines 3 and 7 in Fig. 2) the choices of zero payoffs serve as a reference point. The other parameter values are chosen from the (absolute) range from 0 to 3 in steps of 0.5 and are justified below.

The smallest element (thus $c = -0.5$) accounts for the costs to the West for imposing sanctions and for the small reward offered to Iran in the case of its compliance. Iran's costs from sanctions are much larger ($T = -2$). The harm borne by the West as a result of Iran's nuclear program must of course exceed the costs of sanctions (otherwise sanctions can never be an option), but its level depends on the (unknown) strength of Iran: ($H = -3$) if Iran is strong but only ($h = 1$) if Iran is weak, because then its nuclear force is fairly ineffective (e.g. because of facing Israel's retaliation). Iran's benefit from its nuclear program vary less, between $a = 1$ (if weak) and $A = 2$ (if strong), which assumes that the harm to the West increases sharper with respect to Iran's strength (compare h and H) than Iran's benefits (a versus A). The reason is that a destructive weapon cannot bring prosperity for a country (and we observe this on another example—North Korea) but can serve as a bargain in international negotiations. But the threat to the rest of the world grows rapidly with the number of bombs (or probability to create one). To summarize, the chosen parameters are simple and satisfy the following natural conditions: $|c| < |h| < |T| < |H|$, $a < A < |H|$ and $B < A$, while both cases ($B < a$ and $B > a$) are considered.

Clearly, there exist many other choices for the parameters. We address this by varying the size of the carrot B and the probability p that can induce a threshold.

We denote by n a strategy according to the numbering in Table 3. $W(n)$ denotes the payoff of the West given the strategy pair by number n, while $I(n)$ are the payoffs to Iran. The set of Iranian strategies includes two choices, $\{\text{Exit} = X, \textit{Continue} = C\}$ and the West also has two options, $\{\text{Sanctions} = S, \text{No Sanctions} = N\}$. The payoffs at the end of the game are based on the implicit assumption that the parties stick to

Table 2 Notations of the models and reference parameter values

$p =$ probability of Iran to be strong (not known to the West)
$c = -0.5$; cost of sanctions for the West
$H = -3$; harm to West if Iran continues nuclear program and is strong
$h = -1$; harm to West if Iran continues nuclear program and is weak
$B = 0.5$ or 1.5; benefit to Iran from West's "carrot" if it stops its nuclear program
$A = 2$; benefit to Iran from continuing nuclear program if strong
$a = 1$; benefit to Iran from continuing nuclear program if weak
$T = -2$; cost to Iran from sanctions

Table 3 Little carrot game ($B = ½$): strategies and payoffs

branch	Nature	West	Iran	Payoffs	
n	Iran s/w	N/S	X/C	West W	Iran I
1	s	S	X	-0.5	-1.5
2	s	S	C	-3.5	0
3	s	N	X	0	0
4	s	N	C	-3	2
5	w	S	X	-0.5	-1.5
6	w	S	C	-1.5	-1
7	w	N	X	0	0
8	w	N	C	-1	1

their policies: the West lifts its sanctions and pays out the reward after Iran quits its nuclear program and Iran does not re-start its nuclear program after the sanctions are suspended. The importance of the West's public reputation in future dealings (not only with Iran) is presumably sufficiently high to deter it from withholding the promised carrot. More critical is the assumption that Iran takes the reward and does not continue with its program after some time, after all there is little reputation it can lose. For example, in 1994 the United States and North Korea signed the "Agreed Framework" based on North Korea freezing its plutonium production program for US aid including the construction of two modern light-water nuclear power plants. However, North Korea restarted its nuclear program some time after receiving aid. The extension to a repeated version of this stage game is, however, left for future research.

Incentives can be either in the form of carrots or sticks (Dari-Mattiacci and De Geest 2009) according to Becker's (1968) economic theory of crime (Mendoza and Wielhouwer 2012). The West's strategy includes both punishment (sanctions as a stick) and rewards as additional carrot for Iran stopping the nuclear program. Two different scenarios are considered. In one case the carrot is small and costless for the West (e.g. accepting Iran's request to join the World Trade Organization and other global institutions thus allowing gas exports on a large scale), while in the other case the carrot is big and costly (for example, transfer of technology or even giving out cash, donating nuclear power stations).

4 Solving the Game

We solve our model for two scenarios: Either a small or a big carrot is offered if Iran complies with the IAEA regulations. Table 3 shows the payoffs in the case of a small carrot, while Table 4 illustrates the payoffs for a big carrot. The shaded cells indicate the optimal choices of the players starting at the end with Iran's decisions and the implied subgame perfect equilibrium, which is given by the shaded strategy pair for both players, e.g., strategy $n = 4$, $\{N, C\}$ in case of a strong Iran.

The game is solved by applying backward induction, first for assumption of a small carrot, $B = ½$. Considering the last move of a weak Iran it chooses continuation of its program C since $I(1) = 0 > I(2) = -1.5$. Not surprisingly, the same applies for a strong Iran. Therefore, 'no surrender' is Iran's dominant strategy (irrespective whether Iran is actually weak or strong). Given Iran's dominant strategy of not giving in to the West's demand suggests that the West should have not imposed sanctions in the first place given the costs. Indeed $\{N, C\}$ is the unique subgame perfect equilibrium (see Table 3) irrespective of Iran's strength and the probability p.

Table 4 Big carrot ($B = 1.5$) game: strategies and Payoffs

	Nature	West	Iran	Big carrot (1.5)	
n	Iran s/w	S/N	Stop X/con C	Payoff W	Payoff I
1	*s*	*S*	*X*	-0,5	-0,5
2	*s*	*S*	*C*	-3,5	0
3	*s*	*N*	*X*	0	0
4	*s*	*N*	*C*	-3	2
5	*w*	*S*	*X*	-0,5	-0,5
6	*w*	*S*	*C*	-1,5	-1
7	*w*	*N*	*X*	0	0
8	*w*	*N*	*C*	-1	1

Subgame perfect equilibrium for p < ½

Therefore, a change in parameters is necessary in order to eliminate Iran's dominant strategy of continuation in the last round. This can be done in various ways. We opt for an increase of the reward, $B = 1.5$ but alternatively one may focus on the stick by increasing the harm Iran faces for continuing its nuclear program e.g. due to more severe sanctions as other countries join.

As indicated, the major difference is that Iran's last decision can now be X in order to catch the reward. This is the case of a weak Iran facing sanctions. In all other instances, C remains the optimal strategy. This in turn affects the West's decisions and provides a potential justification for sanctions. Imposing sanctions and accounting again for Iran's rational response, the West obtains −3.5 if Iran is strong and −0.5 if facing a weak Iran. Therefore, the West's expected payoff depends on the probability, either if imposing sanctions on Iran,

$$pW(2) + (1-p)\,W(5) = -3.5p - 0.5\,(1-p) = -0.5 - 3p, \tag{1}$$

or not:

$$pW(4) + (1-p)\,W(8) = -3p - 1\,(1-p) = -1 - 2p. \tag{2}$$

Therefore, the West should impose sanctions if and only if the prior probability about Iran being strong is low, more precisely $p < 0.5$ in the case of the our choice of parameters.

Summarizing, the outcome of this theoretical game confirms the strategy choice of Iran of continuing with its nuclear program in spite of the sanctions. This observation extends presumably beyond the model given the past experience about the ineffectiveness of sanctions in general and those against Iran in particular. However, the West's imposition of sanction is much harder to justify and requires either more effective sanctions and/or the use of carrots. Given this additional incentive, sanctions can be rationalized to achieve the West's target if the probability of a strong Iran is not too large. The down side is that the sole purpose of Iran's nuclear weapon program is then to collect the reward and this is not incentive compatible as it triggers bad behavior as observed in the past with North Korea.

5 Extension: Iran's First Step: Nuclear Program or Not

The above game tries to reflect the current interactions and thus took Iran's violations of the Non-Proliferation Treaty as given. The following extension considers Iran's actions in a period before those covered in Fig. 1 in order to evaluate its choice to go nuclear in the first place. All assumptions about information are as before, i.e., Iran knows about its strength. Nevertheless, ex-post information may differ. The reason is that although the West does not know Iran's strength, Iran's strategy may reveal this implicitly. As we have seen from our analysis, the perfect equilibrium in the following subgame considered in Fig. 1 is (N, C), i.e. no sanctions and continued nuclear development for a strong Iran. In the case of a weak Iran, the outcome depends on the size of the carrot. Assuming the small carrot, then strategy $n = 8$ is played (again (N, C), i.e., no sanctions and continued nuclear development), whereas in the case of the big carrot, strategy $n = 5$ ((S, X), i.e., sanctions work) is the equilibrium.

Now we extend the game for the initial decision of Iran about going nuclear. The basic assumption is collecting the payoffs of the game given in Tables 2 and 3 requires that Iran executes in the first stage the option to embark on this nuclear program. Therefore considering the normalized payoff structure in Tables 2 and 3, forgoing the game analyzed above must impose costs on Iran. The idea is that accepting passively the external pressures and pre-nuclear sanctions (see Table 1) imposes costs, denoted by $z < 0$ and common knowledge, if not to Iran than at least to its politicians. Therefore, Iran will start its nuclear program if it leads to an equilibrium with Iran's payoff being higher than in the no nuclear scenario, which delivers only $z < 0$. We consider the two cases about the size of carrot separately.

In the case of a small carrot, any $z < 0$ is sufficient to induce the game in Fig. 2 and its equilibria (conditional on Iran's strength and also on p). Now consider the case of a big carrot. In the case of sanctions and Iran's decision to continue its nuclear program the size of the carrot does not matter. If Iran stops, its payoff is less than under continuation, except for a weak Iran and a big carrot, where $I(5) = -0.5 > -1 = I(6)$. Thus, for any $z < -0.5$ a weak Iran has an incentive to start its nuclear program but its only objective is then to get the big carrot (and to exit) if $p < 1/2$.

The question remains whether Iran follows the above analyzed truthful policies and implicitly reveal its strength (weak or strong) or prefers to hide it? This depends on the value of the parameters z, p and the carrot size b (by assumption all these parameters are known to both players). Suppose that the West observes the start of Iran's nuclear program. Does this allow the West to infer about the strength of Iran? A strong Iran will always start its nuclear program, while the choice of a weak Iran depends on z and the size of the carrot. In the case of a small carrot, any $z < 0$ is sufficient to lead to the game in Fig. 2. Therefore, only the case of a big carrot is of further interest that leads to the following different cases:

(a) If $-0.5 < z < 0$, a weak Iran does not start its nuclear program if the West considers the probability of a strong Iran small, more precisely, $p < 0.5$ (see Table 3), because this loss is less than what a weak Iran can obtain from going nuclear (-0.5 according to Table 3). However, if $p > 0.5$, a weak Iran benefits from imitating the strategy of a strong Iran and starts its nuclear program, because the West will then not impose sanctions, which allows even a weak Iran to collect the payoff of 2 according to Table 3.
(b) If $z < -0.5$; a weak Iran starts nuclear program for any $p > 0$, because all outcomes along the following sanction game yield higher payoffs.

Summarizing, a weak Iran can have an incentive to hide its nature when taking the first step. In the case of a small carrot, a strong and a weak Iran have an interest to go nuclear. Similarly, if the initial damage to Iran is sufficiently large, $z < -0.5$, then both types have an incentive to go nuclear. Even in the intermediate case of (a), Iran can hide its type when the West perceives the probability of Iran being strong as greater ½. Therefore, only a big carrot and a low prior probability of a strong Iran ($p < 0.5$), lead to separating equilibrium in which Iran reveals its type if it is weak.

6 Conclusions and Policy Implications

This paper addresses the topical issue of the sanctions imposed against Iran in July 2012 as a response to Iran continuing its nuclear program. For this purpose a game was proposed inspired by the analysis of Dixit and Skeath (2004), who studied the Cuban crisis of 1962 and associated brinkmanship strategies. For a wide range of model parameters sanctions do not work since continuation with its program remains Iran's optimal response, which seems to comply with the real world choice (until very recently). One way to trigger Iran's (rational) compliance is to offer Iran a sufficiently big reward; an alternative could be strengthening the stick, e.g., by tightened sanctions if other countries join. And even this outcome requires in addition that its assessment of the probability of a strong and tough playing Iran is not too large. All other subgame perfect equilibria are favorable for Iran (no sanctions, Iran continues its program).

We also consider the influence of old sanctions on Iran on its decision to start a nuclear program. This initial pressure leads to a negative payoff ($z < 0$) for Iran, which always triggers a strong Iran towards the development of nuclear arms. If this pressure is high ($z < -½$), both weak and strong Iran have an incentive to go nuclear such that West cannot infer about Iran's type. Indeed a weak Iran has strong incentives to hide its type over a large range of parameters. Only if the West considers a strong Iran unlikely, $p < ½$, and offers a big carrot, Iran can be induced to abstain from going nuclear. Therefore, it is conceivable that the West's continuous pressure in the past induced even an actually weak Iran to go nuclear.

This paper is only a first shot at modeling the conflict between Iran and the West and there are many ways to extend it. An obvious candidate is to enlarge the game tree. Another is to embed such a stage game into a repeated one in order to justify (or falsify) the implicit commitment that must be assumed in the last round to fix the payoffs ex ante.

Acknowledgements I thank the editors, Lucas Bernard and Unurjargal Nyambuu, for their initiative to honor our friend and colleague Willi Semmler and his wide range of academic work in method, topic and geography covering at least two continents. Moreover, we thank for the opportunity to share our work.

Appendix

See Tables 5 and 6.

Table 5 The history of Iranian nuclear program

Year	IAEA/Iran action
1959	IAEA membership
1970	Iran signed the Nuclear non-proliferation treaty (NPT)
2002	Construction of Iran's first nuclear reactor
2003	IAEA concludes that there is no evidence of a weapons program
2004, June	Iran is rebuked by the IAEA for failing to fully cooperate with an inquiry into its nuclear activities
2004, November	Iran agrees to suspend most of its uranium enrichment under a deal with the EU
2005, August	Tehran insists the program is for peaceful purposes
2005, September	IAEA finds Iran in violation of the nuclear Non-Proliferation Treaty
2006, January	Iran breaks IAEA seals at its Natanz nuclear research facility
2006, February	IAEA votes to report Iran to the UN Security Council over its nuclear activities
2006, February	Iran resumes uranium enrichment at Natanz
2006, April	Iran says it has succeeded in enriching uranium
2006, August	IAEA says Tehran has failed to suspend the program
2007, February	IAEA says Iran failed to meet a deadline to suspend uranium enrichment, exposing Tehran to possible new sanctions
2007, April	President Ahmadinejad says Iran can produce nuclear fuel on an industrial scale
2007, April	IAEA says Iran has begun making nuclear fuel
2007, May	IAEA says Iran could develop a nuclear weapon in 3–8 years if it so chooses
2007, May	Iran agrees to allow inspectors to visit the Arak nuclear plant following talks with the IAEA
2008, May	IAEA says Iran is still withholding information on its nuclear program
2009, October	Five permanent UN Security Council members plus Germany offer Iran proposal to enrich its uranium abroad
2009, November	Iran refuses to accept the international proposal to end the dispute over its nuclear program. UN nuclear watchdog IAEA passes a resolution condemning Iran for developing a second uranium enrichment site in secret
2009, November	Iran denounces the move as "political" and announces plans to create 10 more uranium enrichment facilities
2010, February	Iran says it is ready to send enriched uranium abroad for further enrichment under a deal agreed with the West. The US calls on Tehran to match its words with actions
2010, May	Iran reaches a deal to send uranium abroad for enrichment after mediation talks with Turkey and Brazil; Western states respond with skepticism, saying the agreement will not stop Iran from continuing to enrich uranium

Table 5 (continued)

Year	IAEA/Iran action
2010, August	Iranian engineers begin loading fuel into the Bushehr nuclear power plant
2010, December	Talks in Geneva between Iran and key world powers on Iran's nuclear program
2011, January	Nuclear chief Ali Akbar Salehi says Iran now possesses technology needed to make fuel plates and rods for nuclear reactors
2011, May	Iran's Atomic Energy Organization says the generating unit at the Bushehr nuclear power plant has begun operating at a low level
2011, September	Iran announces that the Bushehr nuclear power station has been connected to the national grid
2011, November	IAEA says Iran is carrying out research that can only be used to develop a nuclear bomb trigger
2011, November	Iran rejects the findings as politically motivated
2012, February	IAEA inspectors leave Iran after being denied access to the Parchin site, south of Tehran
2012, May	UN nuclear inspectors find traces of uranium enriched at 27 % at Iran's Fordo nuclear site
2012, September	IAEA says Iran doubles production capacity at Fordo nuclear site and "significantly hampered" IAEA ability to inspect Parchin military site

Source: "A chronology of key events. Iran Profile", *BBC News*, November, 2012

Table 6 Timeline of Iran's nuclear activities

1950s	1967	1968	1974	1977	1979	1980s	1990	1992	1995	1997	2001
The United States and Iran signed a civil nuclear cooperation agreement as part of the United States Atoms for Peace program	The Tehran Nuclear Research Center, supplied by the United States, opened	Iran signed the Nuclear Non-Proliferation Treaty (NPT)	Iran signed the NPT's Safeguards Agreement with the International Atomic Energy Agency (IAEA)	President Carter reopened negotiations on the shah's quest for a nuclear energy program	After the 1979 revolution, The United States retracts a deal it had made with Iran a year earlier and stops supplying enriched uranium for the Tehran research reactor	Abdul Qadeer Khan, a Pakistani metallurgist sells Iran, North Korea and Libya his uranium enrichment technology, and in Libya's case, a bomb design	Iran decided to rebuild the damaged Bushehr nuclear power plant	Russia and Iran signed a cooperation agreement on the civil use of nuclear energy, including construction of a nuclear power plant	Iran signed a contract with the Russian Ministry of Atomic Energy to build a light water reactor at Bushehr under IAEA safeguards	The IAEA expanded the Safeguards Agreement by adopting the Additional Protocol	Russian President Vladimir Putin and Iranian President Khatami signed nuclear and military cooperation accords

(continued)

Table 6 (continued)

2002	2003	2004	2005	2006	2007	2008	2009	2010	2011	2012
Construction of Iran's first nuclear reactor	IAEA concludes there is no evidence of a weapons program	Iran is rebuked by the IAEA for failing to fully cooperate	Tehran insists the program is for peaceful purposes	IAEA says Tehran has failed to suspend the program	IAEA says Iran could develop a nuclear weapon in 3–8 years	IAEA says Iran is still withholding information on its nuclear program	Iran announces plans to create 10 more uranium enrichment facilities	Iranian engineers begin loading fuel into the Bushehr nuclear power plant	Nuclear chief says Iran now possesses technology needed to make fuel plates and rods	UN nuclear inspectors find traces of uranium enriched at 27 % at Iran
				Iran breaks IAEA seals	IAEA says Iran failed to meet a deadline to suspend uranium enrichment				IAEA says Iran is carrying out research that can only be used to develop a nuclear bomb trigger	

		Iran agrees to suspend most of its uranium enrichment	IAEA finds Iran in violation of the nuclear Non-Proliferation Treaty	Iran says it has succeeded in enriching uranium	President Ahmadinejad says Iran can produce nuclear fuel on an industrial scale		IAEA passes a resolution condemning Iran for developing a second uranium enrichment site in secret	Iran says it is ready to send enriched uranium abroad		IAEA inspectors leave Iran after being denied access to the Parchin site
					Iran agrees to allow inspectors to visit the Arak nuclear plant				Iran rejects the findings as politically motivated	

(continued)

Table 6 (continued)

2012	2013	2014	April, 2015
• Bomb Kills Iranian Nuclear Scientist	• U.S. Bolsters Sanctions	• Negotiators Put Final Touches on Iran Accord	• Iran Agrees to Nuclear Limits, but Key Issues Are Unresolved and are remained for the next meeting at the end of June
• Iran says it is building about 3000 advanced uranium-enrichment centrifuges at the Natanz plant	• US Blacklists an Iranian and Businesses over Violation of Sanctions	• Iran Is Providing Information on Its Detonators, Report Says	
• The P5 + 1 held inconclusive talks with Iran in Baghdad	• U.S. Arms Deal with Israel and 2 Arab Nations Is Near	• Iran Outlines Nuclear Deal; Accepts Limit	
• A European Union embargo on Iranian oil takes effect	• Israeli Officials Stress Readiness for Lone Strike on Iran	• Iran Nuclear Talks Extended, Diplomats Say	
• Israel's 'Red Line'	• Fearing Price Increases, Iranians Hoard Goods	• Role for Russia Gives Iran Talks a Possible Boost	
• Prime Minister Benjamin Netanyahu of Israel tells the United Nations that Iran's capability to enrich uranium must be stopped before the spring or early summer, arguing that by that time Iran will be in a position to make a short, perhaps undetectable, sprint to manufacture its first nuclear weapon	• Iran Nuclear Weapon to Take Year or More, Obama Says	• Iran Altering Arak Reactor in Bid for Nuclear Deal	
• An IAEA report stated that inspectors found traces of uranium enriched to 27 % at Fordow	• U.S. Imposes Sanctions on Those Aiding Iran	• Negotiators Scrambling as Deadline Looms in Nuclear Talks	

(continued)

• The U.S. Treasury Department froze the assets of four Iranian companies and one executive for links to Tehran's missile and nuclear programs	• Iran Is Seen Advancing Nuclear Bid	• U.S. and Allies Extend Iran Nuclear Talks by 7 Months	
• U.S. Bolsters Sanctions	• U.S. Adds to Its List of Sanctions against Iran		
• Iran's Currency Tumbles	• Iran Elects New President		
• After months of harsh, American-led sanctions, Iran's currency, the rial, plunges 40 %	• Iran Slows Its Gathering of Enriched Uranium, Report Says		
	• Iran Said to Seek a Nuclear Accord to End Sanctions		
	• Rouhani, Blunt and Charming, Pitches a Moderate Iran in First U.N. Appearance		
	• First Direct US-Iran Talk Since 1979		
	• Iran Talks Called Substantive		
	• Iran Says It Agrees to 'Road Map' With U.N. on Nuclear Inspections		
	• Obama Calls for Patience in Iran Talks		
	• Deal With Iran Halts Nuclear Program		

Sources: http://iranprimer.usip.org/resource/timeline-irans-nuclear-activities. http://www.nytimes.com/interactive/2014/11/20/world/middleeast/Iran-nuclear-timeline.html?_r=0#/#time243_7169

References

Basuchoudhary, A., & Meredith, R. (2009). War with Iran: Will brinkmanship deter Iran from building the bomb? In MPRA Paper*19764*. Germany: University Library of Munich.

Becker, G. (1968). Crime and punishment: An economic approach. *The Journal of Political Economy, 76*(2), 169–217.

Lacy, D., & Niou, E. M. S. (2004). A theory of economic sanctions and issue linkage: The roles of preferences, information, and threats. *The Journal of Politics, 66*(1), 25–42.

Dari-Mattiacci, G., & De Geest, G. (2009). Carrots, sticks, and the multiplication effect. *Journal of Law, Economics, and Organization, 26*(2), 365–384.

Dixit, A. K., & Skeath, S. (2004). *Games of strategy*. W. W. Norton and Company, ERRATA as of January 6, 2005.

Fayazmanes, S. (2003). The politics of the U.S. economic sanctions against Iran. *Review of Radical Political Economics*.

Fuhrmann, M., & Kreps, S. E. (2010). Targeting nuclear programs in war and peace: A quantitative empirical analysis, 1941–2000. *Journal of Conflict Resolution, 54*(6), 831–859.

Han, B. (2012). *A theory of economic sanctions*. Korea Institute for International Economic Policy (KIEP), WP12-03.

Hufbauer, G. C., Schott, J. J., Kimberley, A. E., & Oegg, B. (2007). *Economic sanctions reconsidered* (3rd ed.). Washington, DC: Peterson Institute for International Economics.

Katzman, K. (2012). *Iran sanctions*. U.S. Congressional Research Service, September 13, fpc.state.gov/documents/organization/198052.pdf

Mendoza, J. P., & Wielhouwer, J. L. (2012). Only the carrot, not the stick: Incorporating trust into the enforcement of regulation.

Mousavi, M. A., & Norouzi, Y. (2010). Iran-US nuclear standoff: A game theory approach. *Iranian Review of Foreign Affairs*.

Shidiqi K. A., & Pradiptyo, R. (2011). A game theoretical analysis of economic sanction. SSRN WP series, http://ssrn.com/abstract=1769423

Tsebelis, G. (1990). Are sanctions effective? A game-theoretic analysis. *Journal of Conflict Resolution, 34*, 3–28.

From the Great Divergence to the Great Convergence

Askar Akaev

1 Introduction

A.V. Korotaev has recently carried out a study which has reasonably demonstrated an evolutionary process of transition from the Great Divergence (Huntington 1996) between the West and the rest of the World, stretching over one and a half centuries (1820–1970), and convergence, which had started in the middle of the twentieth century, but has rapidly manifested in the first decade of the twenty-first century (Korotayev and Zinkina 2014; Sadovnichij et al. 2014). As correctly noted by Angus Maddison, there was extensive economic growth in the period between 1000 and 1820. After 1820, the Western economies began to intensively develop, creating a dynamic world economy (Maddison 2012, p. 111). During the period 1000–1820 actual per capita income had increased 2.8 times in the West; between 1820 and 2003 it grew 20-fold! In the rest of the world per capita income grew at much slower rate—from 1000 to 1820 it had grown by slightly more than a quarter, and, in the subsequent period only sevenfold (Maddison 2012, p. 112).

By analyzing all available data of the early nineteenth century capitalist era, Maddison came to the conclusion that the transition to accelerated growth in the West had started around 1820 (Maddison 2012, p. 467), when the Industrial Revolution was born. The Industrial Revolution started in England around 1750, and simultaneously spread to most Western European countries: Denmark, Netherlands, Norway, Prussia, France, Finland and Sweden. It then quickly spilled over to

Research work was supported by the Russian Science Foundation (project №14-28-00065)

A. Akaev (✉)
M.V. Lomonosov Moscow State University, Institute of Mathematical Studies of Complex Systems, Moscow, Russia
e-mail: askarakaev@mail.ru

L. Bernard, U. Nyambuu (eds.), *Dynamic Modeling, Empirical Macroeconomics, and Finance*, DOI 10.1007/978-3-319-39887-7_3

English-speaking countries in other continents—USA, Canada, Australia and New Zealand. This created the club of industrially developed states with high standards of living where about 1 billion people, i.e., about 15 % of the world's population, live today. The Industrial Revolution is the reason the GDP and per capita income in these countries began to grow rapidly. This growth became sustainable and continued for almost 150 years, until the world economic crisis of the 1970s.

The rest of the countries, which represented 80 % of the world population, continued to develop under the conditions of the traditional model of low growth for more than 100 years, until the end of the World War II. People remained poor and lived mostly in stagnant conditions, both economically and technically. There were, however, exceptions. For example, Japan had joined the industrially developed countries by the end of the nineteenth century due to the Meiji reforms. *Thus, there appeared a global model of divergence between developed countries of the West and developing countries of the rest of the world.* The rapid divergence of the two worlds—developed and developing, rich and poor had started. In 1820, the West's share in world GDP comprised about 28 % while representing 17 % of world population. By the end of the nineteenth century it had almost doubled, having 54 % of world GDP with just 23 % of the world's population. This is graphically displayed in Fig. 1, plotted by A.V. Korotaev (Sadovnichij et al. 2014). In the twentieth century the gap between the West and the rest of the world continued to grow, though insignificantly—the West's share of GDP had reached a 60 % maximum in the 1950s–1960s and subsequently started to decrease. The main reasons for the growth tapering and decline were the two World Wars and the Great Depression of the 1930s as well as the collapse of world colonial system.

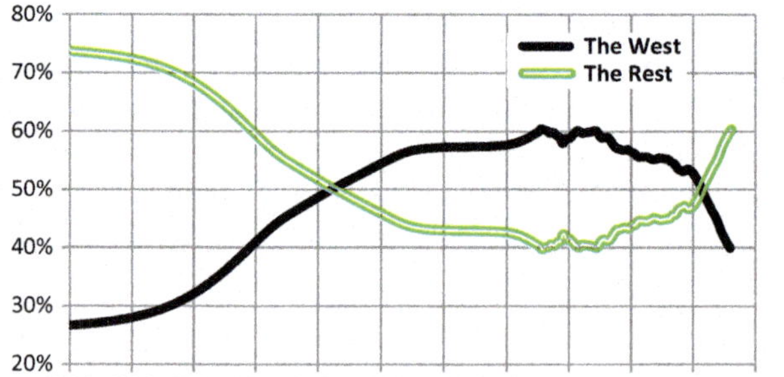

Fig. 1 The dynamics of the West and the rest of the world shares in world GDP after 1800 (*after A. Maddison*). *Data source*: Until 2008 (inclusive)—Maddison (Maddison 2010); after 2008—World Bank 2013: NY.GDP.MKTP.PP.KD. For the sake of compatibility, the World Bank data on GDP after 2008 were recalculated according to Maddison's coefficients of conversion of nominal dollars into dollars after PPP (purchasing power par)

WWII created a unique situation for developing countries. The colonial system had been eliminated and many large countries gained independence. These newly sovereign countries now had an urge to develop. For example, India, which is the largest democratic country, is located on the vanguard of the fastest growing economies. Developing countries aimed to build a new architecture of international relationships based on global security and mutually beneficial collaboration. The United Nations (UN) was established to maintain peace and security. The UN's economic institutions—IMF, WB, GATT, et. al. were established to facilitate the economic development of developing countries and to eradicate pauperism. GATT (*General Agreement on Tariffs and Trade*) was created to decrease and control customs barriers. It has played an important part in lifting unnecessary barriers in international trade and has led to a surge in foreign investment. The growth in international trade was facilitated by the World War II era revolution in transportation and modern means of communication, binding countries in all continents. This enabled developing countries to achieve breakthroughs in science and technology. An unprecedented quest for knowledge began in developing countries and education became a chief priority. "Growth started in the countries of developing world. At first it was relatively slow and was observed only in particular countries but later had embraced other developing countries and its rate began to accelerate" (Spence 2013, p. 19).

Figure 1 clearly shows that the West's share in world GDP started to decrease quite substantially beginning at the end of the 1960s and slowly declined until the end of the 1990s. After 2000, the West's share of world GDP began to rapidly decline while the rest of the world's share increased. A.V. Korotaev suggested that if this rate remains constant, within 15–20 years the ratio between the West and the rest of the world in contribution to world GDP will return to the level observed before the start of the Great Divergence. Obviously, the turn towards accelerated convergence in the beginning of the twenty-first century is largely due to the sustainably accelerated growth of the two largest countries—China and India, as demonstrated in the last 2 decades.

Thus, as early as the 1990s, the GDP gap between developing countries and developed countries had significantly narrowed. However, as Fig. 1 illustrates, the gap had begun a sustainably accelerated decrease after the year 2000. Making a connection, A.V. Korotaev had pointed out a curious fact: in the 1990s, leading economists of the West had carried out extensive research of the convergence problem and concluded that the convergence between developed and developing countries doesn't exist. Instead, the issue is continuing divergence, though not so well pronounced (Sadovnichij et al. 2014). In the first decade of the twenty-first century, convergence occurred so rapidly that it could not go unnoticed. Well-known economist and Nobel Prize winner Michael Spens describes the issue from different points of view as originating from the post-war era (Spence 2013).

A.V. Korotaev discovered a remarkable fact in the evolution of the world system from Great Divergence to Convergence. The surprising similarity of the curve describes the dynamics of the gap between the developed countries of the West and developing countries of the rest of the world, to the curve of world population growth dynamics (Korotayev and Zinkina 2014).

In this paper we aim to model the global process of evolution of the world system from the Great Divergence of the nineteenth to twentieth centuries, to the Convergence of the late twentieth to early twenty-first centuries through the Industrial Age. We do so by means of mathematical models and we build models for long-term forecasting in order to present possible development scenarios of the world system in the first half of the twenty-first century.

2 The Separation of West from the Rest of the World: Forces Facilitating the Great Divergence in the Nineteenth Century

The rise of the West, divergence between the West and the rest of the world, as well as the genesis of modern economic growth are comprehensively and quite fully considered in numerous books. In our study we will be guided by the three sources (Allen 2013; Maddison 2012; Mel'yantsev 1996) which are well known and most credible.

By the beginning of the nineteenth century Europe was the most developed and wealthy continent. This resulted in an unprecedented flourishing of culture, education, science and technologies, which started in the Age of Renaissance and Enlightenment. Per capita income of the European population was twice as much as most other world regions, approximately $1100–1200 per year in prices of 1990. Netherlands was the most prosperous European country by the start of the Industrial Revolution in 1820, with per capita income of about $1800. However, Great Britain became the leader in the rise of the West. From 1750 Great Britain singlehandedly led the path to the Industrial Revolution. It took two generations but the new model of economic growth allowed Great Britain to become the second wealthiest economy of the world by 1820 with per capita income of $1700. Consequently, Great Britain had sustainable growth throughout the entire nineteenth century and became the "*workshop of the world*". It was the economic and technological leader of the world and retained leadership for over 150 years, up to World War II. During this time, capital export and free trade aided acceleration of world development.

As R. Allen writes: "Between 1815 and 1870, the Industrial Revolution spread from Britain to the continent with remarkable success. Not only did the Western European countries catch up to the leader, but they joined the leader in forming a group of innovators that had jointly advanced the world's technology frontier ever since. Of course, North America also industrialized in the 19th century and soon joined the innovation club." (Allen 2013, p. 60) R. Allen included Great Britain, France, Germany, US, Canada, Australia and New Zealand in the first innovation club. R. Lucas divided the first innovation club into two groups, the first group consisting of Great Britain, US, Canada, Australia and New Zealand.

Notably, this English-speaking group of countries demonstrated sustainable growth in per capita income over the last 200 years. The second group included France, Germany, the Netherlands, and Scandinavian countries. (The first and second group of countries had significantly different trajectories of per capita income growth up to World War II and rapidly approaching henceforth) One should note that R. Lucas grouped the countries by the similarity of their per capita growth trajectories, and by contemporary income level (Lucas 2013, pp. 179–180). R. Lucas separated the first innovation club into two groups that would both be considered successful—countries with high levels of per capita income (HI—*high income*) and countries with medium level of per capita income (MI—*middle income*). In our study, we will follow the grouping accepted by R. Lucas, but extending the second group of medium-income countries to include Austria, Belgium, Italy, Spain, Portugal, Switzerland and Japan in the East. Most other countries will be included in the group of developing countries with a low level of per capita income (LI—*low income*). The dynamics of population per capita income for the mentioned groups of countries, including the world average (W) are presented in Fig. 2.

The Industrial Revolution in innovative countries resulted in a five to sixfold general acceleration rate of economic growth as compared to corresponding data from the Renaissance and Enlightenment ages: growth was 0.3–0.4 % per year during the sixteenth to eighteenth centuries and 1.8–2.2 % in the nineteenth to early twentieth century (Mel'yantsev 1996, p. 103). Simultaneously, from 1873 to 1913 there was a considerable increase in growth rate as compared to the period from 1820 to 1870: from 2 to 2.67 % per year (Maddison 2012, p. 114). Per capita income grew in the first group by 2 % annually, doubling every 36–40 years. V.A. Mel'yantsev has also calculated that the rate of gross capital investment grew during the transition to an industrial economy twofold—from 5 to 7 % of GDP in the sixteenth to eighteenth centuries to 12–14 % of GDP (Mel'yantsev 1996, p. 214). He also turned attention to the fact that industrial growth in Western countries was well-

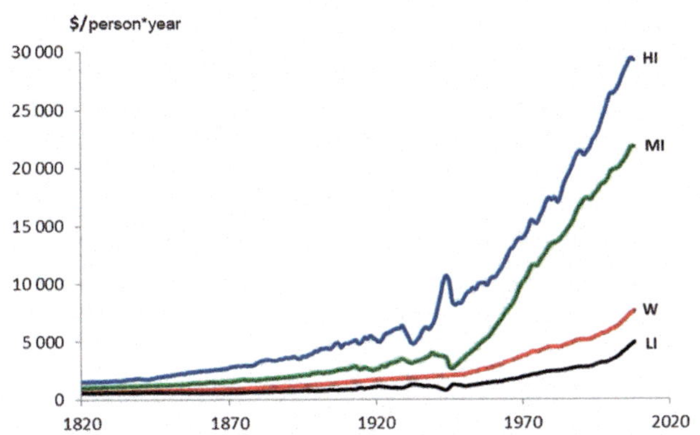

Fig. 2 Per capita GDP for World-system and various groups of countries. *Data source*: (Maddison 2010)

balanced. Indeed, it was sufficiently connected to the development of the agriculture and infrastructure industries. Agrarian reform was implemented conventionally but with an improved crops rotation system, use of selection seeds, more productive breeds of cattle and a wide application of organic fertilizers. There was a parallel revolution in communication, which made transportation cheaper, quicker and more reliable (Mel'yantsev 1996, p. 104). Thanks to all this, the Great Convergence and unseen separation of the West from the rest of the world occurred, and as a result by the end of the nineteenth century the West's share of world GDP was more than 55 % with about 23 % of the world population (see Fig. 1).

A new phenomenon caused by the industrial revolution was a decline in birth rate, which later caused a demographic transition. Thus, sustainable growth of per capita income in developed countries had two sources: technological innovations and demographic transition. As clearly seen from Fig. 3, the fast divergence between the West and the rest of the world in GDP share, observed in the nineteenth century (see Fig. 1) had a powerful demographic component. As is well known, during the first phase of demographic transition (which Western countries went through in the nineteenth century) there is a significant decrease in mortality rate (Vishnevskij 2005). Commeasurable decrease of birth rate is observed only in the second phase (*which Western countries entered only in the very late 19th – early 20th century*). Consequently, there was a significant decrease in mortality rate in Western countries throughout the nineteenth century and, together with a high birth rate, this resulted in a considerable increase of the natural population growth rate, observed in Fig. 4. Due to the time lag in modernization, similar demographic growth took place in most other countries as late as the twentieth century. Thus, in the nineteenth century a considerably higher population growth rate in Western countries compared to other world countries led to a particularly fast GDP share growth.

On the other hand, in the twentieth century Western countries entered the second phase of demographic transition, and the birth rate decreased. This caused demographic growth to increasingly decelerate, culminating in the years preceding World War II (see Fig. 4). R. Lucas believes that limited possibility for the accumulation of human capital has thrown most people into a dilemma of quality and quantity. This has pushed people in favor of decreasing birth rate and increasing the quality of human capital in order to pay more attention and allocate more

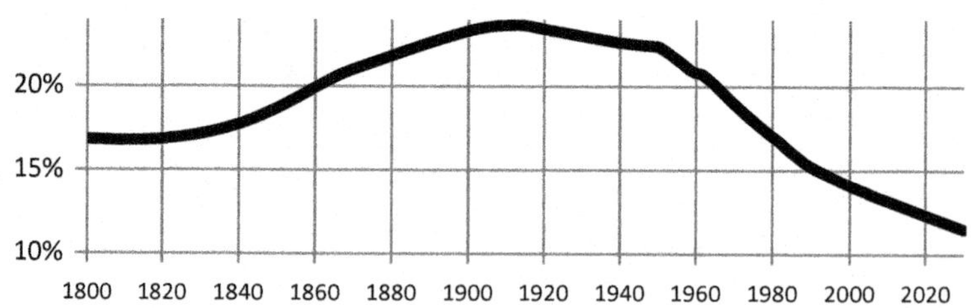

Fig. 3 The dynamics of the population share of the West in world population, 1800–2009, with forecast up to 2030. *Data source*: (Maddison 2010)

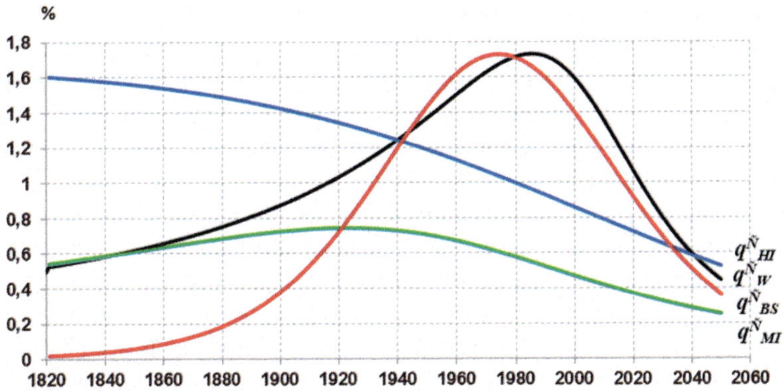

Fig. 4 Population growth rate for different groups of countries in industrial age, with forecast until 2050. *Data source*: (Maddison 2010)

resources to every child. This parental approach, based on economic decisions of investing in children's futures, describes endogenic growth of human capital and, according to R. Lucas, explains demographic transition (Lucas 2013, p. 255).

Most other countries of the world entered the first phase of demographic transition in the twentieth century, which encompassed a significant decrease in the death rate while the birth rate was still high. As a result, West's share in total world population had peaked by the beginning of World War I and started to decline afterwards, though until the early 1950s it progressed at a very slow rate (see Fig. 3). In the 1950s however, when the majority of the Third World countries had entered the first phase of demographic transition, the population outbreak started. This, along with further birth rate decreases in first world countries, resulted in a rapid decrease of the Western population share. The rate of decline had slowed down in the end of the 1980s due to the majority of the countries entering the second phase of demographic transition, as can be seen in Fig. 4. All these facts should be taken into account in considering the dynamics of the GDP per capita gap between the West and the rest of the world.

Let us consider the forces and reasons leading to the separation of the West from the rest of the world countries and factors that facilitated sustained dynamic growth of western economies. Essentially, as reasonably noted by R. Allen, the most important reasons for rapid separation in world income were the industrialization of the West and the de-industrialization of Eastern countries (Allen 2013, p. 15). At the same time, trade revolution had occurred and had a huge impact on the economic development of the East. Explosive development of international trade, according to one of the essential principles of economic science—principle of comparative advantage, facilitated specialization of countries in the output of goods they could produce most efficiently. Since innovator-countries of the West were the first to exploit technological achievements of the Industrial Revolution, they naturally specialized in industrial production, which had high growth potential. The ongoing trade between Europe and Eastern Asia had empowered China, India and other countries of the region to specialize in traditional agriculture, which had low growth potential.

As a result, the West started to separate from Eastern Asia both in production volume and per capita income. Indeed, according to R. Allen's data, in 1750 a major part of industrial production was located in China (*33 % of world output*) and Indian subcontinent (*about 25 %*). In per capita production volume, China and India were only slightly behind wealthiest countries of the West. By the start of the twentieth century, however, the world had drastically changed. By 1913, China's and India's shares went down to 4 % and 1 % respectively. Growing competition from the English enterprises had led to the collapse of manufacturing in Eastern Asia. The reason was the Industrial Revolution in England from 1750 to 1880. During this period, Great Britain's share of industrial production increased from 2 to 23 % (Allen 2013, p. 16).

Two great twentieth century economists—Joseph Schumpeter and Robert Solow have designed an innovation-based theory of economic development (Schumpeter 1982) and a neoclassical theory of economic growth (Solow 1956), which have provided a basis for the most of modern models of economic growth and development. From these theories, it follows that the main sources of economic growth are innovations, accumulation of human and physical capital and productivity. Innovation is new knowledge, new technologies, and new ways of production process engineering, employed for the increase of production potential of the economy. Some innovations save labor while the others—mainly capital. Innovations can be of basis and improving type. Basis innovations are cyclic and give powerful impulse to long-term non-equilibrium economic development, called the long waves of economic development or big cycles of Kondratiev's economic conjuncture (Yakovets 2004) with longevity of 30–40 years. Innovations are usually generated by technology advance.

The Industrial Revolution of late eighteenth to early twentieth centuries produced seminal innovations such as loom, steam engine, electricity and internal combustion engines. It was an absolutely natural result that all these technological innovations were generated and had practical implementation in Europe. Due to the flourishing of education, science, machinery and technology in the era of the Renaissance and Enlightenment (sixteenth to eighteenth centuries), Europe was best prepared for such innovations. The literacy among the European population even at that time exceeded 40–50 %, while in the rest of the world it did not even reach 10 %. Each of the listed inventions had the potential of application and extensive use. T. Bresnahan and M. Trajtenberg have created a term "*general purpose technologies* (GPT)" to describe basis technologies of this type (Bresnahan and Trajtenberg 1995).

Extensive use of steam engines became one of the pacing factors of the nineteenth century revolution in land and naval passenger and cargo transportation. The next impulse to economic growth was provided by the invention of the internal combustion engine, the extensive use of which in the early twentieth century led to considerable growth of international trade. Not without a reason, the period from 1870 to 1913 was called the first globalization wave, which is characterized by the unprecedented growth of international trade, investments and migration. While in the year 1880 international trade comprised only 2 % of world GDP, it increased to 21 % of world GDP in 1913 (Allen 2013, p. 85). The second globalization wave, as it is well known, started after World War II. P. Krugman has shown that international trade does not lead to convergence (Krugman 1987). Even though the trade releases the forces of convergence, it also releases divergence forces. Which one of the two will take over, depends on many factors. Obviously, divergence forces dominated in the nineteenth century.

It is common knowledge that the accumulation of physical and human capital are key factors in defining the growth of production volume and income. Since accumulation of these factors influence the creation of economic stimuli, they take a central part in the theoretical and empirical analysis of economic growth in the twentieth century. In his classical Nobel Prize winning work, R. Solow formulates essential ideas concerning the impact of physical capital accumulation on the economy (Solow 1956). He established that according to the law of decreasing factor productivity, the growth rate of the national economy decreases over the time with growth of capital intensity (capital-labor ratio). It follows, that economic growth should be faster in countries with low capital capacity. On the other hand, it means, that physical capital is used more efficiently in countries where it is in deficit, as they are more attractive for investments.

Therefore, it seemed that under favorable conditions developing countries should enjoy faster accumulation of capital, causing corresponding growth of population and per capita income. Thus, if per capita income were defined only by the accumulation of physical capital and general rate of technical progress, as follows from neo-classical growth theory by R. Solow, the rich and poor countries would converge in per capita income. i.e., the convergence should have occurred due to the faster growth of poor countries. However this didn't happen in the 1960s. In response, he addressed human capital accumulation. R. Solow's advanced model of economic growth has revealed that human capital accumulation is accompanied by the decreasing limit of efficiency, regardless of whether it is taken into account or not (Helpman 2012), pp. 33–34). It has become clear that the model by Solow does not fully explain the convergence. It has also become evident that the accumulation of the factor by itself is insufficient for convergence. Productivity better explains the difference in per capita income and growth rates between different countries.

Productivity is a multifaceted concept. The main source of growth in labor productivity is technological advance. It can increase both labor productivity and capital productivity, as well as other factors. Technology advance can stimulate production output regardless of the composition of factors of production. Hicks. E. Helpman considers this type of technology advance to be neutral and suggests that factor intensity is the value of the conversion coefficient of natural resource units into efficient units (Helpman 2012, p. 39). Such treatment eliminates possible confusion in interpreting the intensity of different factors. Therefore, according to Helpman, the growth of intensity of a certain factor will mean the growth of the coefficient, which is convenient and we will stick to this approach throughout our study.

The neoclassical model by Solow (Solow 1956) became the basis for the evaluation of economic growth and productivity. A key idea is as follows: production growth can be expanded into components, which are explained by the growth production factors and residual growth rate, which cannot be explained by the growth of production factors. The first ones are equal to the share of a given factor in GDP, multiplied by a corresponding growth rate. The second ones, described by a residual term, represent total factor productivity (TFP). For instance, if we consider Solow's production function with technology advance (A) neutral by Hicks $Y = AK^{\alpha}L^{1-\alpha}$ (K—basis capital, L—labor input, α—capital share in GDP) by logarithmic differentiation we obtain the following formula for economic evaluation:

$$q_Y = \alpha q_K + (1-\alpha)\, q_L + q_A, \tag{1}$$

where $q_Y = \frac{\dot{Y}}{Y}$; $q_K = \frac{\dot{K}}{K}$; $q_A = \frac{\dot{L}}{L}$; $q_A = \frac{\dot{A}}{A}$. The q_A here is the residual term, defining TPF.

R. Solow was the first to calculate the TPF growth in the nonagricultural private sector of the U.S. in the first half of the twentieth century, and found out, that it comprises almost 80 % of the production growth rate. He, however, didn't take into account the quality improvement of production factors, so he overestimated. Solow's successor, U.S. economist Eduard Denison, completed a more accurate analysis and presented it in his seminal monograph (Denison 1971). The most accurate TPF estimations with an allowance for the quality improvement of capital and labor were obtained later by D. Jorgenson and E. Yip (Jorgenson and Yip 2001). It turned out that more than 30 % of growth rate is due to TPF growth, which is about 50 % of Japan's growth rate and more than 40 % of German and Italian growth rates. To this we can add that the factor which has the most impact on TPF is technology advance, as R. Solow believed from the beginning. Therefore, for the growth rate to increase, technology changes should accelerate over time, fast enough to overcome diminishing productivity due to accumulation of physical capital.

Underlining the extraordinary importance of TPF in economic development, E. Helpman writes: "There is convincing evidence that total factor productivity plays a major role in accounting for the observed cross-country variation in income per worker and patterns of economic growth." (Helpman 2012, p. 57). Indeed, P. Klenow and A. Rodriguez-Clare have confirmed in their paper (Klenow and Rodriguez-Clare 1997) that differences in TPF play an important role in explaining the income gap. They showed that TPF differences account for more than 60 % of income per worker. The share of TPF in the growth rate of per capita income turned out to be even higher—about 90 % of differences! Moreover, when other resources can't be increased, the only source of growth is TPF.

In summary, we can state that TPF defines productivity of people under given production factors. In other words, TPF is a uniform indicator of all factors used in the production process. Taking into account the special role that TPF plays in contemporary economic growth, we suggest the use the following simple production function for long-term modeling and forecasting of economic growth:

$$Y = \gamma\lambda AN, \quad \gamma = const, \tag{2}$$

where N is the population dynamics of country; $\lambda = \lambda(t)$ is the slow varying function defining the relation between population and the volume of employment (L) in economy, $L = \lambda N$. At certain time intervals, we can take average values $\lambda = \overline{\lambda}$. In the future we will assume $\lambda = \overline{\lambda} =$ **constant** and include it in the coefficient γ.

The production function (2) in rate representation will be as follows:

$$q_Y = q_A + q_N. \tag{3}$$

PF (3) is efficient in two ways. First, as pointed out above, the change of per capita income growth rate is mainly determined by TPF. Second, population dynamics of any country can be modeled accurately enough to forecast for a long period of time by means of adequate mathematical models (Akaev and Sadovnichij 2011).

The most simple and efficient model for the calculation of world population dynamics (N_w) is Kapitsa's model (Kapitsa 2008):

$$N_w = K_w^2 arcctg\left(\frac{T_w - t}{\tau_w}\right), \tag{4}$$

where K_w^2—is Kapitsa's number; T_w—is the year of demographic transition; $2\tau_w$—is the period of demographic transition. All these constant parameters are easily determined from actual data of world population during the industrial era (1820–2009) by the approximation of curve (4) by least squares fit. We obtained the following parameter values: $K_w^2 = 4063$; $T_w = 2003$; $\tau_w = 42$ years by the coefficient of determination equal to 0.997. World population dynamics is presented in graphic form in Fig. 5, which shows a strong relationship between model curve and factual data (Maddison 2010).

Kapitsa's formula is valid at least until year 2050, since it best describes demographic transition, which lasts $2\tau_w$ years in the interval from 1960 until approximately 2045. If one needs to look after year 2050, when world population becomes equal to or exceeding the capacity of the Earth biosphere, it is better to employ the demographic model with cusp and stabilization at the level determined by the capacity of the environment (Akaev and Sadovnichij 2010).

Kapitsa's model (4), as demonstrated by demographic calculations (Akaev et al. 2013a, b) gives a satisfactory description of the demographic dynamic of individual countries—developed and developing, large and small. The main condition is the ability of a state to provide long-term evolutionary sustainable growth. An important additional parameter is the absence of forced birth limitations and significant influence of external migration flow on the social and economic processes. Therefore, we will use Kapitsa's formula (4) to model forecast calculations of demographic dynamics, limiting the time frame of planning to the year 2050.

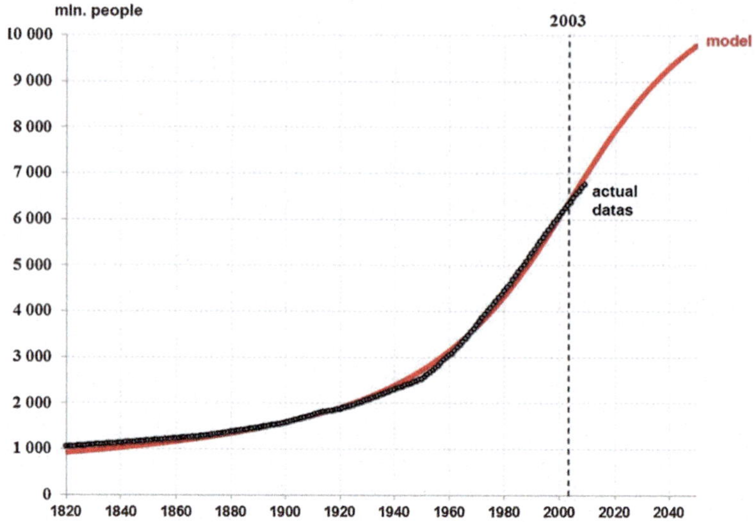

Fig. 5 World population dynamics in industrial era with forecast until year 2050. *Data source*: (Maddison 2010)

For individual countries or groups of countries, extrapolation by logistic function is more efficient:

$$N(t) = \frac{N_0\,(1 + n_m)}{1 + n_m \exp\left[-\vartheta\,(t - T_0)\right]}, \tag{5}$$

where N_0—is population in the initial moment of time T_0; n_m—describes maximum attainable population, $N_{max} = N_0\,(1 + n_m)$; ϑ—is a parameter describing population growth rate. In applied calculations, a shifted logistic function is sometimes more appropriate:

$$N(t) = \overline{N}_0 + \frac{N_0' \cdot (1 + n_m)}{1 + n_m \exp\left[-\vartheta\,(t - T_0)\right]}, \tag{6}$$

where $N_0 = \overline{N}_0 + N_0'$, $\overline{N}_0$—is the basis component of initial population N_0.

Using Kapitsa's formula (4) we have calculated total population dynamics for high-income countries (Fig. 6), middle-income countries (Fig. 7) and low-income countries (Fig. 8). As the figures clearly show, the demographic dynamics of middle-income (MI), and, especially, low-income countries (LI) is approximated by Kapitsa's model (4) much better, than for the high-income (HI) countries. This is due to intense migration to the countries of the first group (USA, Canada, Australia and New Zealand) during the nineteenth and twentieth centuries. Approximation by logistic function (5) yields much better approximation, shown in Fig. 9 (compare it to Fig. 6). Likewise, population dynamics of the BRICS countries is better

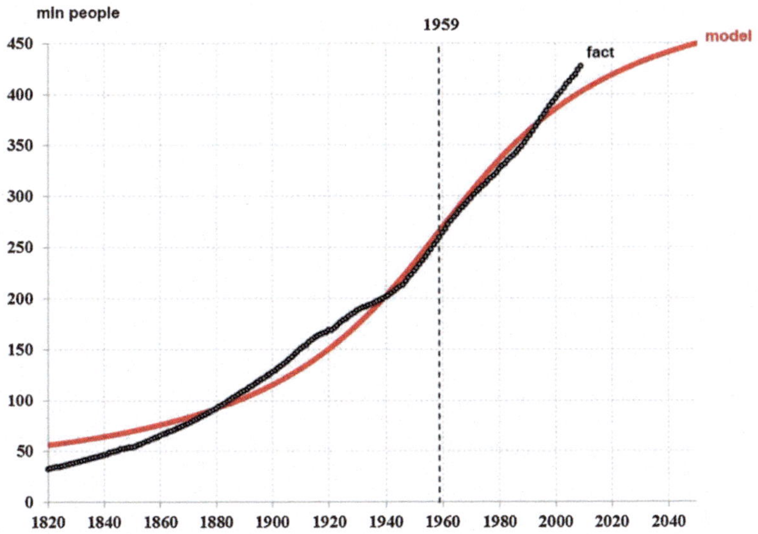

Fig. 6 Population dynamics for high-income countries with approximation by Kapitsa's model. $K_{HI}^2 = 169$, $T_{HI} = 1959$, $\tau_{HI} = 48$ *years*, coefficient of determination is equal to 0.978. *Data source*: (Maddison 2010)

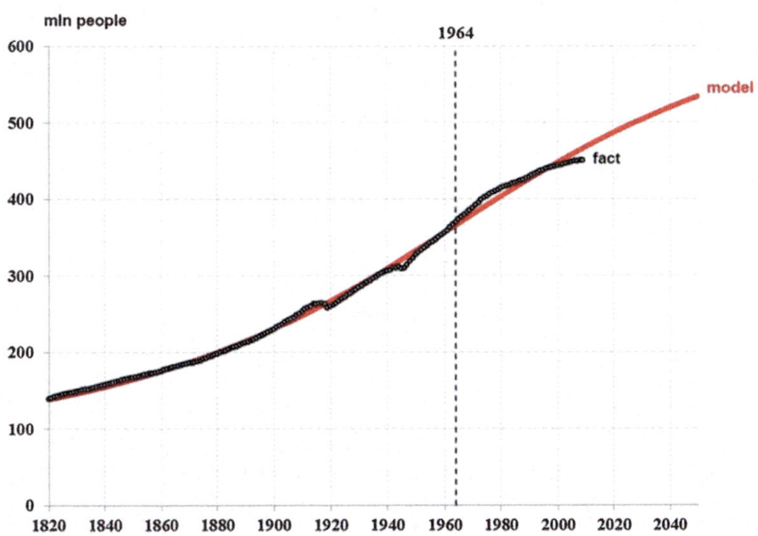

Fig. 7 Population dynamics for middle-income countries. $K_{MI}^2 = 233$, $T_{MI} = 1964$, $\tau_{MI} = 98$ *years*, coefficient of determination is equal to 0.997. *Data source*: (Maddison 2010)

approximated by logistic function, as Fig. 10 shows. It should be pointed out that formula (6) was used for better approximation. From the analysis of Figs. 6, 7 and 8 it is evident that demographic transition occurred in developed countries of the West during 1959–1964, about 40 years ahead of the rest of the world (2002 in Fig. 8).

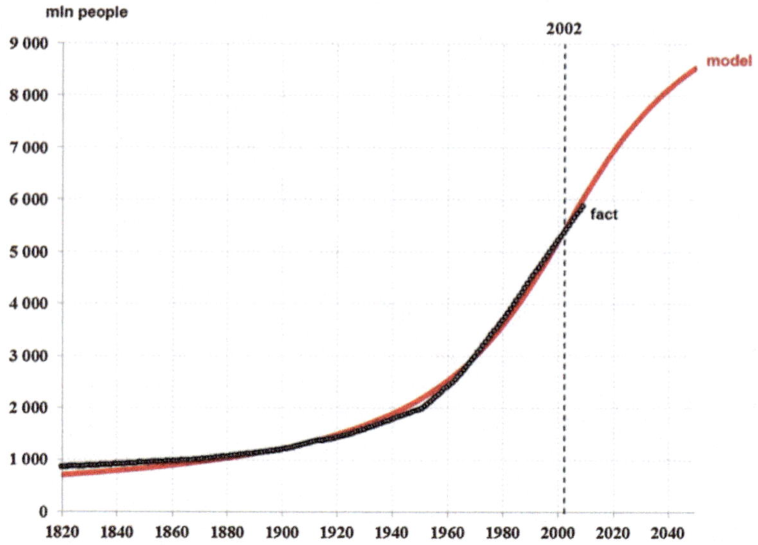

Fig. 8 Population dynamics for low-income countries. $K_{LI}^2 = 3455$, $T_{LI} = 2002$, $\tau_{MI} = 38$ *years*, coefficient of determination is equal to 0.995. *Data source*: (Maddison 2010)

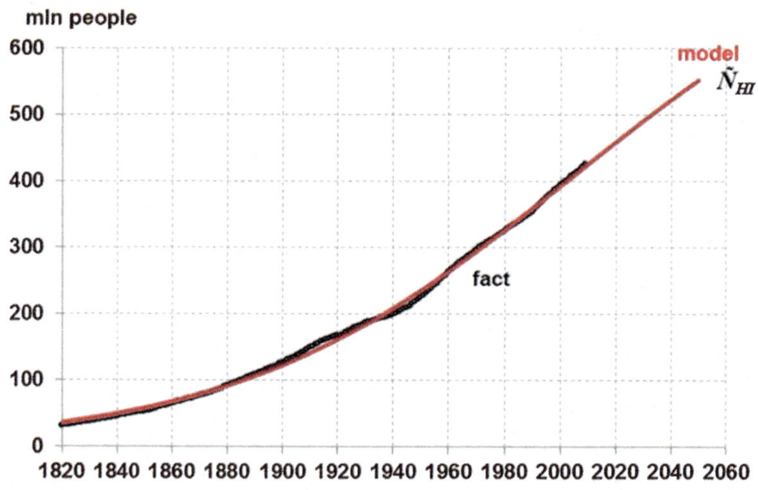

Fig. 9 Population dynamics for high-income countries with approximation by logistic function. $N_0 = 36, 6$; $n_m = 21$; $\vartheta_{HI} = 0.017$, coefficient of determination is equal to 0.992. *Data source*: (Maddison 2010)

The forecasting horizon accepted above coincides with another long wave of economic development (2018–2050s) or 6th BKC (*Big Kondratiev Cycle*), which has a known cluster of basis innovation technologies, with its core made by NBIC—convergent technologies. The effect of NBIC technologies on the rate of economic growth in the developed countries and world economy during the 6th BKC can

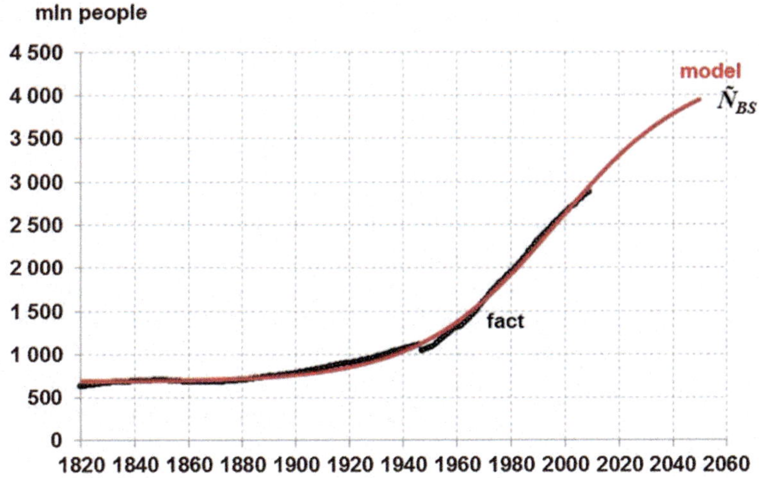

Fig. 10 Population dynamics of the group of BRICS (BS) countries with approximation by logistic function. $\overline{N}_0 = 689$ $(N_0' = 3,2)$; $n_m = 1143$; $\vartheta = 0,04$, coefficient of determination is equal to 0.994. *Data source*: (Maddison 2010)

be estimated quite accurately. Therefore, we can count on a reliable forecast of economic growth until the middle of the twenty-first century. To glimpse further into the future will be quite difficult as we do not know anything about basis technologies of 7th BKC.

To calculate GDP using the AN model (2) an appropriate mathematical model must be selected to describe TPF (A). For the industrial era of the nineteenth to twentieth centuries with domination of general-purpose technologies (GPT), the most appropriate is a basis model by Kremer (Kremer 1993):

$$q_A = \frac{dA}{Adt} = в\frac{dN}{Ndt} = вq_N, \qquad в = const. \tag{7}$$

The coefficient $в$ in the equation is mainly determined by the rate of population literacy—the higher the literacy rate, the higher the value of the coefficient.

The demographic transition started in the West in the early nineteenth century, 40–50 years ahead of the rest of the world. Naturally, the West's population growth rate in the nineteenth century was much higher than in the rest of the world (see Fig. 4), and it facilitated the TPF rate, as follows from Eq. (7). Similar high population growth rate occurred in developing countries almost a century later—by the middle of the twentieth century. As was pointed out above, the rate of literacy in European countries exceeded identical indicators in the rest of the world five to sixfold. Which means the coefficient $в$ in Eq. (7) was 5–6 times higher in developed countries than developing ones. Not surprisingly, the TPF in developed countries in the nineteenth century was much higher than in the rest of the world, accelerating the divergence process.

Let us consider Kremer equation (7) for the group of developed countries (HI), on the condition that population dynamics (N) is described by logistic function (5):

$$q_{A_{HI}}^{(1)} = \frac{dA_{HI}^{(1)}}{A_{HI}^{(1)}dt} = k_{HI}^{(1)} q_{N_{HI}} = \frac{k_{HI}^{(1)} \cdot n_m \cdot \vartheta_{HI} \cdot \exp\left[-\vartheta_{HI}\left(t - T_0\right)\right]}{1 + n_m \exp\left[-\vartheta_{HI}\left(t - T_0\right)\right]}. \tag{8}$$

In this equation, we introduced calibration coefficient $k_{HI}^{(1)}$ identical to the proportionality coefficient in Eq. (7). We also introduced upper index (1), denoting that given formula (8) is valid for the first stage of the industrial era, from year 1820 until 1960. The first group of high-income (HI) developed countries includes USA, Great Britain, Canada, Australia and New Zealand.

By integration of the differential equation (8) with initial value $A_{HI0} = A_{HI}(t = T_0)$ in year 1820, $(T_0 = 1820)$ we obtain the basis growth rate of technology advance:

$$\overline{A}_{HI}^{(1)} = \frac{A_{HI}^{(1)}}{A_{HI0}} = \exp\left\{k_{HI}^{(1)} \int_{T_0}^{t} q_{N_{HI}}(\tau)\, d\tau\right\} =$$

$$= \exp\left\{k_{HI}^{(1)} \cdot \ln\frac{1 + n_m}{1 + n_m \cdot \exp\left[-\vartheta_{HI}\left(t - T_0\right)\right]}\right\} = \left\{\frac{1 + n_m}{1 + n_m \cdot \exp\left[-\vartheta_{HI}\left(t - T_0\right)\right]}\right\}^{k_{HI}^{(1)}}. \tag{9}$$

The only unknown coefficient in the formula is k_{HI}, which can be easily obtained by the least squares fit of data $A_{HI}(t)$ in the interval $t \notin [T_0, T_M]$ by the function (9). Assuming $T_M = 1950$, we obtained $k_{HI}^{(1)} = 0.92$. If the estimates of TPF (A_{HI}) are missing, we can roughly assume $A_{HI} \cong g_{HI}$, where g_{HI} is per capita income. The calculation results of $\bar{A}_{HI}$ from Eq. (9) are presented in Fig. 11. Clearly, the Kremer model (8)–(9) provides a good approximation until the middle of the twentieth century, just as we suggested. We call it the first iteration (1) (see Fig. 11). After 1960, it separates from actual data which go further and further above, which means, that new sources of acceleration of TPF appeared in this time period. The main reason for this is well known today as the Research & Development (R&D) system. Taking R&D into account, we obtain the second iteration, which provides good agreement in modeling, presented in Fig. 11 (2). We will discuss it later.

Let us now get back to consideration of other groups of countries. For these countries, the TPF dynamics is described in an identical way, with the help of Eq. (7), just like for the high-income countries. However, as shown above, population dynamics for this group is described by Kapitsa's (4) formula. Therefore, for example, the Kremer equation (7) for the group of countries with medium per capita income, including Austria, Belgium, Denmark, France, Finland, Germany, Italy, the Netherlands, Norway, Sweden, Switzerland, Spain, Portugal and Japan is written as

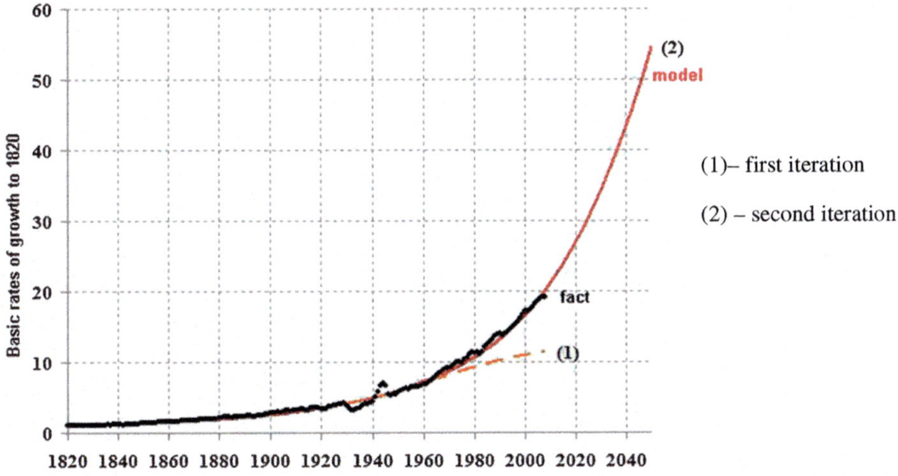

Fig. 11 Dynamics of technology advance in high-income countries. *Data source*: (Maddison 2010)

follows:

$$q_{A_{MI}}^{(1)} = k_{MI}^{(1)} \cdot q_{N_{MI}} = k_{MI}^{(1)} \cdot \left\{ \tau_{MI} \cdot \left[1 + \left(\frac{T_{MI} - t}{\tau_{MI}} \right)^2 \right] \cdot arcctg \left(\frac{T_{MI} - t}{\tau_{MI}} \right) \right\}^{-1}. \tag{10}$$

By integration of a given differential equation, we obtain the function describing the dynamics of technology advance.

Since $q_{A_{MI}}^{(1)} = \frac{dA_{MI}^{(1)}}{A_{MI}^{(1)}dt}$, then $A_{MI}^{(1)} = A_{MI0} \cdot \exp \left\{ \int\limits_{T_0}^{t} q_{A_{MI}}^{(1)}(\tau)\, d\tau \right\}$. Therefore,

$$A_{MI}^{(1)} = A_{MI0} \cdot \exp \left\{ k_{MI}^{(1)} \int\limits_{T_0}^{t} q_{N_{MI}}(\tau)\, d\tau \right\} = A_{MI0} \cdot \exp \left\{ k_{MI}^{(1)} \int\limits_{T_0}^{t} \frac{dN_{MI}}{N_{MI} d\tau} d\tau \right\} =$$

$$= A_{MI0} \cdot \exp \left\{ k_{MI}^{(1)} \cdot \int\limits_{N_{MI0}}^{N_{MI}(t)} \frac{1}{N_{MI}} dN_{MI} \right\} = A_{MI0} \cdot \exp \left\{ k_{MI}^{(1)} \cdot \ln \frac{N_{MI}(t)}{N_{MI0}} \right\} = A_{MI0} \left[\frac{N_{MI}(t)}{N_{MI0}} \right]^{k_{MI}^{(1)}}.$$

Thus, taking into account that $N_{MI}(t) = K_{MI}^2 \cdot arcctg \left(\frac{T_{MI} - t}{\tau_{MI}} \right)$ we obtain:

$$A_{MI}^{(1)}(t) = A_{MI0} \cdot \left[\frac{arcctg \left(\frac{T_{MI} - t}{\tau_{MI}} \right)}{arcctg \left(\frac{T_{MI} - T_0}{\tau_{MI}} \right)} \right]^{k_{MI}^{(1)}}. \tag{11}$$

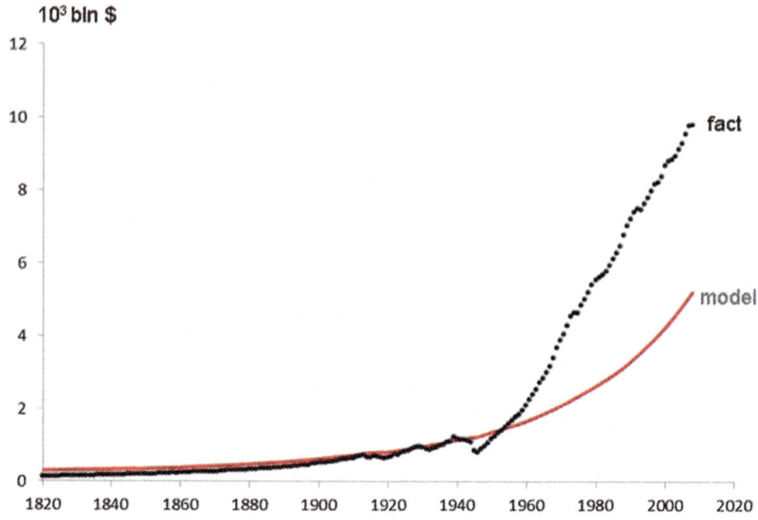

Fig. 12 GDP dynamics of middle-income countries. *Data source*: (Maddison 2010)

In order to find the indefinite coefficient $k_{MI}^{(1)}$ we do not need to know actual values of A_{MI}. Functional expression (11) can be substituted in the AN-model (2) to simultaneously determine two calibration coefficients γ_{MI} and $k_{MI}^{(1)}$ from actual GDP data (Y_{MI}), widely available in reliable databases. By making this, we obtain:

$$Y_{MI}^{(1)}(t) = \gamma_{MI}^{*} \cdot arcctg\left(\frac{T_{MI} - t}{\tau_{MI}}\right) \cdot \left[\frac{arcctg\left(\frac{T_{MI} - t}{\tau_{MI}}\right)}{arcctg\left(\frac{T_{MI} - T_0}{\tau_{MI}}\right)}\right]^{k_{MI}^{(1)}}, \tag{12}$$

where $\gamma_{MI}^{*} = \gamma_{MI} \cdot \overline{\lambda} \cdot A_{MI0} \cdot K_{MI}^{2}$. The best fit estimates of coefficients γ_{MI} and $k_{MI}^{(1)}$ yielded: $\gamma_{MI} = 1.01$; $k_{MI}^{(1)} = 1.59$. Calculation results after formula (12) are presented in Fig. 12. Similar calculations were performed for the group of countries with low per capita income (LI), which are presented in Fig. 13. From the analysis of the figures below, it can be concluded that the Kremer model provides a good approximation of the TPF dynamics, and, therefore, of the GDP dynamics until 1950.

The explosive growth of Western share in World GDP, which had started after 1820, began to slow down by the end of the nineteenth century when the law of decreasing limit of efficiency was in effect, as Fig. 1 shows. The intense accumulation of physical capital in the era of industrial revolution in the nineteenth century resulted in long-term sustainable economic growth, accompanied by a sustainable growth of population per capita income due to outpacing productivity growth. However, with a growing capital-labor ratio, increasing capital stock started to have a diminishing effect on production growth; therefore the stimuli for further

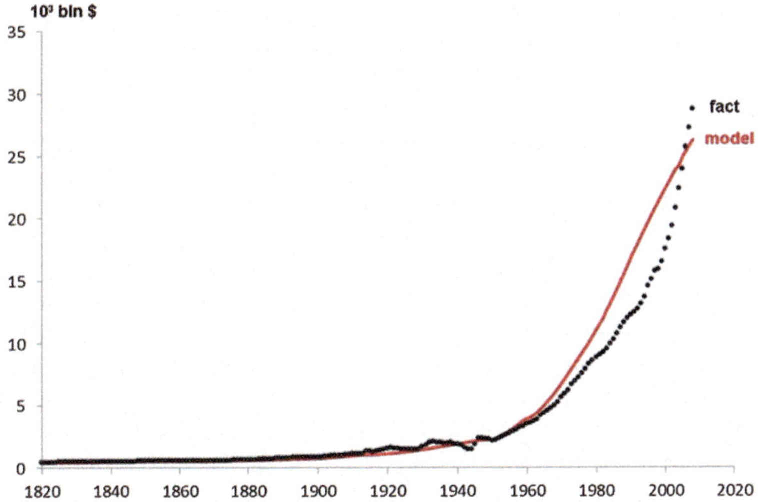

Fig. 13 GDP dynamics of low-income countries. *Data source*: (Maddison 2010)

accumulation began to weaken. Thus, economic growth was limited by the diminishing degree of returns from capital formation. Therefore, average GDP growth of the most developed big countries of the West in 1914–1950 comprised as little as 1.7 %, while in the preceding period of 1870–1913 it reached 2.2 % (Mel'yantsev 1996, p. 152). Per capita income growth was as follows: in 1870–1913—1.2 %; in 1914–1950—1.1 % (Mel'yantsev 1996, p. 152). At the same time, average per capita income growth in China and India in 1870–1933 comprised 0.5–0.6 % (Mel'yantsev 1996, p. 136), i.e. two times lower than in Western countries. There were, of course, some exceptions. For example, Japan achieved great success in the late nineteenth to early twentieth due to open international trade and successful exploration of foreign technologies that provided rapid growth to its economy after the Meiji Revolution. In 1820 Japan was a poor country, but only 100 years later it closed the income gap with the West. Therefore, despite a decrease in the growth rate of developed countries of the West, the divergence went on until 1950, though at a slower rate (see Fig. 1).

3 Post-War Golden Era of World Economic Development. The Origins of Convergence Between the Developing and Developed Countries

A period of unprecedented dynamic world economic growth, which started after WWII and continued for a quarter of a century from 1948 to 1973, had no counterparts in any other periods of human civilization. Outstanding historian of world economy Angus Maddison writes: “The time span from 1950 to 1973 was the “*Golden Era*” of unprecedented flourishing. World per capita GDP grew by about 3 % annually, world GDP – by 5 %, and export – by 8 %. In all world regions, economic results were better than ever before. A distinctive feature of this period was a significant degree of convergence in per capita income and productivity (Maddison 2010, p. 115)”. This was of course due to the new international situation based on global security and cooperation created by the winner countries as well as removal of barriers for international trade and creation of the UN and its economic institutions (IMF, WB, GATT etc.) to facilitate world economic development.

Most significant, however, were epoch-making achievements implemented into efficient active technologies during WWII. Following is an incomplete list of such innovations: creation of fast computers; exploration of nuclear energy; creation of jet and rocket engines which connected continents and allowed breakthrough into space; discovery of coherent light sources—lasers and creation of laser technologies etc. The catalyst for this was a sharp decrease in the time it took to market scientific and technical innovations due to the creation of the modern R&D system which serves the needs of commercialization of science and technology achievements. Investments into R&D today are manifold higher than into machinery and equipment. Between 40 and 66 % of TPF growth is due to R&D development (Helpman 2012, p. 74). Every time investments increase TPF, a higher TPF level causes accelerated capital accumulation. As a result, R&D has a direct and indirect effect on output, with indirect impact being quite significant. Therefore, countries that do not invest into R&D stagnate. The most developed economies require a higher volume of R&D, targeted at maintenance of technologic leadership, creation of innovation products and support of high quality products. It is not surprising, that more than 95 % of R&D is carried out in industry-developed countries. D. Co and A. Helpman have shown that capital invested into national R&D systems in U.S., Germany and Great Britain exceeded 20 % of GDP in 1990, while in Japan and France it was at the 17 % level, and in Italy and Canada—8 % (Coe and Helpman 1995). In most developed countries, this indicator does not exceed 1–2 %.

This raises a natural question: whether R&D is necessary for poor countries? E. Helpman, who a major contributor to the topic, believes investments into R&D, targeted at improvement of foreign technologies can be just as productive in the lagging countries as invention activity in vanguard developed countries (Helpman 2012, p. 129). Besides, the higher the investments of lagging countries into R&D, the quicker the gap diminishes between the backward and vanguard countries. Thus, R&D helps backward countries catch up to technological leaders. When useful knowledge emanating from R&D diffuses to foreign countries at the same speed as it diffuses within the domestic economy, then these knowledge flows provide a potent force of convergence in the world economy. If, however, the international flows are slow relative to the domestic flows, then these knowledge flows provide a potent force of divergence (Helpman 2012, p. 124). It was exactly the case in the nineteenth and twentieth centuries, until WWII.

Thus, R&D forms the basis for contemporary TPF growth, with a direct and indirect impact on final output. Nowadays R&D accounts for up to 90 % of the growth rate. R&D is essentially a modern industry that generates innovative technologies and products. The first R&D model for the calculation of growth rate was suggested by P. Romer (Romer 1990) and upgraded by C. Jones (Jones 1995). These models however, have a scale effect, which of course was not actually observed. A model without scale effect was suggested by authors and co-workers (Akaev et al. 2013a, b), from which it follows:

$$q_A^o = \frac{dA}{Adt} = a\{l_A^2(3l_M - 2l_A) - l_{Ab}^2(3l_M - 2l_{Ab})\}, \tag{13}$$

where q_A^0—is the TPF growth rate, fueled by innovative technologies created in the national R&D system.; $l_A = \frac{L_A}{L}$ is the ratio of scientists, engineers and support staff working in R&D (L_A), to the total number of employees in the economy (L); l_M—is the maximum value of employed in R&D in the saturation mode; l_{AB}—is the initial share of employed in R&D at $t = T_b$; a is a normalizing coefficient. The main variable l_A in Eq. (13) is described by the logistic function (Akaev et al. 2013a, b, pp. 291–292):

$$l_A(t - T_b) = \frac{l_{Ab}(1 + l_m)}{1 + l_m \exp[-\vartheta(t - T_b)]}, \quad l_{Ab}(1 + l_m) = l_M \tag{14}$$

Here l_m, ϑ—are constant parameters, determined by the least squares from fitting actual data l_A; T_b—is the initial moment of time, where $L_A(t = T_b) = L_{Ab}$. . These constants are easily determined, as actual data employees in R&D is readily available (see for example UN data: http://data.un.org/Data.aspx?d=UNESKO&f=series%3aST2001).

3.1 Economic Growth of High-Income Countries

Let us compute the second iteration of TPF for most developed countries with high per capita income, with an allowance for R&D contribution. We have calculated above the first iteration (8)–(9), defined by the Kremer model. Therefore, the second iteration can be determined from the following equation:

$$q_{A_{HI}}^{(2)} = q_{A_{HI}}^{(1)} + q_{A_{HI}}^{(0)} =$$
$$= \frac{k_{HI}^{(1)} \cdot n_m \cdot \vartheta_{HI} \cdot \exp[-\vartheta_{HI}(t - T_0)]}{1 + n_m \cdot \exp[-\vartheta_{HI}(t - T_0)]} + k_{HI}^{(2)}[l_{A_{HI}}^2(3l_{A_{HI}}^{\max} - 2l_{A_{HI}}) - l_{Ab_{HI}}^2(3l_{A_{HI}}^{\max} - 2l_{Ab_{HI}})]. \tag{15}$$

Solving differential equation, we obtain:

$$A_{HI}^{(2)} = \frac{A_{HI}^{(2)}}{A_{HI0}} = \exp\left\{ k_{HI}^{(1)} \cdot \ln \frac{1+n_m}{1+n_m \cdot \exp[-\vartheta_{HI}(t-T_0)]} + \right.$$
$$\left. + k_{HI}^{(2)} \int_{T_0}^{t} [l_{A_{HI}}^2 (3l_{A_{HI}}^{\max} - 2l_{A_{HI}}) - l_{A_{HIв}}^2 (3l_{A_{HI}}^{\max} - 2l_{A_{HIв}})] d\tau \right\}. \quad (16)$$

In this equation we have only unknown coefficient $k_{HI}^{(2)}$, which is determined like $k_{HI}^{(1)}$ in the previous stage, by least squares fit. For $k_{HI}^{(1)}$ we obtained the following estimate: $k_{HI}^{(1)} \cong 1,252$.

However, we should first determine the functional expression for l_A (14) which is equivalent to the determination of parameters $l_{Aв_{HI}}$, l_m and $T_в$, from actual data for $l_{A_{HI}}$. As can be seen in Fig. 14, the reliable evidence for l_A is available only from 1987 to 2008. From this data we obtain the best parameter fit (in terms of deviation square minimum): $l_{M_{HI}} = l_{A_{HI}}^{\max} = 0.008$; $l_m = 7500$; $\vartheta_{HI} = 0.055$; $T_в = 1930$; determination coefficient is equal to 0.793. The curve of TPF growth rate, which is exclusively impacted by national R&D, is presented in Fig. 15. It was calculated by Eq. (13) at optimal value of coefficient $k_{HI}^{(2)} = 41097$ found by the least squares fit of data for A_{HI} over time interval 1950–2013 (see Fig. 11). From Fig. 15 follows, the R&D system in developed countries has had a substantial impact since 1950–1960s, when it increased their economies' growth rate roughly to 0.5–1.2 % annually. In the 2020–2040s this indicator will increase to 2.2 % annually. As can be seen, the

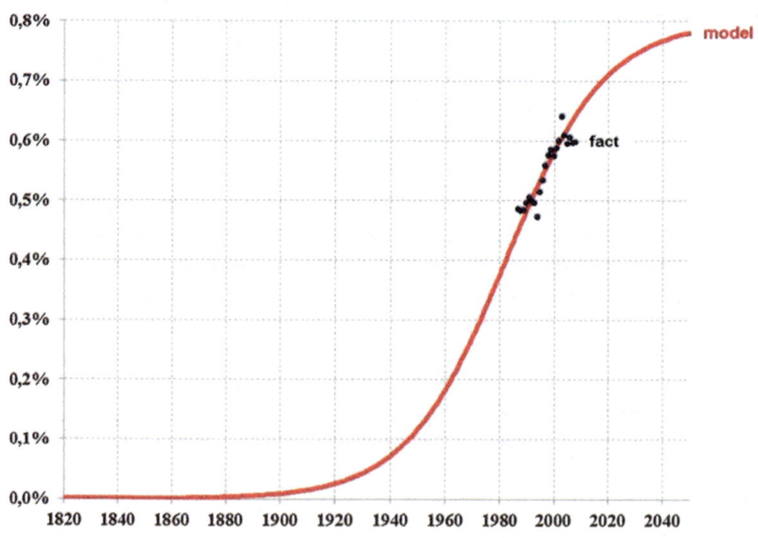

Fig. 14 The percentage of people employed as researchers in high-income countries. *Data source*: http://www.oecd.org/innovation/inno/researchanddevelopmentstatisticsrds.htm

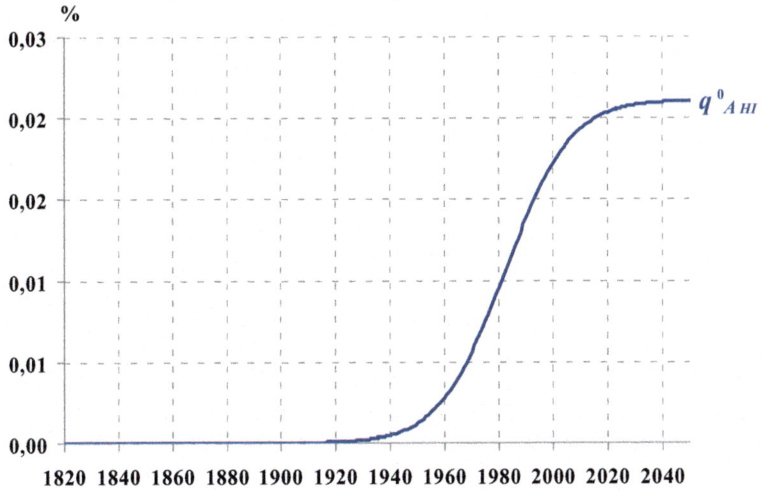

Fig. 15 Technology advance growth rate due to home technologies in high-income countries. *Data source*: http://www.oecd.org/innovation/inno/researchanddevelopmentstatisticsrds.htm

second iteration of $A_{HI}^{(2)}$ presented in Fig. 11 as iteration (2), describes very well actual dynamics of A_{HI} over the industrial era (1820–2013). It can also be used as the basis for the TPF dynamics forecast until year 2050.

The R&D impact on technology advance (TPF) growth rate can be seen in Fig. 16, where the plot of the function $q_{A_{HI}}^{(2)}$ (15) is presented. Traditional ways of promoting technological advance used in the nineteenth and early twentieth century had peaked by year 1940, followed by a declining TPF growth rate. By this time a new R&D system had matured as a result of accelerated technological advance and further increased the TPF growth rate in the 1960s–2000s (see Fig. 18). Figure 18 makes it clear that R&D will soon reach its potential. Therefore, we will need new ways of enhancing the TPF growth to advance technology in order to meet the demands of the twenty-first century.

With TPF calculation results for high per capita income countries $A_{HI}^{(2)}$ (16), and total population dynamics N_{HI} computed after formula (5), presented graphically in Figs. 9 and 11, total GDP growth rate with PF of the type "γAN" (2) can be easily calculated. The corresponding GDP trajectory for the countries with high per capita income is presented in Fig. 17.

The GDP trajectory $Y_{HI}^{(2)}$ was obtained by the optimum value $\gamma_{HI} = 1.556$ (determination coefficient is equal to 0.998). It is evident that the second iteration model trajectory fits well with actual data throughout the industrial era and can also serve as the basis for the forecast until year 2050. Thus, a record high growth rate for most developed economies in the post-war period had two driving sources: epoch-making innovations generated by the technology advance of the twentieth century and creation of modern R&D systems.

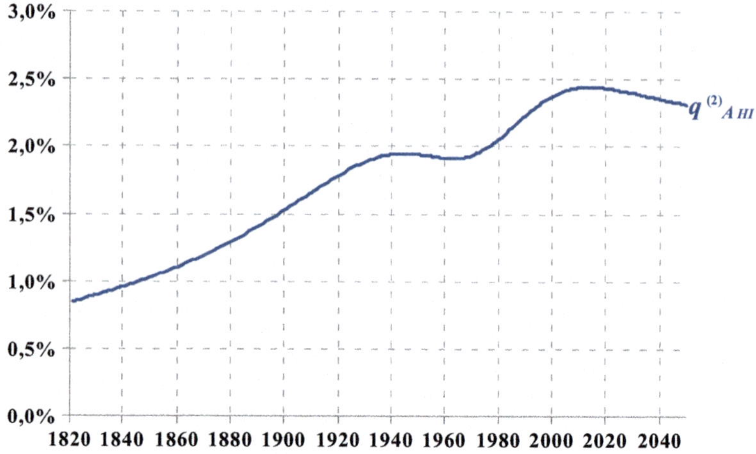

Fig. 16 The growth rate of technology advance in high-income countries

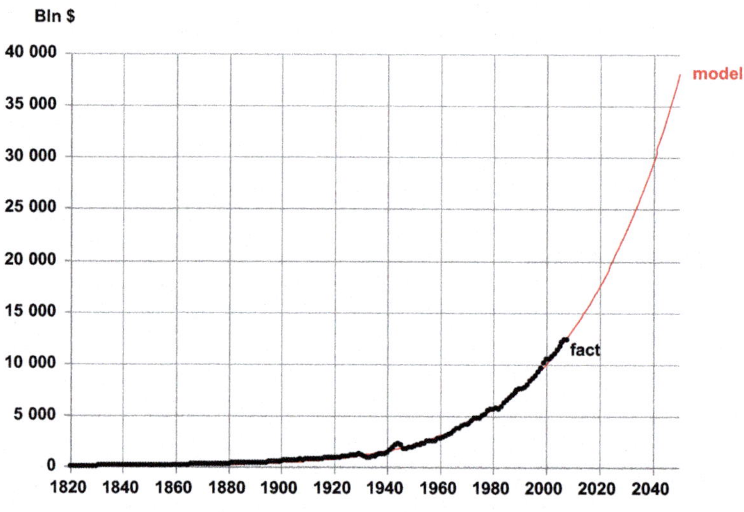

Fig. 17 GDP dynamics for high-income countries (in comparable price in 1990). *Data source*: (Maddison 2010)

3.2 *Economic Growth of Middle-Income Countries*

Let us begin to consider economic development of developed countries with middle level per capita income (*MI*). We address the dynamics of technology advance (TPF) in this group of countries. The first iteration of TPF growth rate was built after Kremer's model in Sect. 2, formula (10). We have also obtained analytical expressions, describing the dynamics of technology advance, $A_{MI}^{(1)}$ (11) and GDP

$Y_{MI}^{(1)}$ (12). Actual dynamics of technology advance for middle-income countries is presented in Fig. 18. As the chart illustrates, the first iteration (1), described by expression (11) gives a good approximation until year 1950, as expected.

We construct the second iteration of TPF growth rate of middle-income economies by adding an additional term, to reflect the national R&D system (12) to the first iteration (10):

$$q_{A_{MI}}^{(2)} = q_{A_{MI}}^{(1)} + q_{A_{MI}}^{(0)} =$$

$$= k_{MI}^{(1)} \left\{ \tau_{MI} \cdot \left[1 + \left(\frac{T_{MI} - t}{\tau_{MI}} \right)^2 \right] \cdot arcctg \left(\frac{T_{MI} - t}{\tau_{MI}} \right) \right\}^{-1} + k_{MI}^{(2)} [l_{A_{MI}}^2 (3 l_{A_{MI}}^{max} - 2 l_{A_{MI}}) - l_{A_{MI0}}^2 (3 l_{A_{MI}}^{max} - 2 l_{A_{MI0}})]. \tag{17}$$

In this equation we need to define a functional expression for $l_{A_{MI}}$ which is described by logistic function (14). Differential equation (17), because of the second term, can no longer be analytically integrated, so we take numerical integration and presented it in graphic form (2) in Fig. 18. Just as expected, the second iteration only slight improves from the first one, due to the small contribution of home innovations.

Above, with reference to E. Helpman, we have pointed out that more than 50 % of output growth in member countries of the Organization for Economic Cooperation and Development (OECD) occurred due to ideas and innovations which had come from abroad. In most OECD countries other than technological leaders (USA, Japan, Germany, France and Great Britain), more than 90 % of TPF growth occurred due to external diffusion of ideas and innovation (Helpman 2012, p. 128). Therefore, the third iteration of TPF growth rate is constructed by adding an additional rate (to reflect the borrowing of innovation technologies that are already in use in

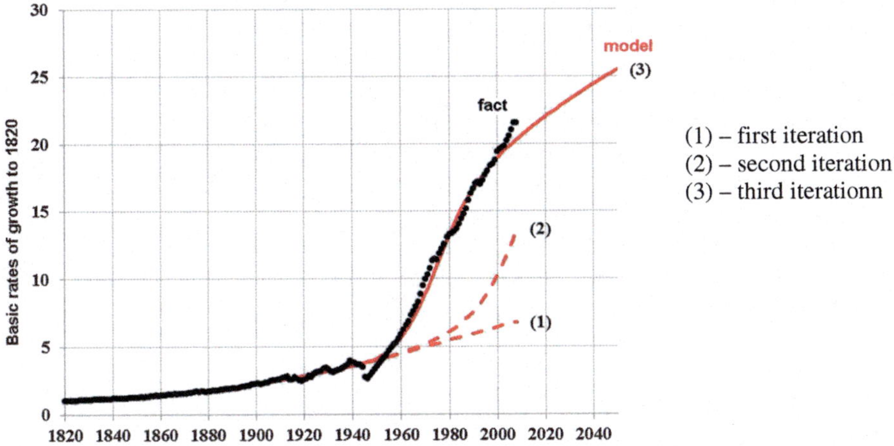

Fig. 18 Technology advance growth rate for middle-income countries. *Data source*: (Maddison 2010)

technologically leading countries) to the second iteration (17). As pointed out above, ideas and technologies obtained from R&D diffuse in OECD countries without particular barriers and lags.

Therefore, additional rates, due to borrowing technologies from vanguard high-income countries, can be written in the following form:

$$q_{A_{MI}}^{(e)}(t) = \frac{s_{MI}\xi_{MI}}{s_{HI}\xi_{HI}} \cdot q_{A_{HI}}^{o}(t) \cdot \frac{[1+1(t-T_{MI\,m})]\cdot \exp[-\vartheta_{MI}^{(e)} \cdot (t-T_{MI\,m})]}{1+\exp[-\vartheta_{MI}^{(e)} \cdot (t-T_{MI\,m})]}. \tag{18}$$

where $q_{A_{HI}}^{(0)}$—is the TPF growth rate in high-income countries due to R&D; $s_{MI}(s_{HI})$—is the rate of saving in middle-income (high-income) countries; $\xi_{MI}(\xi_{HI})$—is the capital productivity in middle-income (high-income) countries. According to V.A. Mel'yantsev data, the rates of saving and capital productivity in high and middle-income developed countries were approximately the same: $s_{MI} \cong s_{HI} \cong 24\%$; $\xi_{MI} \cong \xi_{MI} \cong 0.23$ (Mel'yantsev 2009, p. 17). In the following decades of the twentieth century the situation was unchanged, though the savings rate went down to 21 %, and capital productivity, in contrast, increased. Therefore, the product: $\frac{s_{MI}\cdot\xi_{MI}}{s_{MI}\cdot\xi_{HI}} \approx 1$. The term in the form of a logistic function designates the process of closing down the borrowings, starting from T_{MIm}, when the time of dominant use of own innovation technologies begins. Therefore, there is function $[1 + 1\,(t - T_{MI\,m})]$, equal to 2 at $t \geq T_{MI\,m}$, and equal to 1 at $t < T_{MI\,m}$, where $1\,(t - T_{MI\,m})$—is a unit function, before the multiplier. As the trajectory shows (2), described by the second iteration (see Fig. 18) $T_{MI\,m}$ coincide starting from year 1980. The parameter $\vartheta_{MI}^{(e)}$ is chosen from the condition that the process of closing down technological borrowings from outside lasts one Kondratiev cycle, i.e. about 30 years. That means $\vartheta_{MI}^{(e)} = \frac{\ln 19}{30} \cong 0.098$.

Thus, the third iteration of TPF growth rate for middle-income countries can be written down in final form by summing up $q_{A_{MI}}^{(2)}$ (17) and $q_{A_{MI}}^{(e)}$ (18):

$$q_{A_{MI}}^{(3)} = \frac{dA_{MI}^{(3)}}{A_{MI}^{(3)}dt} = k_{MI}^{(1)} \cdot \left\{ \tau_{MI} \cdot \left[1 + \left(\frac{T_{MI}-t}{\tau_{MI}} \right)^2 \right] \cdot arcctg\left(\frac{T_{MI}-t}{\tau_{MI}} \right) \right\}^{-1} +$$

$$+ k_{MI}^{(2)} \cdot [l_{A_{MI}}^2 (3l_{A_{MI}}^{\max} - 2l_{A_{MI}}) - l_{A_{MIe}}^2 (3l_{A_{MI}}^{\max} - 2l_{A_{MIe}})] +$$

$$+ k_{MI}^{(3)} \cdot q_{A_{HI}}^{(0)}(t) \cdot \frac{[1+1(t-T_{MI\,m})]\cdot \exp[-\vartheta_{MI}^{(e)} (t-T_{MI\,m})}{1+\exp[-\vartheta_{MI}^{(e)} (t-T_{MI\,m})]}, \tag{19}$$

where $k_{MI}^{(3)}$ is a normalization coefficient.

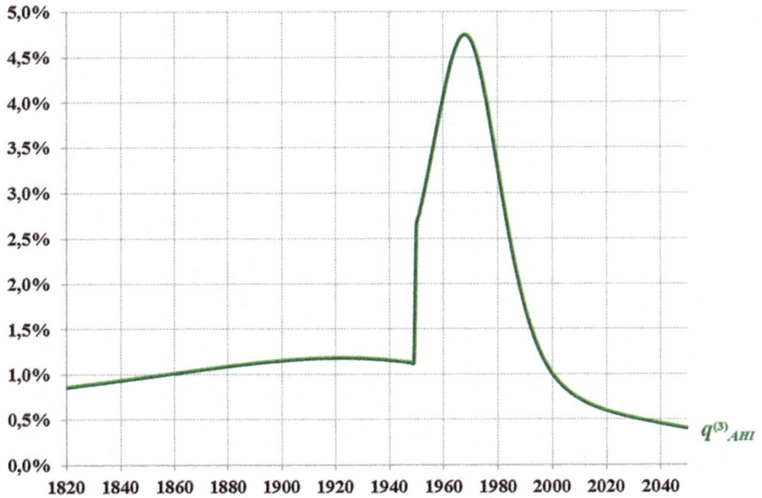

Fig. 19 Technology advance growth rate in middle-income countries

Solving the given differential equation by numerical integration according to the formula $\overline{A}_{MI}^{(3)} = \frac{A_{MI}^{(3)}}{A_{MI0}^{(3)}} = \exp\left\{\int_{T_0}^{t} q_{A_{MI}}^{(3)}(\tau)\,d\tau\right\}$, we obtain curve (3) in Fig. 18, which appropriately reflects actual dynamics of technology advance throughout the Industrial Age. The R&D rates for middle-income countries calculated after expression (19) are also graphically presented in Fig. 19. Figure 19 shows that right after the end of World War II, a group of middle—income countries had demonstrated an explosive growth rate from 1948 to the 1980s, peaking at 4–5 % in the 1960s. Indeed, Japan's contribution to the group is particularly large. This is the result of a significant narrowing of the gap in per capita income. i.e. convergence that had occurred in the group of developed countries. As a whole, the group of developed countries saw a threefold increase in GDP growth—from 1.7 % in the period 1913–1950 to 5.4 % in 1950–1973 and a fourfold increase in per capita income during post-war decades—from 1.1 to 4.4 % (Mel'yantsev 2009, p. 17).

The dynamics for total GPD growth in the countries discussed can now be easily calculated using the AN-model $Y_{MI}^{(3)} = \gamma_{MI} A_{MI}^{(3)} N_{MI}$ by multiplying $A_{MI}^{(3)}$ by N_{MI}, since calculation results for N_{MI} (Fig. 7) and $A_{MI}^{(3)}$ (Fig. 18) are available. Coefficient γ_{MI} is obtained from the least squares fit from GDP data over the interval from 1820 to 2013 ($\gamma_{MI} = 1.01$ was obtained). An alternate calculation can be carried out by the following expression:

$$Y_{MI}^{(3)} = \gamma_{MI} \cdot A_{MI}^{(3)} \cdot N_{MI} = \gamma_{MI}^{*} \cdot arcctg\left(\frac{T_{MI} - t}{\tau_{MI}}\right) \cdot \exp\left\{\int_{T_0}^{t} q_{A_{MI}}^{(3)}(\tau)\,d\tau\right\} \quad (20)$$

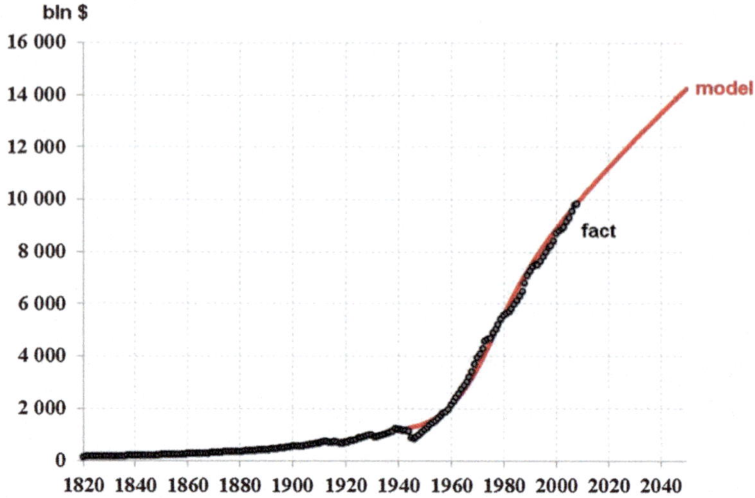

Fig. 20 GDP dynamics for middle-income countries (comparable to prices in 1990). *Data source*: (Maddison 2010)

In our calculations we simultaneously optimized four coefficients γ_{MI}, $k_{MI}^{(1)}$, $k_{MI}^{(2)}$ and $k_{MI}^{(3)}$. We obtained the following estimates: $\gamma_{MI} \cong 1.01$; $k_{MI}^{(1)} \cong 1.59$; $k_{MI}^{(2)} = 1.2 \cdot 10^4$; $k_{MI}^{(3)} = 53.9$; determination coefficient is equal to 0.996. The dynamics of total GDP for middle-income countries with a forecast until year 2050 is presented in graphic form in Fig. 20. Figure 20 shows a solid agreement despite a break in GDP trajectory during the post-war period. Thus, the technique of successive iterations that we suggest to describe the actual technology advance growth rate, or TPF, allows efficient and accurate modeling and forecasting of economic growth in different groups of countries.

The most remarkable phenomenon of the post-war "*golden era*" was undoubtedly the Japanese "*economic miracle*". Japan became the first country in postwar history to experience economic growth at a sustainable high rate of 8–10 % annually for more than 2 decades. Japan's strategies and practical measures for its rapid growth became a model that other Asian countries followed and then spread to countries in different continents. South Korea, Taiwan, and Singapore grew just as fast and demonstrated record longevity of sustainable growth (8–9 % annual GDP growth) for a period exceeding 35 years. These countries contributed 17 % to world production and joined the group of high- and middle-income countries. The surprisingly high growth rate that Japan and other Asian countries exhibited gradually became sustainable. They set targets for developing countries, following the flying geese pattern. Japan, with a population of 120 million people, remained the largest of these countries for a while. Therefore, they could not turn the scales in favor of convergence of the developed and developing worlds.

3.3 The Reasons for the Emergence of New Conditions, Facilitating Rapid Growth of Developing Countries with Low Per Capita Income

Even at a high sustainable growth rate, it takes many decades for a poor developing country to shift to the group of rich countries. For a poor country with per capita income of $1000/person*year to shift to the lower level of rich countries with per capita income of about $20,000/person*year, income would have to double 4–5 times. It is common knowledge that income doubling occurs every 10 years if the economy grows by an average annual rate of 7.2 %. Thus, even at the highest growth rate, the transition from a poor to rich country requires 40–50 years of sustainable growth. Most of this time is required to shift to the group of countries with an average income of $10,000–15,000/ person*year. This is a very important achievement, as the next income doubling would bring this country into the group of developed countries. Therefore, the sustained growth for 2–3 decades in a row is crucial. The last doubling providing the transition from middle-level to high-level income has proven to be extremely difficult, and for many countries insurmountable. For this reason, it is frequently called the "*middle income trap*."

In an attempt to understand process dynamics and reasons to start the model of sustainable expanding growth as well as factors related to the model, Nobel Prize Winner Michael Spence asked two interconnected questions: why did the process of increasingly wide and fast modernization start in the post-war period and what do developing countries do to maintain a high growth rate and cut poverty? His answer to the first question was the global economy, which provides to developing countries a huge potential market and wide access to knowledge and technology (Spence 2013, p. 78). The global market allows developing countries to better specialize in what they do best. Such specialization increases productivity. Knowledge, technologies and practical skills required for modernization of a backward economy are better imported from a global economy, especially from forward countries. There are many channels for the transfer of knowledge and technology, some of which have already been discussed above. Processes, connected to foreign direct investments (FDI), involvement in global economy via multinational supply chains, are probably the most important of these (Spence 2013, p. 81).

Even though a global economy is probably a necessary condition it is definitely not sufficient. It only creates opportunities for sustained high growth. Here M. Spence answers the second question: " Recipes of high growth include many critically important ingredients, but the most important of them is education, i.e. obtaining already existing knowledge, important for production. Besides, people obtaining new knowledge should be able to accept it in the transfer process" (Spence 2013, p. 81). A literate population has much more capability to accept and use modernization achievements, and also actively implements innovations, facilitating further modernization and economic growth. Therefore, a poor developing country has to channel available resources into critical assets, mainly education and infrastructure. This became the priority for developing countries in the postwar

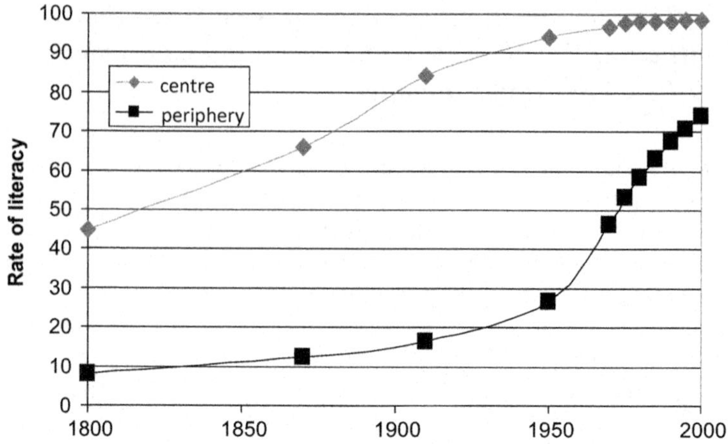

Fig. 21 The dynamics of literacy of Western countries (center) and the rest of the World (periphery) in nineteenth and twentieth centuries. *Data source*: (Mel'yantsev 1996)

development period. The results were not long in coming. Figure 21 shows the explosive percentage rise in the literate population among the developing countries in the 1950s and in the following decades of the twentieth century. It is also evident that by the start of the twenty-first century the gap between the literate populations in developed and developing countries decreased threefold as compared to 1950.

Thus, high growth of developing countries during the postwar period was possible due to the transfer of knowledge and technologies from developed countries as well as reduced barriers of trade, services and capital. M. Spence believes that GATT (*General Agreement on Tariffs and Trade*) was an important part of this process and insists that GATT started the transformation of the global economy (Spence 2013, p. 46). GATT became known as the World Trade Organization (WTO) in the 1990s. One of the main functions of the WTO today is to enable small and poor countries to participate in changing regulations of global trade in order to take into account the interests and needs of developing countries when making decisions. Surprisingly, countries of East and South-East Asia, the poorest and most deprived of natural resources as compared to Africa and Latin America, which made use of the new possibilities that appeared in post-war times. Asia was the first to find a new way to be included in the global economy and to develop at an unprecedented rate. Asia invested in its main treasure—people, enabling acceleration and a high sustained growth.

M. Spence described this period of third world country development: "It was the start of a century-long journey to a global economy. The end point is likely to be a world in which perhaps 75 percent or more of world population live in advanced countries with all that it entails: increasing income levels, with likewise increasing patterns of consumption and energy use. In addition to the spreading pattern of growth, the remarkable feature of the postwar modern era is speed. In the high-growth developing countries, there had been sustained periods (a quarter of century or more) of growth at 7 percent and more. To put this in perspective, high-speed growth in the first 200 years of the Industrial Revolution would have been between 2 and 2.5 percent" (Spence 2013, pp. 19–20).

Let us construct a mathematic model of economic growth for low-income poor countries. As shown above, the total population of low-income developing countries is described by Kapitsa's formula (4) and graphically presented in Fig. 8. Just like in the case of middle- and high-income countries, the first iteration of TPF growth is built after Kremer model (7):

$$q_{A_{LI}}^{(1)} = \frac{dA_{LI}^{(1)}}{A_{LI}^{(1)}dt} = k_{LI}^{(1)} \cdot \left\{ \tau_{LI} \cdot \left[1 + \left(\frac{T_{LI} - t}{\tau_{LI}} \right)^2 \right] \cdot arcctg\left(\frac{T_{LI} - t}{\tau_{LI}} \right) \right\}^{-1}. \quad (21)$$

The equation has the following solution [see solution (11) of Eq. (10)]:

$$A_{LI}^{(1)}(t) = A_{LI0} \cdot \left[\frac{arcctg\left(\frac{T_{LI}-t}{\tau_{LI}}\right)}{arcctg\left(\frac{T_{LI}-T_0}{\tau_{LI}}\right)} \right]^{k_{LI}^{(1)}}. \quad (22)$$

The curve of actual relative technology advance A_{LI}/A_{LI0} over the interval from 1820 to 2013 is presented in Fig. 22. Theoretical dependence of technology advance dynamics $A_{LI}^{(1)}/A_{LI0}$ obtained after (23) is also presented. The best fit of $k_{LI}^{(1)}$ obtained after (22) is equal to $k_{LI}^{(1)} \cong 0.51$. Just as expected, curve (1) provides good approximation until 1960.

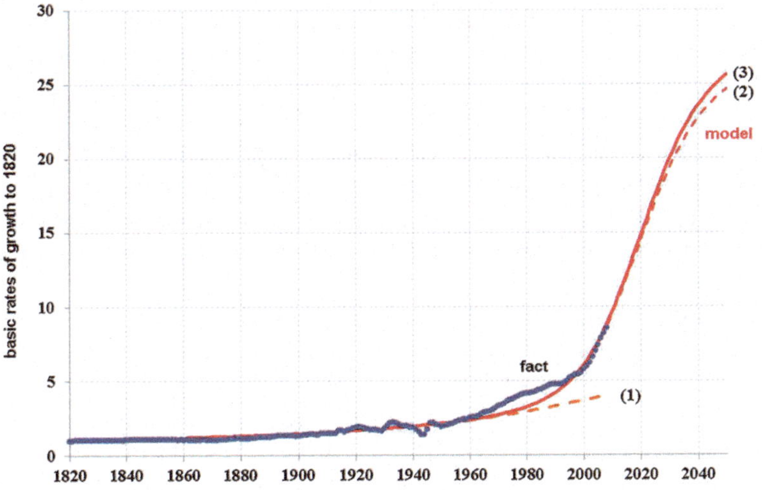

Fig. 22 Technology advance rate in low-income countries. *Data source*: (Maddison 2010)

In the 1950s developing countries did not yet have their own R&D system and their economies grew almost completely by borrowing innovation technologies in developed countries. Therefore the second iteration of TPF growth rate can be directly obtained by adding a term which is due to borrowing the technologies from highly developed countries (similar to expression (18) for middle-income countries):

$$q_{A_{LI}}^{(в)} = \frac{s_{LI}\xi_{LI}}{s_{HI}\xi_{HI}} \cdot q_{A_{HI}}^{(o)}(t - T_{LID}) \frac{[1 + 1(t - T_{LIm})] \cdot \exp[-\vartheta_{LI}^{(в)} \cdot (t - T_{LIm})]}{1 + \exp[-\vartheta_{LI}^{(в)} \cdot (t - T_{LIm})]}. \tag{23}$$

where T_{LIm} is the year of the start of closing-up of production of borrowed production.

Here, additionally (compare to (18)), we get delay time in borrowing the technologies from vanguard countries (T_{LID}). While developed countries with middle per capita income obtain new mode innovation technologies without delay, developing countries obtain old mode technologies with a big lag. In the 1950s the lag was equal to about 50 years, i.e. $T_{LID} = 30$ years.

Therefore, we obtain for the second iteration:

$$q_{A_{LI}}^{(2)} = q_{A_{LI}}^{(1)} + q_{A_{LI}}^{(в)} = k_{LI}^{(1)} \cdot \left\{ \tau_{LI} \cdot \left[1 + \left(\frac{T_{LI} - t}{\tau_{LI}} \right)^2 \right] \cdot arcctg\left(\frac{T_{LI} - t}{\tau_{LI}} \right) \right\}^{-1} +$$
$$+ k_{LI}^{(2)} \cdot \frac{s_{LI}^{(k)}\xi_{LI}^{(k)}}{s_{HI}^{(k-1)}\xi_{HI}^{(k-1)}} \cdot q_{A_{HI}}^{(o)}(t - T_{LID}) \cdot \frac{[1 + 1(t - T_{LIm})] \cdot \exp[-\vartheta_{LI}^{(в)} \cdot (t - T_{LIm})]}{1 + \exp[-\vartheta_{LI}^{(в)} \cdot (t - T_{LIm})]}. \tag{24}$$

In the equation above we introduced the calibration coefficient $k_{LI}^{(2)}$ which is determined by the least squares technique. It is assumed, that active borrowing of technologies started in the 1980s and will last until 2010s, therefore $T_{LI\,m} = 2010$, and parameter $\vartheta_{LI}^{(в)}$ is still defined by the expression: $\vartheta_{LI}^{(в)} = \frac{\ln 19}{30} \cong 0.098$. By using V.A. Mel'yantsev' s estimates, we also obtain $s_{LI}^{(k)}/s_{HI}^{(k-1)} \cong 1.2$ and $\xi_{LI}^{(k)}/\xi_{HI}^{(k-1)} \cong 3$ (Mel'yantsev 2013, p. 18). Upper indices denote the numbers of BKC and take into account the effect of lag by one Kondratiev cycle and borrowing the technologies in most developed countries. We obtained the following best estimate of the calibration coefficient: $k_{LI}^{(2)} \cong 14.56$.

By integration of Eq. (24) in the following expression

$$\overline{A}_{LI}^{(2)}(t) = \frac{A_{LI}^{(2)}(t)}{A_{LI0}} = \exp\left\{ \int_{T_0}^{t} q_{A_{LI}}^{(2)}(\tau)\, d\tau \right\}, \tag{25}$$

we obtain curve (2) in Fig. 22, which approximates actual data over the industrial era (1820–2013) and can form the basis for the forecast of economic development until 2050. Observed deviation of the actual curve from the simulated one in the 1960s–1990s is explained by the fact that we don't separate the USSR and socialist countries in a specific group with its own properties. The latter had dynamic growth in the given period which waned by the 1990s. Considering that developing countries borrowed technologies not only from high-income countries (23) but also from middle-income countries, we built the third iteration in a way similar to the second one. The result of the iteration is presented in Fig. 22 as curve (3). Just as expected, the correction is small and impacts only forecast part.

TPF growth rate, calculated by formula (24) is graphically presented in Fig. 23. It is evident that TPF growth in developing countries is currently at its peak and it will decrease to the level of the 1960s by year 2040. We have also calculated GDP dynamics for low-income countries using production function $Y_{LI}^{(3)} = \gamma_{LI} A_{LI}^{(3)} N_{LI}$ (we obtained $\gamma_{LI} = 0.58$ by determination coefficient equal to 0.873) by multiplication on numerical values of $A_{LI}^{(3)}$ (see Fig. 22) and N_{LI} (see Fig. 8). The third iteration of GDP dynamics of low-income countries $Y_{LI}^{(3)}$ is presented in Fig. 24. The coefficient of determination is low only because of the difference in the actual and calculated data in the period from 1960 to 1990, where the contribution of the USSR countries of the socialist camp is underestimated. As one can see from Fig. 24, the model provides a good fit for the majority of the industrial epoch and can provide a good forecast until year 2050. Of course, if we separate developing countries into three groups and single out the group of countries which actively implement the catching-up development strategy, the USSR and the countries of socialist camp would get a more accurate model. Interested readers should try our technique to prove it. We confine ourselves by what is listed, as we are more interested in long-term trends rather than in particular details and deviations observed in different periods of industrial era.

We should emphasize an exceptionally important part of the world economic revolution that occurred because of the collapse of the world socialist system and

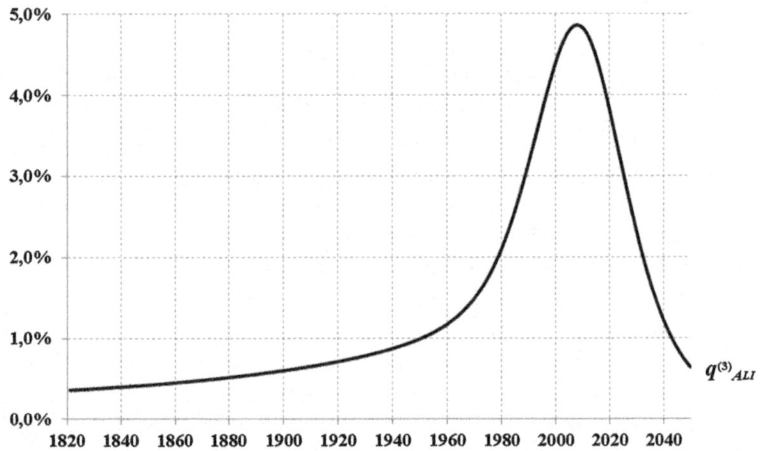

Fig. 23 Technology advance growth rate in low-income countries. *Data source*: (Maddison 2010)

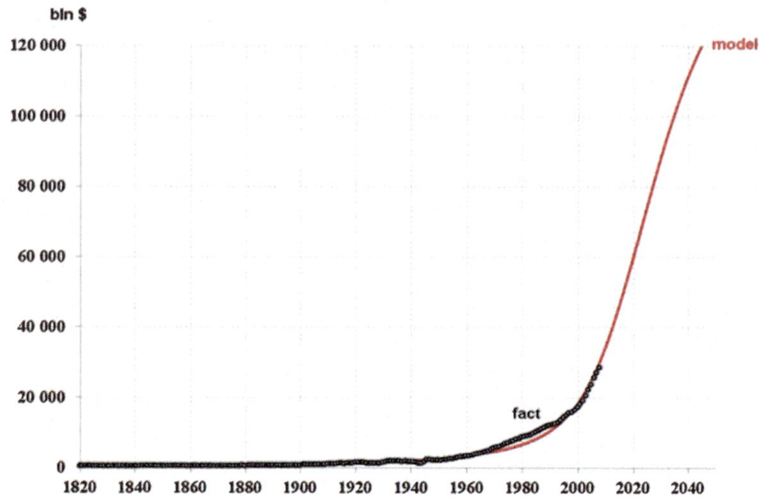

Fig. 24 GDP dynamics for low-income countries (in comparable price in 1990)

breakup of the USSR-acceleration of economic growth in developing countries and acceleration of convergence between rich and poor countries. Eventually, virtually the whole economy of the socialist camp with USSR shifted to capitalist development. Especially successful were market reforms in socialist China with a population of a billion. Therefore, if the statement about the market being the best distribution mechanism of vital resources is true, it increases the chances of developing countries to nail down the success of convergence.

4 Transition to Fifth BKC and the Start of Real Convergence

The downturn of the fourth Big Kondratiev Cycle (*approximately 1946–1982*), which had brought record high growth to the world economy, had started in the early 1970s and began a slowdown of economic growth in all in developed countries. World structural economic crisis caused by the depletion of the 4th technological mode and need for transition to a new-fifth technology mode was approaching. Oil price shocks of 1973–74 and 1979–80 had aggravated the crisis mode of the 4th BKC, turning it into a lengthy depression. After 1973 came the economic crisis of 1975–76 which caused a significant slowdown of world economic growth. However, the growth rate remained high by historical standards. Developed countries had started large-scale energy saving programs, substitution of liquid fuel by alternative resources (natural gas, coal, biofuel), and increased end user energy efficiency. All of these measures provided very fast and efficient results. After the second oil price shock in 1979–1980, when oil prices skyrocketed to $41/barrel, both oil prices and the share of oil consumption expenses in GDP of developed countries had dropped. As early as 1986, consumption stabilized in the range of \$15–\$25/barrel in comparable prices and stayed at the level until the start of the 2000s.

Unlike developed countries, developing countries seriously suffered from the oil crisis, as they did not have access to energy-saving and energy-efficient technologies. Developed countries experienced a noticeable slowdown of per capita growth, and developing countries experienced decrease in per capita growth. E. Helpman numerically estimated the impact of the 1970s oil crisis on world economic development. He has divided these decades into two periods, one prior to the oil crisis of 1973 and one following the oil crisis: 1960–1972 and 1974–1990. He has shown that the average rate of growth in 104 countries was 3.0 % in the former period; it dropped to 1.1 % in the latter (Helpman 2012, p. 19). Moreover, he demonstrated that the difference in growth rate became significant in the second period, with an insignificant decline in the average growth rates of the 21 richest economies. Their average growth rate declined from 4 % per annum to 2 %. But in each period it was higher than the average growth rate for the larger sample of 104 countries (Helpman 2012, p. 20). Thus, the average growth rate of the rich countries declined by a factor of 2, while in the larger sample it declined by a factor of 3. This explains why there were no noticeable results of convergence observed from the 1960s until the 1990s.

Accelerated economic growth reappeared in the 1980s–1990s as a result of intense use of innovation technologies of the 5th technology mode, with microelectronics, personal computers and information technologies being the most efficient. These technologies enabled a significant increase of labor productivity but insufficient to bring back the record high rates of the 1960s. However, by the end of the 1990s and early 2000s there was another slowdown of economic growth in developed countries. According to V.A. Mel'yantsev this slowdown was the result of a 95–100 % deindustrialization. He has shown that slowdown manifested itself in the decline of the growth rate in the industrial sector from 2.2 % in 1973–1990 to 1.7 % in 1990–2006 and its GDP share (in current prices) from 33 % in 1990 to 26 % in 2006 (Mel'yantsev 2009, p. 18).

In developing countries, average per capita income growth first decreased from 2.7–2.9 % in 1950–1973 to 2.1–2.2 % in 1973–1990, exceeding the value for developed countries, and, then in 1997–2007 increased to 3.7–3.9 % per annum, being twice as high as in the developed countries (Mel'yantsev 2009, p. 66). This means real convergence had started in the 1990s. Available evidence shows that the group average for developing countries of annual per capita GDP growth rate in 1950–2007 was about 5–6 times higher than in 1900–1938, when many of them were in the state of colonies and semi-colonies (Mel'yantsev 2009, p. 68).

It should not be forgotten that the growth of developing countries is substantially dependent upon the economic growth of the two giants of the developing world—China and India. Up to date modernization of economic and market-oriented reforms had started in China after 1978 due to new economic policy announced by Deng Xiaoping. China soon after entered a period of sustained growth. In 1978–2003 the average growth rate was 7.9 % per annum. At the same time, growth of the Chinese population slowed down and per capita income growth rate reached

6.6 % per annum. Since the start of the 2000s, Chinese economic growth was at an exceptionally high level of 8–10 % per annum. The most important growth factors were accelerated industrialization and an exceptionally high share of investments in GDP. Per capita GDP grew from $400 in 1978 to $3500 at present, doubling three times. India started economic reforms in the 1980s. It successfully overcame the financial crisis of the 1990s, and the economic growth rate started to exceed 6 % per annum. Right before the financial-economic crisis of 2008–2009, India reached a growth rate of 9 % per annum. After China and India, which account for about 40 % of world population, entered the path of sustained growth, the situation in the world economy had radically changed for the best. It was the rapid economic growth of China and India in the beginning of the twenty-first century that turned the tide with slowly progressing convergence in the 1990s (see Fig. 1) and since 2000 it accelerated the pace.

Thus, South-East Asian countries made significant contributions to the reversal of the Great Divergence, which lasted almost 150 years, and to convergence and acceleration of its pace in the beginning of the century. E. Maddison wrote: "Undoubtedly, the best economic results in 1973-2003 were shown by the countries of South-East Asia (except for Japan), which cover more than a quarter of the world GDP and about half of the world population. The phenomenon of Asian revival was an unprecedented success. After 1973 per capita income there grew faster than in "golden era". The gap between Eastern Asia and the vanguard group of capitalist countries had fallen substantially. We witnessed another step forward, now of the whole group of countries, which emulated what Japan had done in 1950-1973. Had the world been comprised of only developed countries and revived Asia, its development model since 1973 could have been interpreted as a vivid demonstration of possibilities of the concept of conditional convergence, suggested by the proponents of neoclassical growth theory" (Maddison 2012, p. 116). It cannot be better said. For the sake of comparison, he continues: "In all other regions of the world the results of economic growth in 1973-2003 got worse. The strongest decline of growth rate occurred in Africa and Latin America. The decrease of growth rate in developed countries was a powerful shock for the economies of these regions" (Maddison 2012, p. 116).

Postwar explosive growth in the world and the start of convergence between developed and developing world countries was accompanied by unprecedented poverty reduction all over the world. For example, between 1970 and 1998 the number of people who lived on less than $2 a day (*poverty level*) declined by 350 million; the number of people who lived in extreme poverty, on less than $1 a day, declined by 200 million. The number of people affected by this reduction in poverty is staggering. In 1970 1323 million people lived on less than $2 a day and 554 million lived on less than $1 a day (Helpman 2012, p. 168). Rapid economic growth in China and India had made a quintessential contribution to poverty reduction in the world. According to D. Quah's estimates, the fraction of people who lived on less than $2 a day declined in China from 37–54 % in 1980 to 14–17 % in 1992, and it declined in India from 48–62 % to 12–19% during the same period. The number of poor people declined in China from 360–530 million in 1980 to 158–192 million, though the population increased from 981 million to 1162 million people. In India during the same period, the number of poor people declined from 326–426 million to 110–166 million people (Helpman 2012, p. 170). Thus, we conclude, that postwar dynamic economic growth has, on average, increased the income of poor people all over the world.

Thus, by the beginning of the 2000s the result of convergence grew measurable. In 2003, the countries of the West still accounted for 43 % of world GDP, while their total population share did not exceed 12 %. While in the Western countries per capita income was equal to $23,710 (*after PPP in 1990*), the average income in other countries of the world, with 88 % of world population reached $4172 (Maddison 2012, p. 112), i.e. the gap was less than six. As one can observe from Fig. 1, after 2000 the convergence had significantly accelerated. As pointed out above, it occurred largely due to an extremely high economic growth rate, as exhibited in China and Japan. The accelerated growth of the two largest world economies has become the driving force of promoting the global economy in the last 2 decades. In the last decade, the aggregate contribution of the two giants in the developing world to global economic growth is one and a half times higher than that of the leaders of developed world. Soon enough, the economies of the two countries will overcome the economies of the US and the EU. Therefore, further advance of convergence in the twenty-first century will be determined by economic development of China and India.

4.1 Developed Economies Shift to Evolutional Development

Generally, in the last quarter of the twentieth century and beginning of the twenty-first century, developed countries managed to maintain sustained growth on the up wave of the 5th BKC (1982–2006), though the growth was only half as powerful as during the up wave of the 4th postwar BKC (1946–1970). The economic development of high- and middle-income countries became evolutionary during this period. Let us consider this period in detail, taking middle-income countries (MI) as an example. As Fig. 18 shows, the third iteration trajectory (curve 3) separates from the actual trajectory of technological advance by the end of the 1980s. It is quite natural, because expression (19), describing the dynamics of middle-income countries does not take into account the contribution of borrowed technologies during 5th BKC, starting from the 1980s.

Therefore, adding one additional term similar to (18) but shifted by 30 years, to the right part of expression (19) we obtain the fourth iteration:

$$q_{A_{MI}}^{(4)}(t) = q_{A_{MI}}^{(3)}(t) + k_{MI}^{(4)} \cdot 1(t - T_{MI\,e}) \cdot q_{A_{HI}}^{(0)}(t) \cdot \frac{[1 + 1(t - T_{MI\,e})]\exp[-\vartheta_{MI}^{(e)}(t - T_{MIf})]}{1 + \exp[-\vartheta_{MI}^{(e)}(t - T_{MIf})]}. \quad (26)$$

Here, as stated above, $T_{MIв} = 1980$ and $T_{MIf} = 2010$, while still $\vartheta_{MI}^{(в)} = \frac{\ln 19}{30}$. The best fit $k_{MI}^{(4)}$ yields: $k_{MI}^{(4)} \cong 0.11$. In comparison, we should note that we obtained $k_{MI}^{(3)} \cong 80$ from Eq. (19). Meaning, the efficiency of borrowing the technologies had sharply declined as the technological level of middle-income countries in the 1980s became virtually equal to the level of high-income countries.

Let us consider the forecast calculation of technology advance in 6th BKC (2018–2050). As shown above, due to practical implementation of NBIC technologies, by 2018–2020, developed countries will be out of depression and the up wave of 6th BKC (2018–2040) will continue sustainable growth. As shown, the convergence of NBIC technologies would create a powerful synergetic effect. This would result in an increase in the technology growth rate from 1.2 % per annum during 1980–2000 to 1.6 % by 2030. Thus, we should add an additional rate, equal to the difference i.e. 0.4 % per annum achieved by 2030 and increasing by logistic law, to the existing curve.

Finally, the fifth iteration provides the forecasted growth rate of technological advance until 2050, taking into account additional synergetic effect generated by the convergence of NBIC technologies during the up wave of the 6th BKC,

$$q_{A_{MI}}^{(5)}(t) = q_{A_{MI}}^{(4)}(t) + \frac{\Delta q_{NB}^{syn}}{1 + c_{NB} \cdot \exp\left[-\vartheta_{NB}\left(t - T_{NB}\right)\right]}, \tag{27}$$

where T_{NB} is the year production of innovative products on the basis of NBIC starts—technologies ($T_{NB} = 2010$); $\Delta q_{NB}^{syn} = 0.004$. The parameters c_{NB} and ϑ_{NB} are determined from the condition that saturation has to be reached by 2030: $c_{NB} = 19$; $\vartheta_{NB} = 0.294$. All parameter values were taken from Hirooka's innovation paradigm for NBIC-technologies.

By numerical integration of Eqs. (26) and (27) after the formula

$$\overline{A}_{MI}^{(n)}(t) = \frac{A_{MI}^{(n)}(t)}{A_{MI0}} = \exp\left\{\int_{T_0}^{t} q_{A_{MI}}^{(n)}(\tau)\, d\tau\right\}, \tag{28}$$

where $n = 4;5$, we obtain the 4th and 5th iterations of basis rate of technology advance, presented in graphic form in Fig. 25. Figure 25 shows that the fourth iteration provides a good approximation of actual data throughout the industrial era (1820–2010). In addition, the allowance for synergetic effect of NBIC technologies allows us to refine the forecast until 2050 with respect to the trend defined by the 4th iteration.

Next, using the AN-model $Y_{MI}^{(n)} = \gamma_{MI} \cdot A_{MI}^{(n)} \cdot N_{MI}$ we subsequently calculate $Y_{MI}^{(4)}$ and $Y_{MI}^{(5)}$, because N_{MI}, is also known in numerical and graphic representation (see Fig. 7). Total GDP trajectories of middle-income countries in the 4th and 5th iterations are presented in Fig. 26. As Fig. 26 shows, the 4th iteration provides a quite accurate approximation of actual for the period 1820–2010. The fifth iteration

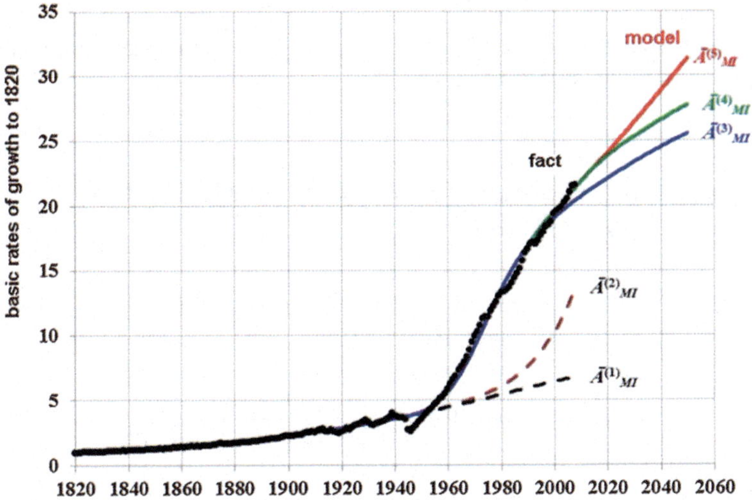

Fig. 25 Modeling and forecast of basis technology advance growth rate for the group of middle-income countries. *Data source*: (Maddison 2010)

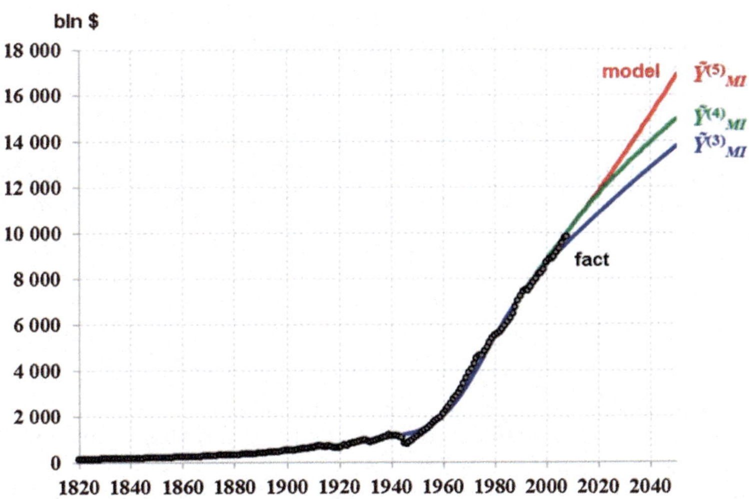

Fig. 26 Modeling and forecast of GDP for the group of middle-income countries. *Data source*: (Maddison 2010)

takes into account the additional contribution NBIC technologies made to GDP during this period.

Thus, forecast computations performed with the help of the AN-model unambiguously show that both high-income and middle-income developed economies will have evolutionary development in the forthcoming 6th BKC (2018–2050), without revolutionary breakthroughs. Hence, the fate of convergence will mainly depend on the pattern of economic growth in the developing world, most significantly in the vanguard BRICS countries. Therefore, we will now consider the dynamics of development in these countries.

4.2 *Vanguard Countries of Developing World (BRICS) Perform Breakthrough*

In Sect. 2 we have shown that the total population of BRICS countries is well described by a logistic function of the type (6):

$$N_{BS}(t) = \overline{N}_{BS0} + \frac{N'_{BS0}\,(1 + n_m)}{1 + n_m \cdot \exp\left[-\vartheta_{BS}\,(t - T_0)\right]}. \tag{29}$$

We have obtained numerical estimates of function' parameters: $\overline{N}_{BS0} = 689$; $N'_{BS0} = 3.2$; $\vartheta_{BS} = 0.04$; $n_m = 1143$. An expression can be derived from this equation, to describe population growth rate in BRICS countries:

$$q_{N_{BS}}(t) = \frac{N'_{BS0} \cdot n_m \cdot \vartheta_{BS} \cdot \exp\left[-\vartheta_{BS}\,(t - T_0)\right]}{\{N_{BS0} + \overline{N}_{BS0} \cdot n_m \cdot \exp\left[-\vartheta_{BS}(t - T_0\,]\right\}\{1 + n_m \exp[-\vartheta_{BS}\,(t - T_0)]}. \tag{30}$$

Hence, the first basis iteration for the chain growth rate of technology advance for BRICS countries, according to Kremer's equation (7) is written as follows:

$$q^{(1)}_{A_{BS}}(t) = k^{(1)}_{BS} q_{N_{BS}}(t). \tag{31}$$

The plot of the function is presented in Fig. 27. We obtained $k^{(1)}_{BS} \cong 0.71$ as the estimated calibration coefficient. As one can see from the plot, the first iteration works well until 1920.

We obtain the second iteration by adding supplementary rates to expression (31), generated by the national R&D system (13):

$$q^{(2)}_{A_{BS}}(t) = q^{(1)}_{A_{BS}}(t) + k^{(2)}_{BS} \cdot [l^2_{A_{BS}} \cdot (3l^{\max}_{A_{BS}} - 2l_{A_{BS}}) - l^2_{A_{BS0}} \cdot (3l^{\max}_{A_{BS}} - 2l_{A_{BS0}})]. \tag{32}$$

From the UN database [*UNdata. A world of information*] we obtained the following values of the parameters: $l^{\max}_{A_{BS}} \cong 0.008$; $l_{A_{BS0}} \cong 0.0$. We have obtained the best fit estimate for the calibration coefficient $k^{(2)}_{BS} \cong 5 \cdot 10^4$. The plot of the function (32) is also presented in Fig. 27. It shows the second iteration starts to improve the first one from the 1970s, with its contribution intensifying over time. This is a logical outcome as the BRICS countries' (except Russia) R&D projects began to yield results in this period.

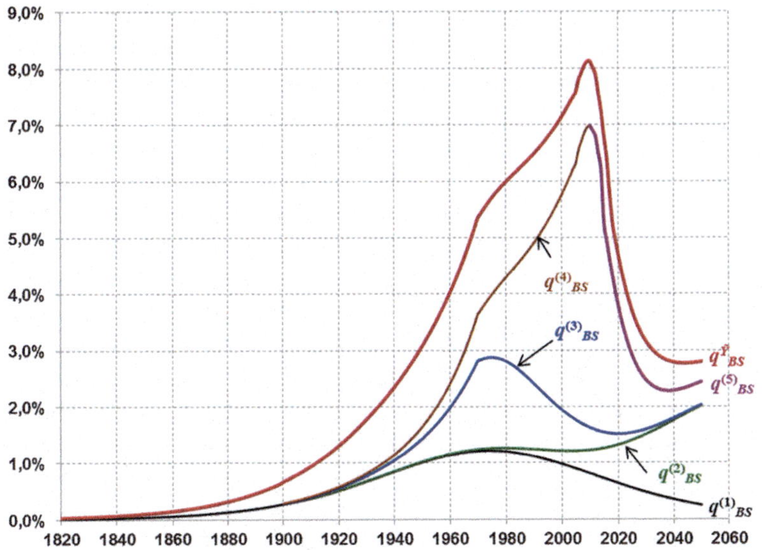

Fig. 27 The modeling and forecast of chain growth rate of technology advance and GDP for BRICS countries (BS)

In order to build the third and fourth iteration, the contribution of borrowed technologies should be considered during the fourth (1950–1980s) and fifth (1980–2010) BKC, respectively, in a way, similar to middle-income (18) and low-income (23) countries. Therefore, the third iteration as written as follows:

$$q_{A_{BS}}^{(3)}(t) = q_{A_{BS}}^{(2)}(t) + k_{BS}^{(3)} \cdot q_{A_{HI}}^{(0)}(t - T_{BSD}^{(4)}) \cdot \frac{[1 + 1(t - T_{BSf}^{(4)})] \cdot \exp[-\vartheta_{BS}^{(\text{в})}(t - T_{BSf}^{(4)})]}{1 + \exp[-\vartheta_{BS}^{(\text{в})}(t - T_{BSf}^{(4)})]}. \quad (33)$$

where $T_{BSD}^{(4)}$ is the time lag in borrowing technologies; $T_{BSD}^{(4)} = 30$ years; $T_{BSf}^{(4)}$—is the year production of innovative goods borrowed in the upswing of the fourth BKC begins to cease, $T_{BSf}^{(4)} = 1980$. Assuming it will take 30 years to completely stop production, we obtain $\vartheta_{BS}^{(\text{в})} = \frac{\ln 19}{30} \cong 0.098$. Further, $q_{A_{HI}}^{(0)}\left(t - T_{BSD}^{(4)}\right)$ represent the chain growth rate of technological advance, generated by the R&D system of high-income countries (15), shifted by $T_{BSD}^{(4)}$; $1\left(t - T_{BSf}^{(4)}\right)$ unit function, starting in $T_{BSf}^{(4)}$.

The fourth iteration is written as follows:

$$q_{A_{BS}}^{(4)}(t) = q_{A_{BS}}^{(3)}(t) + k_{BS}^{(4)} \cdot 1(t - T_{BS\text{в}}^{(5)}) q_{A_{HI}}^{(0)}(t - T_{BSD}^{(5)}) \cdot \frac{[1 + 1(t - T_{BSf}^{(5)})] \cdot \exp[-\vartheta_{BS}^{(\text{в})}(t - T_{BSf}^{(5)})]}{1 + \exp[-\vartheta_{BS}^{(\text{в})}(t - T_{BSf}^{(5)})]}, \quad (34)$$

where $T_{BS\theta}$—is the year that marks the start of borrowing advanced technologies in the upswing of the 5th BKC, $T^{(5)}_{BS\theta} = 1980$; $T^{(5)}_{BSD}$—is the delay time in borrowing at 5th BKC; $T^{(5)}_{BSD} = 20$ years (as we can see, less than in previous BKC); $T^{(5)}_{BSf}$—is the year production of innovative goods borrowed in the upswing of fifth BKC starts to cease; $T^{(5)}_{BSf} = 2010$; $\vartheta^{(\theta)}_{BS}$ is still equal 0.098. We have now fixed the start of borrowing advanced technologies, while in the previous cycle it was left open (see (33)). Note that active borrowing of advanced technologies dates back to the pre-war era in the 1920s and 1930s (for example, by the USSR).

The plots of the chain growth rate of technology advance in 3rd and 4th iterations, calculated by the expressions (33) and (34) are presented in Fig. 27. The best fit estimates of the coefficients $k^{(3)}_{BS}$ and $k^{(4)}_{BS}$ yielded the following values: $k^{(3)}_{BS} = 14.83$ and $k^{(4)}_{BS} = 7.8$. As Fig. 27 shows, during the pre-war (1920–1940s) and, in particular, the post-war (1950–1980s) periods, the growth rate was largely supported by borrowing technologies from advanced developed countries. The process became especially efficient in 1970–2010s, because it was accompanied by the dynamic development of national R&D systems in BRICS countries. Chain growth rate of technology advance in BRICS countries during this period achieved 4–6 % per annum, and, sometimes, a record high 7 % per annum (see Fig. 27).

The fifth iteration forecasts and refines the trend of the fourth iteration until 2050. The iteration is obtained by adding the contribution of additional synergetic effect, generated by convergence of NBIC technologies to the fourth iteration, similarly to middle-income countries (27):

$$q^{(5)}_{A_{BS}}(t) = q^{(4)}_{A_{BS}}(t) + \frac{k^{(5)}_{BS}\Delta q^{syn}_{NB}}{1 + c_{NB} \cdot \exp\left[-\vartheta_{NB}\left(t - T_{NB}\right)\right]}. \tag{35}$$

In this equation we introduced a new calibration coefficient $k^{(5)}_{BS}$, while all other parameters describing the dynamics of synergetic contribution remain unchanged (see 27). We estimated the coefficient $k^{(5)}_{BS}$ with the help of a gain multiplier (see 18), assuming $s^{(6)}_{BS} = 26.5\%$; $s^{(6)}_{HI} = 22\%$; $\xi^{(6)}_{BS} = 0.42$; $\xi^{(6)}_{HI} = 0.34$, and obtained: $k^{(5)}_{BS} = 0.95$. The plot of the fifth equation, which contains a forecasting part until 2050, is also presented in Fig. 27. Analysis of the plot shows that the synergetic effect of NBIC technologies is manifested in the 0.4 % increase in the chain growth rate of technological advance in the 2030s and 2040s. The figure also illustrates that the growth rate of technology advance had reached 1 % only by the start of the World War II, and then grew rapidly in the post-war period until 2010 when the growth stopped due to the financial-economic crisis of 2008–2009. In the future, the rate will gradually decline until the end of the 2030s, as demonstrated by the forecasting part of the curve.

The resultant plot of total GDP of BRICS countries is also presented in Fig. 27; it is equal to $q^{(5)}_{Y_{BS}} = q^{(5)}_{A_{BS}} + q_{N_{BS}}$ since $Y^{(5)}_{BS} = \gamma_{BS} \cdot A^{(5)}_{BS} \cdot N_{BS}$, which follows from the AN-model. As one can see from the plot, the GDP growth rate in BRICS countries was quite low during the nineteenth century, typical of Western countries

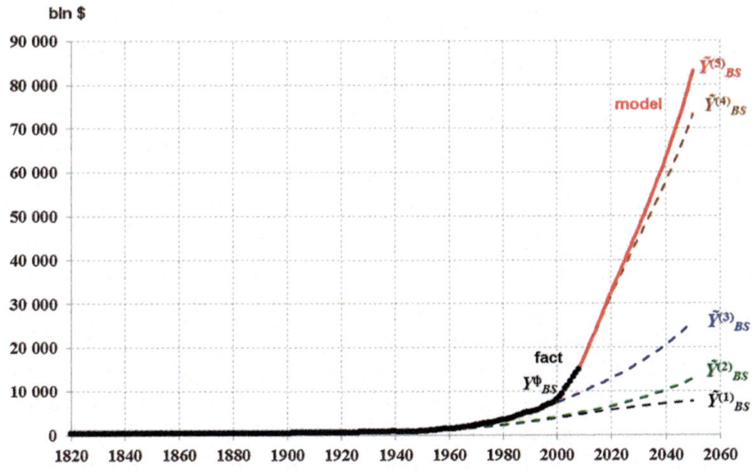

Fig. 28 Modeling and forecast of the GDP of BRICS countries. *Data source*: (Maddison 2010)

in the eighteenth century. The GDP growth rates in these countries had reached 1 % around 1912, almost a 100 years after the West achieved this rate. However, in the eighteenth century, especially in its second half, the economies of BRICS countries accelerated and became the locomotives of world economic development, exhibiting GDP growth rates ranging from 6 to 8 % during 1980–2010. As shown in the forecast (see Fig. 27), GDP growth rates in BRICS countries will decrease slowly and stabilize at the level of 3 % per annum, a rate typical for developed countries. Naturally, the structure of BRICS economies transforms into classic economies of developed countries.

The AN-model has an amazing property—it enables calculation of the GDP trajectory (Y_{BS}) without having to calculate the trajectory for technological advance (A_{BS}). Since the AN-model in rate representation is written as $q_Y = q_A + q_N$, to calculate GDP trajectory in n[th] iteration we can use the following expression:

$$Y_{BS}^{(n)} = \gamma_{BS} \cdot \exp\left\{ \int_{T_0}^{t} q_{Y_{BS}}^{(n)}(\tau)\, d\tau \right\} = \gamma_{BS} \cdot \exp\left\{ \int_{T_0}^{t} \left[q_{A_{BS}}^{(n)}(\tau) + q_{N_{BS}}(\tau) \right] d\tau \right\}, \tag{36}$$

where n = 1,2,3,4,5. All retrospective calibration coefficients $k_{BS}^{(1)}, \ldots, k_{BS}^{(4)}$, including γ_{BS}, are estimated on the basis of historical data of actual aggregate GDP of BRICS countries—$\tilde{Y}_{BS}$ [*Historical Statistics of the World Economy: 1-2008 AD (Copyright Angus Maddison)* http://www.ggdc.net/MADDISON/oriindex.htm].We obtained the following calibration coefficient by the least squares fit: $\gamma_{BS} = 0.577$; $k_{BS}^{(1)} = 0.71$; $k_{BS}^{(2)} = 5 \cdot 10^4$; $k_{BS}^{(3)} = 14.83$; $k_{BS}^{(4)} = 7.8$. The GDP dynamics of BRICS countries for all five iterations are presented in Fig. 28.

The plot shows that, just as expected, the first two iterations are valid only until 1960. The third iteration provides good approximation until the 1990s, while the fourth one provides a good approximation of GDP trajectory over a complete historical period (1820–2010) and creates the main forecasting trend (2015–2050). The fifth iteration $Y_{BS}^{(5)}$ is obtained by adding the contribution of additional synergetic effect generated via convergence of NBIC technologies. It reveals that in the remaining 35-year forecasting period until 2050 the GDP of BRICS countries will increase roughly fourfold, which is possible at an average growth rate of 4 % per annum. Now we have a chance to compare the ratio of economies of developed countries and BRICS countries in the forecasting period (2015–2050). While the economies of BRICS countries today is about 1.5 times smaller in size (see Fig. 28) than the aggregate size of economies of high-income (see Fig. 17) and middle-income (see Fig. 26) developed countries, by 2050 the aggregate size of BRICS economies will be 1.5 the total size of developed economies. BRICS countries, both in population and in the total size of economies, comprise about half of developing world. Thus, forecast computations reveal that the vanguard of the developing world—BRICS countries, could nail down the success of convergence at the global level.

5 Possible Results of the Ongoing Global Convergence

Before summarizing possible results of the forthcoming Global Convergence, we complete the modeling and forecasting of the development trajectory for high-income countries. We have already constructed two iterations, which provide a perfect description of the historical period (1820–2010). The computation results according to the above iterations were presented in graphic form in Figs. 16 and 17. In order to obtain the third iteration, we need to add the synergetic effect, generated by NBIC technologies, to the second iteration (15). The computation can be the same as we did for the fourth iteration for middle-income developed countries (27):

$$q_{A_{HI}}^{(3)}(t) = q_{A_{HI}}^{(2)}(t) + \frac{k_{HI}^{(2)} \cdot \Delta q_{NB}^{syn}}{1 + c_{NB} \cdot \exp\left[-\vartheta_{NB}\left(t - T_{NB}\right)\right]}. \tag{37}$$

Naturally, $k_{HI}^{(3)} = 1$, because NBIC technologies were developed first in high-income developed countries, which will be the first to explore the production and sale of innovative goods and services. Therefore, continuing (15) and (37) we obtain:

$$q_{A_{HI}}^{(3)}(t) = \frac{k_{HI}^{(1)} \cdot n_m \cdot \vartheta_{HI} \cdot \exp[-\vartheta_{HI}(t - T_0)]}{1 + n_m \cdot \exp[-\vartheta_{HI}(t - T_0)]} +$$

$$+ k_{HI}^{(2)} \cdot \left\{ l_{A_{HI}}^2(t)[3l_{A_{HI}}^{\max} - 2l_{A_{HI}}(t)] - l_{A_{HI*}}^2(3l_{A_{HI}}^{\max} - 2l_{A_{HI*}}) \right\} + \frac{\Delta q_{NB}^{syn}}{1 + c_{NB} \cdot \exp[-\vartheta_{NB}(t - T_{NB})]}, \tag{38}$$

where $l_{A_{HI}}(t) = \dfrac{l_{A_{HI}}^{\max}}{1 + c_{HI} \cdot \exp[-\vartheta_{HI}(t - T_{HI\mathit{e}})]}$; $l_{A_{HI}}^{\max}$ is the saturation mode where the maximum share of people are employed in R&D; $T_{HI\mathit{e}}$—is the start time of practical innovation activity of the R&D system in high-income countries; c_{HI} and ϑ_{HI}—are constant parameters, determined from available actual data $l_{A_{HI}}(t)$. The numerical values of parameters in expression (38) were presented above. Hence, by calculating $q_{A_{HI}}^{(3)}(t)$ according to (39), and building the dependence in Fig. 29, we obtain the complete view of the technology advance growth rate over the whole industrial era (1820–2050).

Taking into account the GDP growth rate $q_{Y_{HI}}^{(3)} = q_{A_{HI}}^{(3)} + q_{N_{HI}}$, as $q_{N_{HI}}$ is already known (8), we have also calculated $q_{Y_{HI}}^{(3)}(t)$ and plotted it in Fig. 29. It shows that throughout the industrial era, highly developed countries, as technological leaders, could maintain a GDP growth rate at the level 2.5–3 % per annum. This was sufficient to achieve the highest per capita income, equal to $35,000 in the prices of 1990–further evidence of the importance and advantage of long-term sustained growth of an economy as compared to volatile development. Next, without difficulty we calculate the trajectory of total GDP of high-income countries forecasting until 2050 with a similar expression (36):

$$Y_{HI}^{(3)} = \gamma_{HI} \cdot \exp\left\{ \int_{T_0}^{t} \left[q_{A_{HI}}^{(3)}(\tau) + q_{N_{HI}}(\tau) \right] d\tau \right\} \tag{39}$$

The plot of this function is presented in Fig. 30 together with the first two iterations. It is clear that the third iteration does not differ much from the second one, while

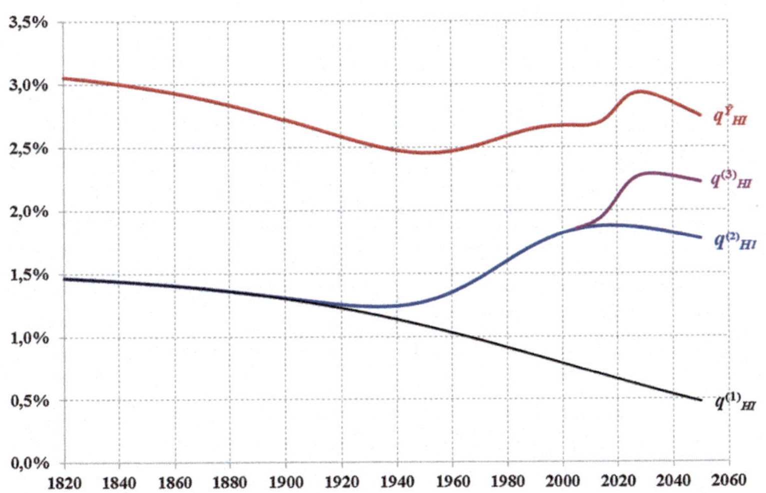

Fig. 29 The modeling and forecasting of chain growth rate of technology advance and GDP for the group of high-income countries (HI)

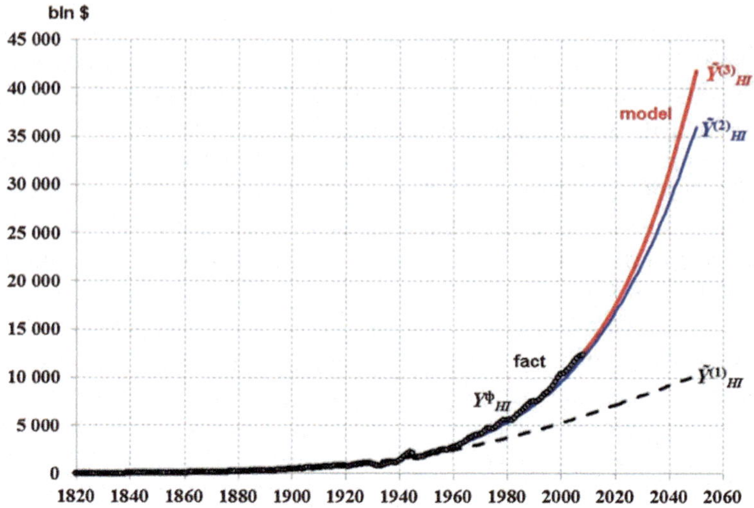

Fig. 30 Modeling and GDP forecast for high-income countries (*HI*)

Table 1 Population forecast until 2050 (million people)

Years	2010 (actual)	2020	2030	2040	2050
World	7028	7919	8679	9294	9782
High-income countries	427	460	492	524	553
Middle-income countries	468	487	504	520	534
BRICS countries	2979	3299	3569	3784	3948
The rest of the world	3154	3673	4114	4467	4747

the second is quite different from the first. This is evidence that further development of highly developed countries will be exclusively evolutionary. Unsurprisingly, the final trajectory of GDP of highly developed countries is reminiscent of exponential behavior. This is due to the fact that GDP growth rates over the interval at issue change smoothly and are close to a constant value of 2.75 % per annum.

Let us now summarize possible results of the Great Convergence, in case it will take place, as shown by the forecast calculations of the AN-model. First of all, we consider dynamic properties of the total population for various groups of countries, which are the most predictable variables. With actual data and forecasting calculations of the total population of various groups of countries at hand (see Fig. 7, 8, 9 and 10), we can easily calculate the dynamic share of those groups in the total population of the world (see Fig. 5). Some forecasted values for world population and populations of various groups of countries until 2050 are presented in Table 1.

Plots illustrating dynamics of the population shares of the four groups of countries discussed are presented in Fig. 31. The figure shows the population of high-income (HI) developed countries of the West will slowly decrease and by 2050 will comprise about 5 %, as in 1820 when the era of industrialization was starting.

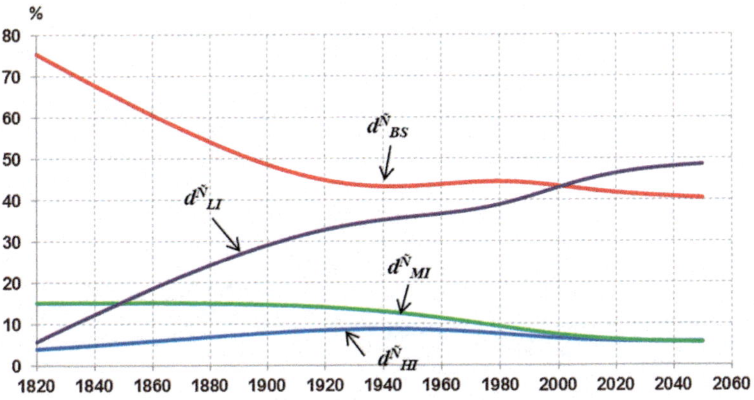

Fig. 31 The dynamics of various groups of countries in world population

The share of middle-income countries will also decrease to 5 %, shrinking threefold from 15 % in 1820. Thus, the total share of high- and middle-income countries in world population will decrease twofold, as compared to 1820, and will comprise only 10 % of the world population.

BRICS countries population share will also decrease and comprise by 2050 about 40 % of world population, as compared to 75 % in 1820, i.e. ¾ of world population. It is a quite natural result, as per capita income in these countries grew faster than in other groups of countries over the last 75 years, resulting in the acceleration of demographic transition and a relative decrease in birth rate. The share of the remaining part of the developing world, most of which is the African continent, will continue its growth and achieve by 2050 almost half of the world population, as compared to 6 % in 1820. This part of the world will represent the majority of poor people by the middle of the twenty-first century.

We have shown above, that the most important determinant of economic growth is population growth rate. Therefore, we calculated population growth rate for all the main groups of countries, as well as for the whole world and presented them in graphical form in Fig. 4. The figure shows that population growth rates in highly developed countries of the West throughout the nineteenth and twentieth centuries were persistently high, exceeding 1 % per annum. This population growth was critical to their technological leadership and it follows directly from basis Kremer's equation (7). The same applies to the group of middle-income countries, though population growth rates were half as high and stayed at the level of 0.6–0.7 % per annum.

Population growth rates in BRICS countries in the nineteenth century were extremely low, which defined their technological backwardness. Population growth rate in BRICS countries became equal to the corresponding values in middle- and high-income countries in 1920 and 1940, respectively (see Fig. 4). We have seen above, that it was during these years, that the GDP growth rates in BRICS countries exceeded 1 % and 2 % per annum, respectively (see Fig. 27). In the postwar period,

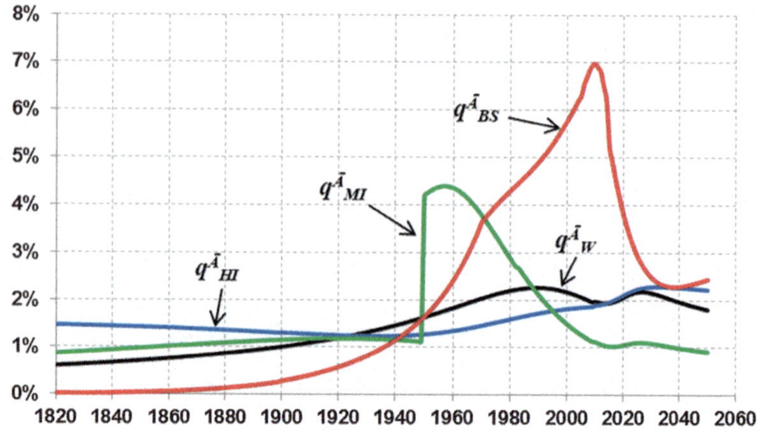

Fig. 32 Chain growth rate of technology advance for groups of countries

the population growth rates of BRICS countries were much higher than in developed countries, enabling high growth rates of technological advance of 4–6 % per annum for 2 or 3 decades, and, sometimes, 7 % per year. We should also point out the following empirical law: while in the nineteenth century world population growth was determined by the population growth in developed countries, starting from 1960 it was defined by the population growth rate in BRICS countries (see Fig. 4).

We will now consider the dynamics of technology advance and GDP growth for the countries of interest, already calculated and analyzed above. For visual convenience, we will put them in one plot. The plots of the chain growth rate and GDP for the main groups of countries are presented in Figs. 32 and 33. As Fig. 32 shows, the rates of technology advance in both high- and middle-income countries during most of the Industrial Era until WWII were almost equal at about 1 % per annum. However, the GDP growth rates in high-income developed economies were about 1 % and greater (see Fig. 33) than in middle-income economies due to higher population growth (Fig. 4)

After WWII, leading Western countries' growth rates of technology advance were permanently growing and stabilized at 2 % in the early 2000s. Middle-income countries actively borrowed technologies from the leading countries of the West in the postwar period, and supported by the Marshall Plan, they step-wise increased the growth rates of technology advance to 3–4 % per annum and maintained them at this level up to the world economic crisis of the 1970s (see Fig. 32). Because of the step-wise increase of the technology advance growth rate, they managed to achieve high GDP growth rates and maintain them at the level of 4–5 % per annum throughout the third quarter of the twentieth century (see Fig. 33). It was then that Germany and Japan demonstrated a new type of "*economic miracle*". Japan's achievements were especially impressive: it had attained unprecedented growth rates of technology advance of almost 8 % per annum for 2 decades, and turned into a technological and economic world leader. After the world economic

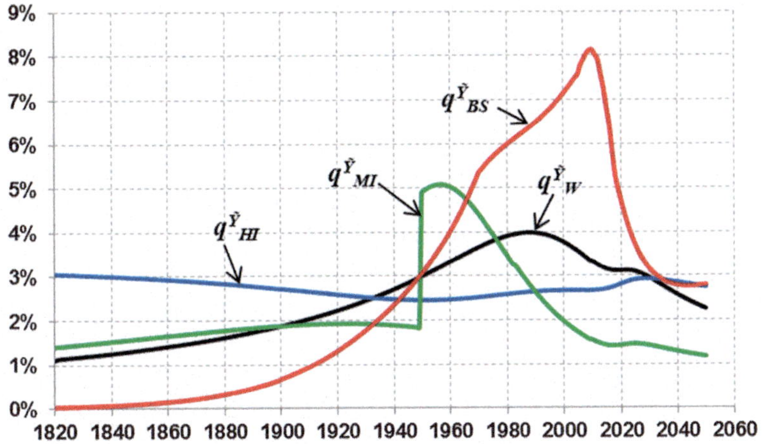

Fig. 33 Chain growth rate of GDP for groups of countries

Table 2 Forecast of the annual average GDP growth rates (%)

Years	2010–2020	2020–2030	2030–2040	2040–2050
World	3.2	3.1	2.8	2.4
High-income countries	2.8	3.0	3.0	2.8
Middle-income countries	1.5	1.5	1.4	1.2
BRICS countries	6.2	3.8	3.0	2.8
The rest of the world	−0.1	2.9	2.9	1.3

crisis of the 1970s, technology advance growth rates in middle-income countries started to decrease and, as indicated by forecast calculations, will stabilize at 1 % per annum level (see Fig. 32) during 6th BKC (2018–2050s).

Since after the 1970s BRICS countries (using Japan's expertise in harmoniously borrowing from abroad and creating its own innovation technologies through its national R&D system), achieved an even higher rate of technology advance, reaching 5–6 % per annum and maintaining this rate for decades (see Fig. 32). However, after the financial—economic crisis of 2008–2009, a natural decrease of technology advance had started in this group as well and will finish by the end of 2030s, stabilizing at approximately 2.5 % per annum, which is still higher than in developed countries of the world. From the analysis of Figs. 32 and 33 it follows that the decrease of the technology advance growth rate and GDP will continue throughout the 6th BKC (2018–2050), though for the high-income and BRICS countries some stabilization is observed in the end of the forecast period. This can be seen in Table 2, which presents annual average GDP growth rates calculated over decades for the various groups of countries.

Finally, we shift to consider the dynamics of the main parameter, describing the presence of convergence and its rate—the GDP of selected groups of countries. Forecast GDP values for the groups at issue are presented in Table 3.

Table 3 GDP forecast until 2050 ($billion)

Years	2010 (actual)	2020	2030	2040	2050
World	55,067	75,424	102, 264	134, 187	170, 021
High-income countries	13,330	17,608	23, 629	31, 611	41, 826
Middle-income countries	10,211	11,792	13, 620	15, 576	17, 591
BRICS countries	17,660	32,343	46, 886	62, 805	83, 098
The rest of the world	13,866	13,681	18, 130	24, 194	27, 506

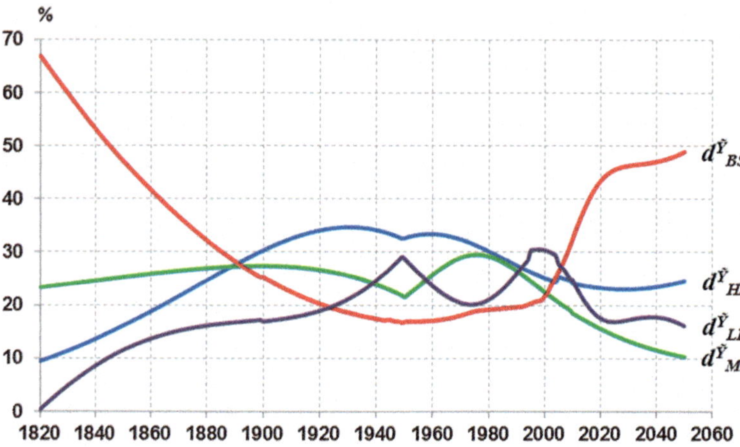

Fig. 34 Share of dynamics of various groups of countries in world GDP

The dynamics of shares of different groups of countries in world GDP is presented in Fig. 34. It shows that in the forecast period only one group of countries—BRICS will significantly increase its share in world GDP from 30 % to almost 50 % by 2050. High-income countries will stabilize their share at about 24 %, while middle-income countries will cut their share from 30 % in 1980 to 10 % in 2050. Developing countries other than BRICS will produce only about 17 % of world GDP by 2050 while accounting for almost 50 % of world population.

Naturally, the comparison of joint shares of the developed states ($HI + MI$) and developing countries ($BS + LI$) in world GDP, presented graphically in Fig. 35 is of main interest. The plots' analysis shows that in the time interval at issue, GDP obtained by means of AN-models with subsequent stage wise refinement of technology advance provide a good approximation of actual data both in qualitative and quantitative terms. We can see that the Great Divergence had taken place at a swift rate within a century (1820–1920). Then, a certain balance was kept until the 1970s when convergence started. Convergence began slowly and accelerated only after the 2000s (see Fig. 35).

The forecast part of Fig. 35 shows that the process of convergence is completed by the middle of the twenty-first century. The developing countries share of world GDP becomes 65 %, a bit less than the 70 % share in 1820. Developed countries'

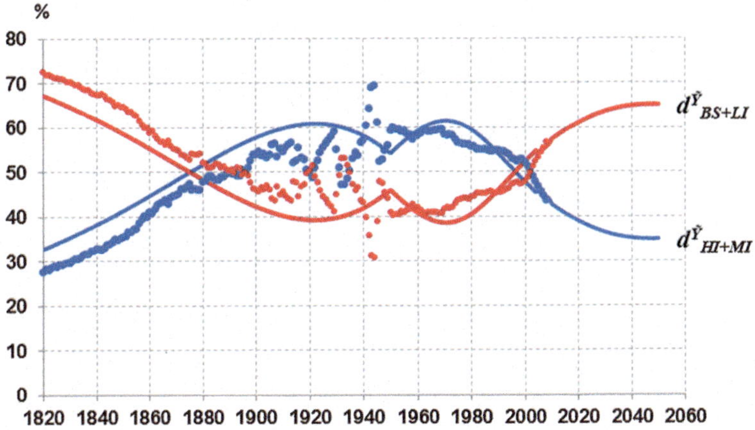

Fig. 35 Dynamics of shares of the developed and developing countries in world GDP. *Actual data source*: (Maddison 2010)

share decreases to 35 %, still higher than 28 % in 1820. Thus, the wheel of world history will come a full circle of 300 years (1750–2050) and, again, the lion's share of real goods production will be in the East—in the Asian part of the world, like it was in the middle of the seventieth century before the start of the Industrial Revolution in England.

The most outstanding result of the convergence will be that BRICS countries, the vanguard of the developing world with 40 % of world population, will shift by the middle of the twenty-first century to the group of middle-income developed countries. Hence, about 5 billion people, slightly more than half of the Earth's population will live in developed countries with a middle level of per capita income. It is quite possible that developing countries will by that time mature for long-term sustainable growth at high rates and convergence will occur between the two halves of humankind—present day developed countries HI, MI, and the group of BRICS countries that has joined them and the remaining world of poor developing countries.

In the beginning of the second part of that paper we mentioned that A.V. Korotaev noted a striking similarity between the dynamics of the gap between developed countries of the West and the rest of developing world in per capita income (times) and the dynamics of population growth. Let us consider this phenomenon with the example of the gap in per capita income in developed countries (HI + MI) and BRICS countries. This gap (times) is presented in graphic form in Fig. 36, together with the plot of world population growth rate—q_{N_W} (in %). The plot shows that similarity mentioned above has merit, and is, of course, not accidental.

An empirical dependence had been formulated earlier by the analysis of the population growth rate for various groups of countries (see Fig. 4): the Western population growth rate determined the world population growth rate during the first lengthy stage of industrialization (1820–1950). Starting in 1960, the developing

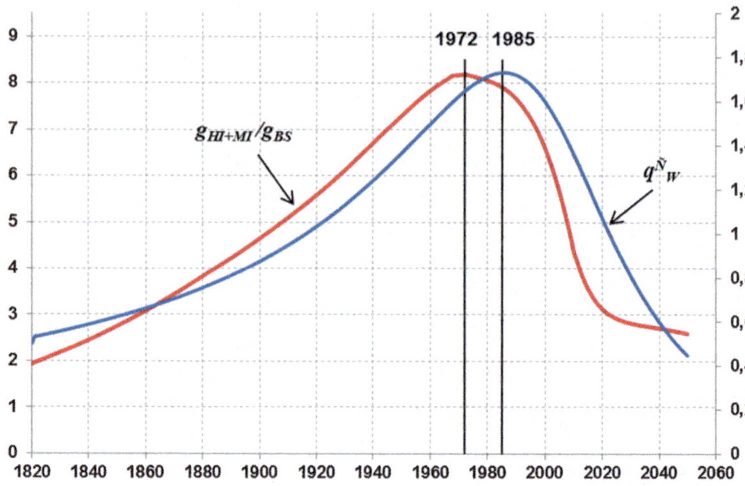

Fig. 36 The ratio of per capita GDP in the developed and BRICS countries and world population growth rate

world's vanguard BRICS countries' growth in population had been governing the world population growth. Additionally, rapid growth of per capita income in the developing countries of the West was observed at the first stage of industrialization, while BRICS countries per capita income grew very slowly until 1960 when it significantly outpaced growth in developed countries. Combining the two empirical dependencies clarifies the correlation between the two curves in Fig. 36.

Further, Fig. 36 shows that as early as the 2020s, the ratio of average per capita incomes in the developed and BRICS countries will decrease to 2:3 and stabilize at the pre-industrial level. This provides evidence that the BRICS countries will transform into countries with middle per capita income. Nobel Prize Winner R. Lucas predicted that this phenomenon would certainly occur in the twenty-first century. He wrote, "It is difficult to predict the future on the basis of data between 1960 and 1990, but I believe the incredible inequality of the post-war time was a global maximum and will continue to fall until it returns to the level of 1800 (Lucas 2013, p. 255)". Considering the world as it is today, it is evident that this transformation will probably occur by the end of the twenty-first century.

References

Akaev, A. A., Anufriev, I. E., & Akaeva, B. A. (2013a). Vanguard world countries in the XXI century under the conditions of converging development. Long-term forecasting of economic growth [Авангардные страны мира в XXI веке в условиях конвергентного развития. Долгосрочное прогнозирование экономического роста] (p. 144). LIBROKOM: Moscow.

Akaev, A. A., Anufriev, I. E., & Akaeva, B. A. (2013b). Mathematic modeling of world development. Demography, economy, energy production, technologies/projects and future risks. [Математическое моделирование мирового развития. Демография, экономика, энергетика, технология / Проекты и риски будущего] (pp. 278–311). KRASAND: Moscow.

Akaev, A. A., & Sadovnichij, V. A. (2010). *The mathematical model of demographic dynamics with stabilization of world population around stationary level* [Математическая модель демографической динамики со стабилизацией численности населения мира вокруг стационарного уровня] (Vol. 435, #3, pp. 317–321). DAN, Moscow.

Akaev, A. A., & Sadovnichij, V. A. (2011). Global demographic models as basis for strategic forecast/Projects and future risks. Concepts, models, instruments, forecasts [Глобальные демографические модели как основа для стратегического прогноза / Проекты и риски будущего. Концепции, модели, инструменты, прогнозы.] (pp. 17–44). KRASAND: Moscow.

Allen, R. C. (2013). *Global economic history: A very short introduction, Translated from Eng* [Глобальная экономическая история] (p. 224). Gaidar Institute Publishers: Moscow.

Bresnahan, T., & Trajtenberg, M. (1995). General purpose technologies: "Engines of growth". *Journal of Econometrics, 65*, 83–108.

Coe, D. T., & Helpman, E. (1995). International R&D spillovers. *European Economic Review, 39*, 859–887.

Denison, E. (1971). *Study of differences in economic growth, Short. Translated from English* [Исследование различий в темпах экономического роста] (p. 648). Progress: Moscow.

Helpman, E. (2012). *The mystery of economic growth, Translated from English* [Загадка экономического роста]. Gaidar Institute Publishers: Moscow.

Huntington, S. (1996). *The clash of Civilizations and the Remaking of World Order*. Newyork, NY: Simon and Shuster.

Jones, C. I. (1995). R&D-based models of economic growth. *Journal of Political Economy, 103*(4), 759–784.

Jorgenson, D. W., & Yip, E. (2001). *Whatever happened to productivity analysis*. Chicago: University of Chicago Press.

Kapitsa, S. P. (2008). *Outline of growth theory of humankind* [Очерк теории роста человечества] (Vol. 33). Nikitskij klub: Moscow.

Klenow, P. J., & Rodriguez-Clare, A. (1997). The neoclassical revival in grown economics: Has it gone too Far? *NBER Macroeconomics Annual, 12*, 73–103.

Korotayev, A., & Zinkina, J. (2014). On the structure of the present-day convergence. *Campus-Wide Information Systems, 31*(2–3), 139–152.

Kremer, M. (1993). Population growth and technological change: On million B.C. to 1990. *Quarterly Journal of Economics, 108*, 681–716.

Krugman, P. R. (1987). The narrow moving band, the Dutch Disease, and the competitive consequences of Mrs. Thatcher: Notes on trade in the presence of dynamic scale economies. *Journal of Development Economics, 27*, 41–55.

Lucas, R. E. (2013) *Lectures on economic growth, Translated from Englist* [Лекции по экономическому росту] (p. 288). Gaidar Institute Publishers: Moscow.

Maddison, A. (2010). World population, GDP and per capita GDP, A.D.1-2008. URL: http://www.ggdc.net/maddison.

Maddison, A. (2012). *Contours of the World Economy 1-2030 AD, Essays on Macroeconomic History, Translated from English* [Контуры мировой экономики в 1-2030 гг. Очерки по макроэкономической истории]. Gaidar Institute Publishers: Moscow.

Mel'yantsev, V. A. (1996). *The East and West in the second millennium: Economy, history and modernity* [Восток и Запад во втором тысячелетии: экономика, история и современность]. MGU Publishing: Moscow.

Mel'yantsev, V. A. (2009). *Developed and developing countries in the era of transformations* [Развитые и развивающиеся страны в эпоху перемен]. Kluch-S Publishing House.

Mel'yantsev, V. A. (2013). *Analysis of the most important trends of global economic growth* [Анализ важнейших трендов глобального экономического роста]. Kluch-S Publishing House.
Romer, P. M. (1990). Endogenous technical change. *Journal of Political Economy, 98*(5), 71–102.
Sadovnichij, V. A., Akaev, A. A., Korotaev, A. V., & Malkov, S. Y. (2014). *Complex modeling and forecast of development of BRICS countries in the context of world dynamics* [Комплексное моделирование и прогнозирование развития стран БРИКС в контексте мировой динамики]. Nauka: Moscow.
Schumpeter, J. (1982). *Theory of economic development* [Теория экономического развития]. Progress: Moscow.
Solow, R. M. (1956). A contribution to the theory of economic growth. *Quarterly Journal of Economics, 70*, 65–94.
Spence, M. (2013). *The next convergence: The future of economic growth in a multispeed world, Translated from English* [Следующая конвергенция. Будущее экономического роста в мире, живущем на разных скоростях] (p. 336). Gaidar Institute Publishers: Moscow.
Vishnevskij, A. (2005). *Selected demographic works. In 2 volumes* [Избранные демографические труды.]. Nauka: Moscow.
Yakovets, Y. V. (2004). *Epoch-making innovations of XXI century* [Эпохальные инновации XXI века]. Ekonomika: Moscow.

Borrowing Constraints and Monetary Policy: The Inflation Tax-Net Worth Channel

Tiziana Assenza and Domenico Delli Gatti

JEL codes: E31, E32, E44, E52

1 Introduction

During the 1990s a remarkable body of literature has been developed under the name of *financial accelerator* or *net worth channel* to explore the transmission of monetary policy on investment and production through the impact on firms' financial structure.

In the early 1990s two alternative frameworks have been proposed to model the Financial Accelerator hypothesis. Bernanke and Gertler (1989, 1990) and Bernanke et al. (1999) emphasized the role of agency costs in the decision to supply credit while Greenwald and Stiglitz (1988, 1993) focused on the role of bankruptcy costs in limiting investment and production. In the late 1990s Kiyotaki (1998) and Kiyotaki and Moore (1997, 2002a,b) put forward a new framework based on the

---

T. Assenza (✉)
Complexity Lab in Economics (CLE), Department of Economics and Finance, Università Cattolica del Sacro Cuore, Largo Gemelli 1, 20123 Milano, Italy

Center for Non-linear Dynamics in Economics and Finance (CeNDEF), Amsterdam School of Economics, University of Amsterdam, Amsterdam, The Netherlands
e-mail: tiziana.assenza@unicatt.it

D. Delli Gatti
Complexity Lab in Economics (CLE), Department of Economics and Finance, Università Cattolica del Sacro Cuore, Largo Gemelli 1, 20123 Milano, Italy

CESifo Group, Munich, Germany
e-mail: domenico.delligatti@unicatt.it

L. Bernard, U. Nyambuu (eds.), *Dynamic Modeling, Empirical Macroeconomics, and Finance*, DOI 10.1007/978-3-319-39887-7_4

idea that borrowers face a financing constraint because lenders extend credit up to the present value of borrowers' collateralizable wealth, i.e. "land", which proxies real assets. The novel and appealing feature of their model is the linkage of asset price changes and borrowing constraints: booming asset prices relax borrowing constraints and boost economic activity; the upswing, in turn, affects asset prices.[1]

Thanks to this feature this model is the natural vehicle to explore the balance-sheet channel of the monetary transmission mechanism. In the original model, however, all the variables, credit included, are in real terms. Therefore the model as such is not suitable to explore the effects of monetary policy. In order to assess the impact of a such a policy, the model must be reformulated to fit a *monetary economy with borrowing constraint.*[2] An attempt in this direction has been proposed by Cordoba and Ripoll (2004a,b), who introduce a cash in advance constraint for consumption and investment to analyze the role of collateral constraints as a transmission mechanism of monetary shocks.

In our paper we follow a well known but different route. We model a monetary economy with financing constraints *adopting the Money In the Utility function (MIU) approach.*[3]

As in Kiyotaki and Moore (KM) population consists of financially constrained farmers and non constrained gatherers. There are two types of goods, a non storable consumption good (fruit) and a collateralizable, durable asset/input (land). The total supply of land is constant.

By assumption the farmer and the gatherer have different rates of time preference[4] and have access to different technologies. The farmer is endowed with a linear technology while the gatherer has a well behaved production function.

In the present framework, however, the utility function is non linear and has two arguments: consumption and money holdings. since preferences are non-linear in consumption, we must not assume that part of production is non-tradable and that consumption is limited to bruised fruit as in KM.

In this setting, we study the effects of a monetary injection on the asset price and the allocation of land to the farmers and the gatherers. Notice that, in equilibrium

---

[1]An interesting extension is presented in Iacoviello (2005). The author applies the Kiyotaki and Moore framework to a general equilibrium model to study the effects of houses' prices fluctuations on the business cycle and possible monetary policy interventions.

[2]In footnote 18 of the 1997 paper KM recognize that a possible transmission channel of a monetary shock in their framework, once the role of money is specifically taken into account, would take place through a redistribution of wealth between borrower and lenders. In our framework this is indeed the case. Redistribution is induced by an inflation tax effect on net worth.

[3]The introduction of money in the utility function is a controversial but well known and largely adopted modelling procedure in the monetary literature. A simple argument in favour of such a procedure is that agents allocate time to shopping to purchase consumption goods and that the amount of shopping time needed to purchase a certain level of consumption goods is negatively related to agents' holdings of real money balances.For description of a shopping time monetary economy see Ljungqvist and Sargent (2000, Chap. 17).

[4]Another way of characterizing the framework put forward by Kiyotaki and Moore, in fact, is in terms of *preference heterogeneity.*

the real interest rate is anchored to the rate of time preference of the gatherer. Since this is exogenous, the real interest rate is given and constant. As a consequence, a change of the inflation rate brings about the same change in the nominal interest rate. Finally, in the steady state the inflation rate is pinned down to the rate of growth of money supply. In the end therefore, in the steady state a change of the rate of growth of money supply brings about a change of the same sign and the same magnitude of the nominal interest rate.

In the present context, therefore, by construction an expansionary monetary policy cannot affect real output through the *liquidity effect*. This channel of monetary transmission is not active because, in the end, an increase in the rate of growth of money supply raises the inflation rate and pushes the interest rate up, not down. This does not necessarily mean, however, that superneutrality always holds. A change in the growth rate of money supply can affect real output through *the impact of inflation on the farmer's net worth, landholding and output*. In a sense the monetary transmission mechanism we are focusing on consists in a combination of the *inflation tax* effect and the net worth channel. Contrary to the traditional view, at least for some parameter restrictions, an increase of the inflation tax can bring about an increase of aggregate output.

We can go as far as to gauge the possibility to determine an optimal inflation rate. For instance we can determine an output maximizing inflation rate, i.e. a rate of growth of money supply (and inflation) which equals the marginal productivity of land of the farmer and of the gatherer. In our context, we can also determine an inflation rate which maximizes social welfare.

The paper is organized as follows. In Sect. 2 we discuss the optimization problems of the farmer and the gatherer. In Sect. 3 we present the equilibrium conditions and in Sect. 4 we discuss the dynamics generated by the model. In particular we explore the effects of a monetary shock and discuss the criteria to determine optimal inflation. In Sects. 5 and 6 we present the effects on individual and aggregate variables of unexpected and temporary shocks to the productivity parameter and to the rate of growth of money supply respectively. Section 7 concludes.

2 The Model Set Up

Differently from KM we assume that preferences of both the farmer and the gatherer are defined over consumption and real money balances. Moreover, while KM adopt a linear utility function, we consider a Cobb-Douglas specification:

$$U^i = \sum_{s=0}^{\infty} (\beta^i)^s \left[\gamma \ln c^i_{t+s} + (1-\gamma) \ln m^i_{t+s}\right] \qquad i = F, G \tag{1}$$

The superscript F (G) denotes the farmer (gatherer); $0 < \gamma < 1$; c^i_{t+s} is real consumption and $m^i_{t+s} = \frac{M^i_{t+s}}{P_{t+s}}$ are real money balances. Following Ljungqvist and Sargent (2000) we assume that *at the beginning of period t* agent i decides the amount of money he wants to hold *at the beginning of period* $t+1$. Therefore the Flow of Funds (FF) constraint in real terms is

$$c^i_t + (1+\pi_{t+1})\, m^i_{t+1} + q_t \left(K^i_t - K^i_{t-1}\right) \leq y^i_t + (1-\varphi^i) m^i_t \pm (b_t - Rb_{t-1})$$

where $\pi_{t+1} = \frac{P_{t+1}}{P_t} - 1$ is the inflation rate, $(1+\pi_{t+1})\, m^i_{t+1}$ are real money balances the agent decides to carry from period t to period $t+1$, $q_t := \frac{Q_t}{P_t}$ is the real price of land, $q_t \left(K^i_t - K^i_{t-1}\right)$ represents investment in land, $b_t := \frac{B_t}{P_t}$ is credit in real terms and $R = 1 + r := \frac{1+i_t}{1+\pi_{t+1}}$ is the real interest rate so that Rb_{t-1} is debt reimbursement. In the FF of the lenders (borrowers) the term $(b_t - Rb_{t-1})$ shows up with a negative (positive) sign. $\varphi^i m^i_t$, $0 \leq \varphi^i < 1$ are reserves that the agent keeps as a buffer stock. At the beginning of period t the nominal money balances available to the lender will be $M^i_t \left(1 - \varphi^i\right)$.

As in KM we assume that the farmer has a constant return to scale production function, i.e. $y^F_t = \alpha K^F_{t-1}$, with $\alpha > 0$.[5] The gatherer has a decreasing returns to scale technology $y^G_t = G\left(K^G_{t-1}\right)$ that fulfills the Inada conditions, i.e. $G' > 0$ and $G'' < 0$, $G'(0) = \infty$, $G'(\infty) = 0$.

In order to establish the way in which liquidity changes over time in our framework, let's assume that each agent gets a transfer in money in $t+1$ from the public sector proportional to his money holdings in t: $T^i_{t+1} = g_M M^i_t$. The supply of money, therefore, follows the law of motion $M^i_{t+1} = M^i_t + T^i_{t+1} = M^i_t (1 + g_M)$. In words, money holdings grow at the (exogenous) rate g_M. In order to keep the analysis as simple as possible we assume that the rate of change of money supply g_M is uniform across agents. This means that the allocation of money to the agents, i.e. the ratio of money of the farmer to money of the gatherer—which will be denoted by σ in the following—is constant. Therefore real money balances of the agent in $t+1$ are $m^i_{t+1} := \frac{M^i_{t+1}}{P_{t+1}} = \frac{M^i_t}{P_t}\frac{1+g_M}{1+\pi_{t+1}} = m^i_t \frac{1+g_M}{1+\pi_{t+1}}$.

Substituting this relation into the FF constraint we get:

$$c^i_t + \left(g_M + \varphi^i\right) m^i_t + q_t \left(K^i_t - K^i_{t-1}\right) \leq y^i_t \pm (b_t - Rb_{t-1}) \tag{2}$$

[5]Since preferences are not linear in consumption we must not distinguish between tradable output and bruised fruit as KM do.

The term $(g_M + \varphi^i)\, m_t^i = \dfrac{M_{t+1}^i - M_t^i\,(1-\varphi^i)}{P_t}$ is the increase in the agent's money holdings between t and $t+1$ in real terms, i.e. at prices of period t. For lack of a better term we will refer to this magnitude as the increase in real money balances.[6]

2.1 *The Farmer/Borrower*

The farmer, as in KM, will play the role of the borrower in our economy. He maximizes the utility function (1) subject to the FF (2). For the sake of simplicity we assume the farmer does not hold reserves, i.e. $\varphi^F = 0$. After substitutions the farmer's FF specializes to:

$$c_t^F + g_M m_t^F + q_t\left(K_t^F - K_{t-1}^F\right) + Rb_{t-1} \leq \alpha K_{t-1}^F + b_t \tag{3}$$

The term $g_M m_t^F = \dfrac{M_{t+1}^F - M_t^F}{P_t}$ is the increase in the farmer's money holdings between t and $t+1$ in real terms. In the light of this remark, (3) can be interpreted as follows: "resources" of the farmer, of internal or external origin (αK_{t-1}^F and b_t respectively), can be employed to consume (c_t^F), "invest" ($q_t\left(K_t^F - K_{t-1}^F\right)$), reimburse debt ($Rb_{t-1}$) and increase money balances ($g_M m_t^F$). The increase in desired money balances is always equal to the increase of liquidity engineered by the central bank by means of money transfers. In other words, we are ruling out the mismatch between desired and actual increase in money holdings.

The farmer is also financially constrained, the financing constraint can be expressed as:

$$b_t \leq \frac{q_{t+1}}{R} K_t^F \tag{4}$$

From the FOCs we get the following relation (see Appendix 1 for details):

$$\frac{c_t^F}{m_t^F} = \frac{\gamma}{1-\gamma} g_M \qquad \forall t \tag{5}$$

Given the rate of growth of money supply g_M, the equality above states that the ratio of consumption to real money balances is constant over time.

[6] We assume that agents are aware of the way in which the public sector carries out money transfers, i.e. that money holdings grow at the (exogenous) rate g_M. Therefore, this transfer mechanism is incorporated in the budget constraint and will be taken into account in the optimization problem.

Moreover the financing constraint is binding if the following condition holds (see appendix for details):

$$\frac{1 + g_M}{1 + \pi_{t+1}} > \beta^F R$$

In the steady state real money balances are constant, i.e. the rate of growth of money is equal to the inflation rate $g_M = \pi$. The inequality above, therefore, boils down to:

$$R < \frac{1}{\beta^F}$$

The same condition holds also in the original KM framework.

Given the binding conditions for the FF and the financing constraints we can write

$$c_t^F + g_M m_t^F = \alpha K_{t-1}^F - \mu_t K_t^F \tag{6}$$

where $\mu_t = q_t - \dfrac{q_{t+1}}{R}$ is the downpayment. In other words, the output produced by the farmer (αK_{t-1}^F) is employed to consume (c_t^F), hold money balances ($g_M m_t^F$) and provide the downpayment ($\mu_t K_t^F$). Given output, the higher the rate of growth of nominal money balances, the smaller consumption and/or downpayment, the smaller therefore the investment in landholding.

The optimal level of consumption and real money balances are:

$$c_t^F = \gamma \left(\alpha K_{t-1}^F - \mu_t K_t^F\right) \tag{7}$$

$$g_M m_t^F = (1 - \gamma) \left(\alpha K_{t-1}^F - \mu_t K_t^F\right) \tag{8}$$

Thanks to the Cobb-Douglas specification of preferences, consumption and the increase in real money balances are a fraction γ and $1 - \gamma$ respectively of the resources available to the farmer, which in turn are equal to output less downpayment.

After trivial substitutions from the flow of funds constraint one gets:

$$K_t^F = \frac{1}{\mu_t} \left[\alpha K_{t-1}^F - \delta g_M m_t^F\right] \tag{9}$$

where $\delta = \dfrac{1}{1 - \gamma}$. Equation (9) is the law of motion of the land of the farmer. Notice that it differs from the law of motion obtained by KM due to the term that represents the increase of real money balances. In particular there exists a negative relation between the demand for land and the demand for money: the higher the real money balances demanded by the farmer the lower landholding and viceversa.

Since both the FF and the financing constraints are binding, from the very definition of net worth in the present context we can conclude that net worth is equal to *saving net of the increase in real money balances*:

$$n_t^F = y_t^F - c_t^F - g_M m_t^F - Rb_{t-1}^F + q_t K_{t-1}^F = s_t^F - g_M m_t^F \tag{10}$$

In KM $n_t^F = s_t^F$. The increase in real money balances therefore is a negative component of net worth.

Making use of (5), after some algebra we can rewrite the equation above as:

$$n_t^F = \alpha K_{t-1}^F - \delta g_M m_t^F$$

Other things being equal the higher the increase in real money balances, the lower net worth.

Finally notice that since the sum of consumption and the increase in real money balances is equal to the resources available to the farmer, net worth is devoted completely to downpayment:

$$n_t^F = y_t^F - (c_t^F + g_M m_t^F) = y_t^F - (y_t^F - \mu_t K_t^F) = \mu_t K_t^F \tag{11}$$

The same condition holds in KM. Considering (10) and (11) simultaneously we infer that in a KM economy with money

$$n_t^F = s_t^F - g_M m_t^F = \mu_t K_t^F$$

In the steady state $g_M = \pi$ so that

$$n^F = s^F - \pi m^F \tag{12}$$

Eq. (12) states that an increase in the rate of growth of money (which translates into an increase of inflation) has an *inflation tax effect* on the accumulation of net worth.

2.2 The Gatherer/Lender

The gatherer, as in KM, will play the role of the lender in our economy. He maximizes the utility function (1) subject to the FF (2). Since the gatherer is carrying on lending activity, he has to hold reserves as a buffer stock (i.e. $0 < \varphi^G < 1$). Therefore we can think of φ^G as a policy parameter, possibly established by the central bank in his role of regulator/supervisor of the banking system or as a rough measure of transaction cost due to "financial frictions". After substitutions, the gatherer's FF specializes to:

$$c_t^G + q_t\left(K_t^G - K_{t-1}^G\right) + b_t + (g_M + \varphi^G)m_t^G \leq G\left(K_{t-1}^G\right) + Rb_{t-1} \tag{13}$$

The term $(g_M + \varphi^G)m_t^G = \dfrac{M_{t+1}^G - \left(1 - \varphi^G\right) M_t^G}{P_t}$ is the increase in the gatherer's money holdings between t and $t+1$ in real terms. According to (13) the resources of the gatherer, i.e. the sum of output $G\left(K_{t-1}^G\right)$ and interest payments Rb_{t-1}, can be employed to invest $(q_t\left(K_t^G - K_{t-1}^G\right))$, increase money holdings $((g_M + \varphi^G)m_t^G)$, consume (c_t^G) and extend credit to the farmer (b_t).

Solving the maximization problem for the gatherer we find that the FF constraint is always binding (see Appendix 2 for details).

From the FOCs and $\dfrac{m_{t+1}^G}{m_t^G} = \dfrac{1+g_M}{1+\pi_{t+1}}$ follows:

$$\frac{1+g_M}{1+\pi_{t+1}} = \beta^G R$$

In the steady state $g_M = \pi$ so that:

$$R = \frac{1}{\beta^G} \tag{14}$$

i.e. the real interest rate is pinned down to the rate of time preference of the gatherer (as in KM).

Since the rate of time preference of the gatherer is exogenous, the real interest rate is given and constant. As a consequence, a change of the inflation rate brings about the same change in the nominal interest rate. In the steady state the inflation rate is pinned down to the rate of growth of money supply. In the end therefore, in the steady state a change of the rate of growth of money supply brings about a change of the same sign and the same magnitude of the nominal interest rate.

Let's recall now that from the maximization problem of the farmer in the steady state the financing constraint is binding if:

$$R < \frac{1}{\beta^F} \tag{15}$$

From (14) and (15) follows:

$$\beta^G > \beta^F \tag{16}$$

Therefore we can conclude that the farmer is the more impatient agent as in KM.

Substituting the financing constraint into the flow of funds constraint and recalling that both constraints are binding we get:

$$c_t^G + (g_M + \varphi^G)m_t^G = G\left(K_{t-1}^G\right) + \mu_t K_t^F \tag{17}$$

Equation (17) states that the resources of the gatherer, i.e. output ($G\left(K_{t-1}^{G}\right)$) and the downpayment ($\mu_t K_t^F$) received from the farmer, can be employed to consume (c_t^G) and increase money balances ($(g_M + \varphi^G)m_t^G$).

Moreover from the FOCs we obtain (see the appendix for details):

$$G'\left(K_t^G\right) = R\mu_t \tag{18}$$

We get the same condition obtained by KM which equates the present value of the marginal productivity of the gatherer to the downpayment.

Finally we determine the optimal level of consumption and real money balances for the gatherer:

$$c_t^G = \gamma\left[G\left(K_{t-1}^G\right) + \mu_t K_t^F\right] \tag{19}$$

$$\left(g_M + \varphi^G\right) m_t^G = (1-\gamma)\left[G\left(K_{t-1}^G\right) + \mu_t K_t^F\right] \tag{20}$$

Thanks to the Cobb-Douglas specification of preferences, consumption and the increase of real money balances are a fraction γ and $1-\gamma$ respectively of the *resources* available to the gatherer, which in turn are equal to the sum of the output and the downpayment of the farmer.

3 Equilibrium

Total consumption in t is equal to the sum of the optimal consumption of the farmer and the gatherer. Recalling (7) and (19) we can write:

$$c_t = c_t^F + c_t^G = \gamma\left[\alpha K_{t-1}^F + G\left(K_{t-1}^G\right)\right]$$

In words, consumption in t is a share γ of total production available in the same period,

$$y_t = \alpha K_{t-1}^F + G\left(K_{t-1}^G\right) \tag{21}$$

We assume that Government expenditure is $g_t = (1-\gamma)\, y_t$ i.e. the public sector plays the role of "buyer of last resort" purchasing all the output not consumed by the private sector. The aggregate resource constraint $c_t + g_t = y_t$ is always satisfied.

Since by construction there are no taxes, g_t represents also the public sector deficit, which we assume is financed by means of money. The change of money supply between beginning-of-period t and beginning-of-period $t+1$, i.e. $M_{t+1} - M_t$, is equal to nominal Government expenditure $P_t g_t$. Hence $g_t = g_M m_t^F + \left(g_M + \varphi^G\right) m_t^G$. Agents end up "saving" a portion $(1-\gamma)\, y_t$ of total income in the form of real money balances.

In fact from (5) follows:

$$c_t^F + c_t^G = \frac{\gamma}{1-\gamma}\left[g_M m_t^F + \left(g_M + \varphi^G\right) m_t^G\right]$$

from which we get:

$$\left[\alpha K_{t-1}^F + G\left(K_{t-1}^G\right)\right](1-\gamma) = g_M m_t^F + \left(g_M + \varphi^G\right) m_t^G$$

or

$$(1-\gamma)\, y_t = g_M m_t^F + \left(g_M + \varphi^G\right) m_t^G \tag{22}$$

For the sake of simplicity we assume that the ratio between the real money balances of the gatherer and the farmer is constant over time: $\frac{m_t^G}{m_t^F} := \sigma$. Substituting into (22) we get:

$$(1-\gamma)\, y_t = \left[g_M + \left(g_M + \varphi^G\right)\sigma\right] m_t^F$$

After trivial algebra we get:

$$m_t^F = \frac{1-\gamma}{g_M(1+\sigma) + \varphi^G\sigma}\left[\alpha K_{t-1}^F + G\left(K_{t-1}^G\right)\right] \tag{23}$$

4 Dynamics

In a monetary KM economy, the maximization problems of the farmer and the gatherer yield a system of dynamic equations (consisting of (9), (18), (23) and the definition of the downpayment) which we rewrite for convenience of the reader:

$$\begin{cases} \mu_t K_t^F = \alpha K_{t-1}^F - \delta g_M m_t^F \\ G'\left(K_t^G\right) = R\mu_t \\ m_t^F = \dfrac{1-\gamma}{g_M(1+\sigma) + \varphi^G\sigma}\left[\alpha K_{t-1}^F + G\left(K_{t-1}^G\right)\right] \\ \mu_t = q_t - \dfrac{q_{t+1}}{R} \end{cases} \tag{24}$$

We search for the steady states and explore their dynamic properties in Sect. 4.1). In Sect. 4.2 we describe the transmission mechanism of a monetary shock. Section 4.3 is devoted to the properties of this system from the point of view of aggregate output while we make some welfare considerations in Sect. 4.4.

4.1 Steady States and Their Properties

Substituting the third and fourth equations of system (24) into the first one we get:

$$\left(q_t - \frac{q_{t+1}}{R}\right) K_t^F = \alpha K_{t-1}^F - A\left[G\left(\bar{K} - K_{t-1}^F\right) + \alpha K_{t-1}^F\right] \tag{25}$$

where

$$A := \left[1 + \sigma\left(1 + \frac{\varphi^G}{g_M}\right)\right]^{-1} \tag{26}$$

is a polynomial of policy parameters g_M and φ^G and σ. Moreover $0 < A < 1$ and it is increasing with g_M and decreasing with φ^G and σ. Recalling Eq. (6) it is easy to verify that A is the share of total output Y_t that goes to the farmer and that he can consume or employ to increase real money balances. An expansionary monetary policy—i.e. an increase of g_M—therefore, yields a redistribution of income in favour of the farmer.

In the steady state $q_{t-1} = q_t = q$ and $\mu = q\eta$ where $\eta = 1 - \frac{1}{R}$. Therefore Eq. (25) becomes:

$$q = \frac{\alpha}{\eta}\left[(1 - A) - Ah\left(K^F\right)\right] \tag{27}$$

where $h\left(K^F\right) = \dfrac{G\left(\bar{K} - K^F\right)}{\alpha K^F}$; $h'\left(K^F\right) < 0$.

Equation (27) is the farmer's steady state *landholding* equation.

Substituting the fourth equation into the second one and adapting the time index we get:

$$q_t = Rq_{t-1} - G'\left(\bar{K} - K_{t-1}^F\right) \tag{28}$$

In the steady state equation (28) becomes:

$$q = \frac{G'\left(\bar{K} - K^F\right)}{R - 1} \tag{29}$$

Eq. (29) is the steady state *asset price* equation.

In a sense (27) and (29) are demarcation curves in discrete time. In principle the system consisting of these equations can be solved for the steady state values of $K^F$and q. Equation (27) yields an increasing relationship between q and K^F. It crosses the x-axis when $K^F = h^{-1}\left(\frac{1-A}{A}\right) = \hat{K}^F$. If $K^F \rightarrow 0$ then $q \rightarrow -\infty$,

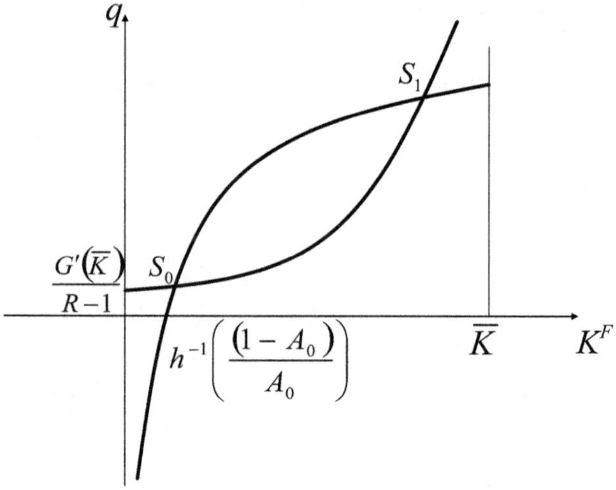

Fig. 1 Steady state loci

while if $K^F = \bar{K}$ then $q = \frac{\alpha}{\eta}(1-A)$. Therefore we conclude that, in the domain of interest, Eq. (27) yields an increasing and concave relationship between q and K^F (see Fig. 1).

On the other hand, Eq. (29) yields an increasing and convex relationship between q and K^F. It crosses the y-axis when $q = \frac{G'(\bar{K})}{R-1}$. When $K^F \to \bar{K}$ then $q \to \infty$.

We can explore different scenarios. In fact, given the properties of the two relations above we can obtain zero, one or two steady states depending on the level of A. We will assume the following:

Assumption A1 *A is greater than a threshold $\hat{A}$ defined as follows:*

$$\hat{A} := \frac{-G''(\bar{K}-K^F)\left(K^F\right)^2}{R\left[G\left(\bar{K}-K^F\right)+K^F G'\left(\bar{K}-K^F\right)\right]}$$

This assumption allows for multiple steady states (see Fig. 1).[7,8]

[7]If $A = \hat{A}$ the demarcation curves are tangent. If $A < \hat{A}$ there will be no steady state.

[8]In order to obtain a closed form solution for the steady state we should specify the production function $G(.)$. We could obtain very complicated closed form solutions without much insight. Therefore we do not proceed on this route.

In order to assess the properties of each of the two steady states, first of all we rewrite (25) using the second equation of the system (24) to get:

$$\frac{G'\left(\bar{K}-K_t^F\right)}{R}K_t^F = \alpha K_{t-1}^F - A\left[G\left(\bar{K}-K_{t-1}^F\right)+\alpha K_{t-1}^F\right] \tag{30}$$

which is an implicit function of K_t^F and K_{t-1}^F, independent of q.

Then we linearize the system (28) and (30) around each steady state and compute the jacobian matrix:

$$J_i = \begin{bmatrix} R\frac{\alpha(1-A)+AG'\left(\bar{K}-K_i^F\right)}{-G''\left(\bar{K}-K_i^F\right)K_i^F+G'\left(\bar{K}-K_i^F\right)} & 0 \\ G''\left(\bar{K}-K_i^F\right) & R \end{bmatrix} \tag{31}$$

with $i = 0, 1$.

J_i is a lower triangular matrix. Therefore the eigenvalues coincide with the elements on the main diagonal. One of the eigenvalues is $R > 1$. The second eigenvalue $R\frac{\alpha(1-A)+AG'\left(\bar{K}-K_i^F\right)}{-G''\left(\bar{K}-K_i^F\right)K_i^F+G'\left(\bar{K}-K_i^F\right)}$ is smaller than one in the higher steady state S_1 ($i = 1$). In this case therefore the steady state is a saddle point. This means that there is only one trajectory on the $\left(K_t^F, q_t\right)$ phase space which is converging to the steady state S_1, i.e. the saddle path. All the other trajectories diverge.

In Fig. 1 there are two steady states. The "low" steady state S_0 is unstable, while the "high" steady state S_1 is a saddle point. Therefore in the discussion that follows we will focus on steady state S_1.

Notice that A is an increasing function of the rate of growth of money supply. In Fig. 2 the convex solid line represents Eq. (29), that is independent from A.

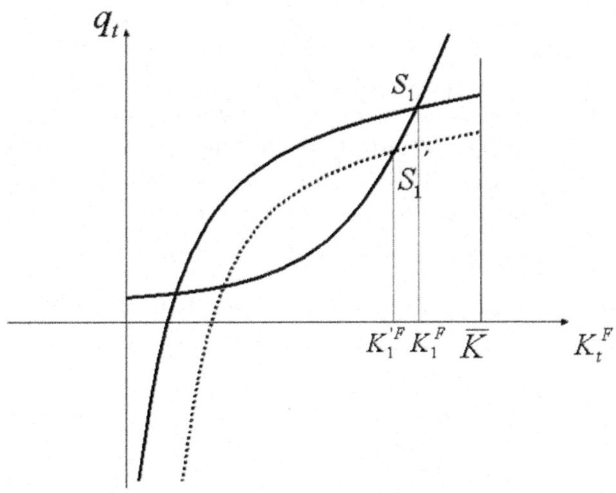

Fig. 2 Effects of an expansionary monetary policy when the initial condition is S_1

The concave solid line represents Eq. (27). Let's assume that the economy lies in the higher steady state (S_1). Suppose that an expansionary monetary policy increases A. The concave curve shifts down: the new situation is represented by the dotted line. Assuming that q_t jumps down to the new saddle path, the economy follows a trajectory which converges to the new saddle point ($S_1^{'}$). In other words in the end an increase of the rate of growth of the money supply yields a decrease of the farmer's landholding and of the asset price.

4.2 Monetary Policy, Net Worth and the Inflation Tax

"In the long run" $g_M = \pi$, i.e. a change in the rate of growth of money supply translates into a change in the rate of inflation of the same magnitude.

In order to assess the overall impact of this policy move, we recall that aggregate output is defined as $Y_t = f\left(K_{t-1}^F\right)$ where $f\left(K_{t-1}^F\right) = \alpha K_{t-1}^F + G\left(\bar{K} - K_{t-1}^F\right)$. Therefore it is a non monotonic function of the farmer's landholding. In particular Y_t is an increasing (resp. decreasing) function of K_{t-1}^F if $\alpha > G'\left(\bar{K} - K^F\right)$ ($\alpha < G'\left(\bar{K} - K^F\right)$). In other words, defining the farmer's landholding which maximizes aggregate output as $K_m^F = \bar{K} - G'^{-1}(\alpha)$, Y_t is an increasing (resp. decreasing) function of K_{t-1}^F if $K_{t-1}^F < K_m^F$ ($K_{t-1}^F > K_m^F$).

Let 's refer to the aggregate output in the higher steady state (S_1) as $Y_1 = f\left(K_1^F\right)$ (see Fig. 3)

Assume the economy is in S_1, where $K_{t-1}^F = K_t^F = K_1^F$ and $Y_1 = f\left(K_1^F\right)$. There are two scenarios.

(1) Suppose that $K_1^F < K_m^F$, then, an expansionary policy move (i.e. an increase of g_M) has a negative long run effect on aggregate output.
(2) Suppose now that $K_1^F > K_m^F$. In this case things become more complicated. Due to non-monotonicity, there are two points on the x-axis such that output is equal to Y_1. One of them is K_1^F.Let's indicate the other point as $\underline{K}_1^F$.By construction, $\underline{K}_1^F < K_1^F$. In this scenario, an expansionary policy move (i.e. an increase of g_M) has a positive long run effect on aggregate output if the reduction in the farmer's landholding is smaller in absolute value than $\Lambda := K_1^F - \underline{K}_1^F$. If this is the case, in fact, the new after shock steady state, say $K_1^{'F}$, would be such that $\underline{K}_1^F < K_1^{'F} < K_1^F$ so that $Y_1' = f\left(K_1^{'F}\right) > Y_1 = f\left(K_1^F\right)$.

Example 1 We can be a little more specific on this issue assuming, for the sake of discussion, that $G\left(\bar{K} - K_1^F\right) = \sqrt{\bar{K} - K_1^F}$. Solving (27) and (29) for K_1^F, by means of implicit differentiation one gets:

$$\frac{dK_1^F}{dA} = \frac{2K_1^F\left(\bar{K} - K_1^F\right)Y_1}{2A\bar{K}\sqrt{\bar{K} - K_1^F} - (1-A)\,\alpha\left(K_1^F\right)^2}$$

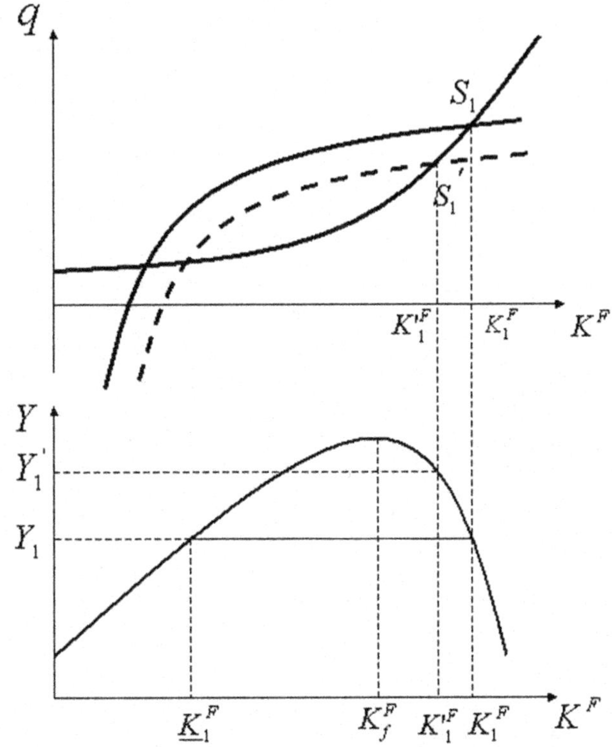

Fig. 3 Positive effect of an expansionary policy move on aggregate output

The expression above measures the impact on the steady state farmer's land of an infinitesimal increase of A. We assume that A is "sufficiently small" so that the expression is negative, as one would expect since the steady state we are considering is the saddle point.

For the sake of discussion, let's assume that this evaluation of the impact extends to discrete changes, i.e.

$$\Delta K_1^F = \frac{2K_1^F\left(\bar{K} - K_1^F\right)Y_1}{2A\bar{K}\sqrt{\bar{K} - K_1^F} - (1 - A)\,\alpha\left(K_1^F\right)^2}\Delta A$$

As to Λ, after some algebra we get the following equation:

$$\Lambda = \frac{1}{\alpha}\left(\frac{1 - 4\alpha Y_1}{\alpha^2} + 4\bar{K}\right)^{1/2}$$

Therefore, if the absolute value of ΔK_1^F is smaller than Λ, then $\dfrac{\Delta Y_1^F}{\Delta A} > 0$. After some algebra, we reach the conclusion that this is indeed the case if A is smaller than a threshold $\bar{A}$

$$A < \bar{A} = \frac{\alpha \left(K_1^F\right)^2}{\alpha \left(K_1^F\right)^2 + 2\bar{K}\sqrt{\bar{K} - K_1^F}} \left\{ 1 - \frac{2\left(\bar{K} - K_1^F\right) Y_1}{K_1^F \sqrt{\frac{1-4\alpha Y_1}{\alpha^2} + 4\bar{K}}} \right\} \tag{32}$$

In Fig. 3 we represent the effects of an expansionary policy move as described above. An increase of the rate of growth of money supply—i.e. an increase of the inflation tax—has a positive effect on aggregate output. This result is due to the fact that following the decrease in the steady state farmer's landholding from K_1^F to $K_1'^F$, the economy moves along the downward sloping branch of the aggregate output function: The loss of output due to the reduction in farmer's landholding is more than offset by the increase of output due to the increase of the gatherer's landholding.

In Fig. 4 we represent the negative effect of an expansionary monetary policy move as described in scenario (1). In this case, in fact, the economy moves along the upward sloping branch of the aggregate production function.

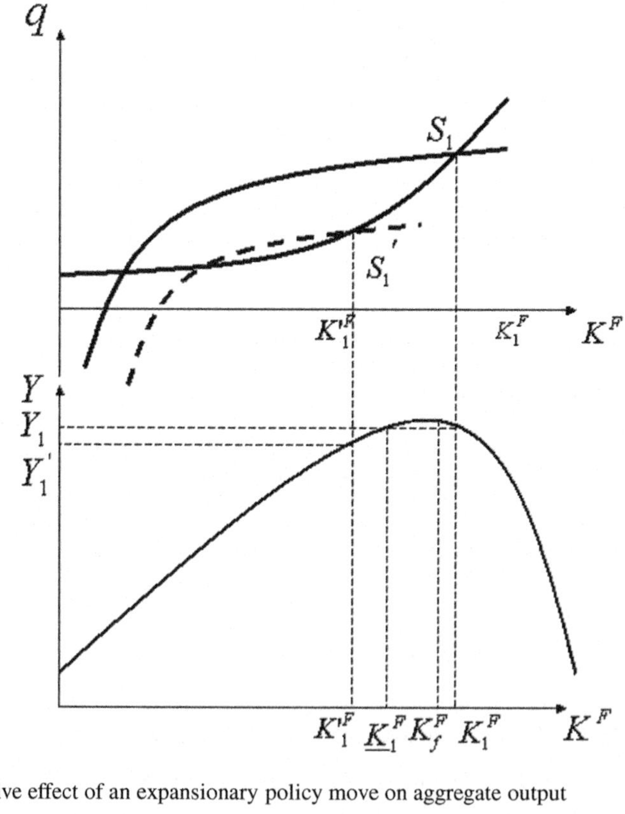

Fig. 4 Negative effect of an expansionary policy move on aggregate output

4.3 Is There an Output Maximizing Inflation Rate?

The allocation of land in the steady state depends on the distribution of total income. An increase of A, i.e. a redistribution of income in favour of the farmer engineered, for instance, through an increase of money growth and inflation, will entail a reallocation of land in favour of the gatherer because $K_1^F = K(A), K' < 0$ as shown above. Therefore the long run impact of a monetary expansion (i.e. an increase of the rate of growth of the money supply, g_M, that translates into an increase of A) can be attributed to *an inflation tax effect.*

If a higher share of total income is employed by the farmer to consume and increase money holdings, net worth—and therefore the availability of resources to be used as downpayment—will be lower and therefore also landholding will be smaller. The effect on output of this downsizing of the farmer's landholding is uncertain. If the farmer is relatively poor and more productive than the gatherer (which is the case in scenario (1) above), an increase of π which brings about a decrease of K_1^F will unequivocally imply a decrease of aggregate output. In this case, the redistribution of land will reduce aggregate output. If the farmer is relatively rich and less productive than the gatherer (which is the case in scenario (2) above), an increase of π which brings about a decrease of K_1^F may lead to an increase of aggregate output if the shock and/or the impact on land are not "too big". In this case, the redistribution of land will increase aggregate output. This line of reasoning leads to conjecture that one can indeed find an inflation rate such that output is maximized.

In order to verify this conjecture notice that, imposing the steady state condition, from (30) follows:

$$G'\left(\bar{K} - K^F\right) = R\alpha\left(1 - A\right) - RA\frac{G\left(\bar{K} - K^F\right)}{K^F} \tag{33}$$

In Fig. 5 the upward sloping convex curve represents the Marginal Productivity of Land of the Gatherer (MPL^G) as a function of the steady state land of the farmer, i.e the LHS of (33) while the upward sloping concave curve represents the RHS, which we label NW curve. A is a shift parameter of this curve. In the figure, we compare two types of income distributions, characterized by A_H and A_L, $A_H > A_L$. As a consequence, $K_{1L}^F = K(A_H)$ and $K_{1H}^F = K(A_L)$. Once the allocation of land is determined, one can compute total output: $Y = \alpha K_1^F + \int_{K_1^F}^{\bar{K}} G'\left(\bar{K} - K_1^F\right) dK^F$.

We will use M as a reference point. In M, the marginal productivity of the farmer and the gatherer are equal $\left(G'\left(\bar{K} - K^F\right) = \alpha\right)$ so that total output is at a maximum:

$$K_m^F = \arg\max G\left(\bar{K} - K^F\right) + \alpha K^F$$

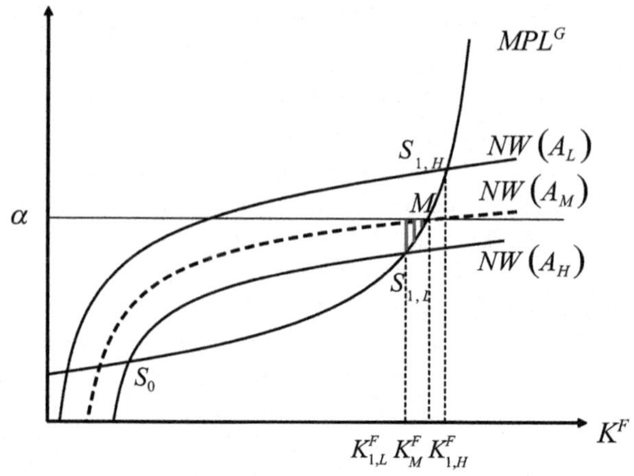

Fig. 5 Allocation of land between the farmer and the gatherer

The allocation $K^F_m = \bar{K} - G'^{-1}(\alpha)$ yields maximum output. In KM it also yields maximum welfare because preferences are linear and defined only on consumption. In the present framework, however, maximizing output is not equivalent to maximizing welfare because money balances enter the utility function (see Sect. 4.4 below).

By construction, in the figure, $K^F_{1L} < K^F_m < K^F_{1H}$. If "too much income" is appropriated by the farmer, then "too little land" is allocated to the farmer, i.e. $K^F_1 = K^F_{1L}$, and there will be a loss with respect to maximum output represented by the shaded area. In KM inefficiency emerges exclusively in this scenario, i.e. too little land is allocated to the farmer.[9] In our framework, inefficiency may arise also in the case of "too much land" allocated to the farmer, i.e. $K^F_1 = K^F_{1H}$. This is the case if "too little income" is appropriated by the farmer. It is easy to determine geometrically the loss also in this case (not shown).

Following this line of reasoning, one can determine the distribution of income which maximizes total output, let's call it A_m. The NW schedule consistent with maximum output is represented by the dashed line in the figure. It intersects the marginal productivity schedule at point M. In order to determine A_m the marginal productivity of the farmer must equal the marginal productivity if the gatherer ($\alpha = G'(\bar{K} - K^F)$) i.e., we have to solve for A the following equation: $\alpha = R\alpha(1 - A) - RA\frac{G(\bar{K}-K^F_m)}{K^F_m}$. The solution is:

$$A_m = \frac{\eta}{1 + \frac{G(\bar{K}-K^F_m)}{\alpha K^F_m}}; \quad \eta := \frac{R-1}{R}$$

[9]In KM, in the steady state $G'(\bar{K} - K^F) = Ra$ while maximum output is obtained when $G'(\bar{K} - K^F) = a + c$ with $Ra < a + c$ (assumption 2 in KM).

Notice that the share of total output appropriated by the farmer is an increasing function of the rate of growth of money supply, which, in the steady state, coincides with the inflation rate [see Eq. (26)]. Therefore, we can determine the inflation rate which maximizes total output:

$$\pi_m = \frac{\varphi^G}{\left[\frac{G(\bar{K}-K_m^F)}{\alpha K_m^F} + 1\right]\frac{1}{\eta\sigma} - \frac{1+\sigma}{\sigma}}$$

Example 2 If we specify the production function of the gatherer as $G\left(\bar{K} - K^F\right) = \sqrt{\bar{K} - K^F}$, the steady state condition becomes:

$$\frac{1}{2\sqrt{\bar{K} - K^F}} = R\alpha\left(1 - A\right) - RA\frac{\sqrt{\bar{K} - K^F}}{K^F} \tag{34}$$

The solutions of this equation can be evaluated numerically assigning values to the parameters of the model. We choose the following parameterization: $R = 1.05, \alpha = 3, \bar{K} = 1$. The farmer's share of output A can be determined setting σ, φ^G and π. We fix $\sigma = 0.1$, $\varphi^G = 0.5$ and choose 10 levels of π (from 1 to 10 %) to determine 10 values of A. Hence we can compute 10 steady state (saddle point) solutions associated to these rates of inflation.

They are shown in Fig. 6. The relationship between the steady state allocation of land and inflation is quasi-linear and obviously decreasing: higher inflation yields a redistribution of income in favor of the farmer and therefore reduces the farmer's landholding in the steady state.

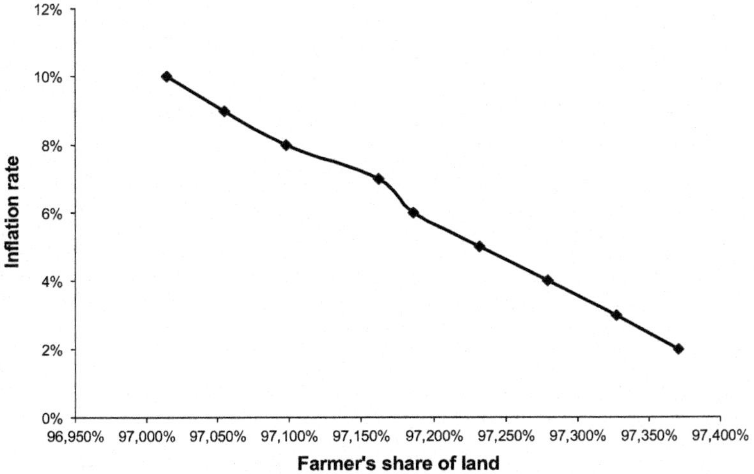

Fig. 6 Farmer's share of land and inflation

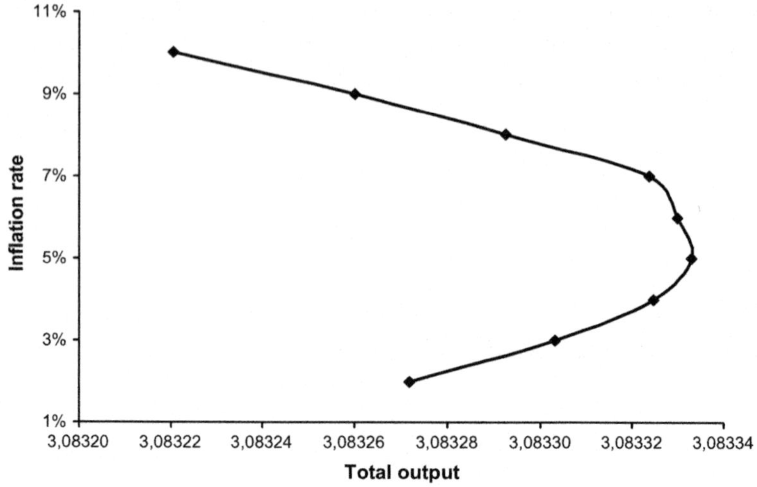

Fig. 7 Total output and inflation rate

The reduction of landholding, however, has a non-monotonic impact on total output because the latter is a non monotonic function of K^F. In Fig. 7 we report the levels of total output associated to the rates of inflation chosen parametrically. The relationship between the steady state level of total output and inflation is hump shaped. There is a rate of inflation (approximately 4.5 % in our example) which maximizes output.

It is easy to see that this inflation rate is consistent with the following definition

$$\pi_m = \frac{\varphi^G}{\left[\frac{2}{4\alpha^2\bar{K}-1} + 1\right]\frac{1}{\eta\sigma} - \frac{1+\sigma}{\sigma}}$$

which determines the farmer's share of output:

$$A_m = \frac{\eta}{1 + \frac{2}{4\alpha^2\bar{K}-1}}$$

and in turn determines the output maximizing land allocation $K_m^F = \bar{K} - \left(4\alpha^2\right)^{-1}$.

4.4 *Inflation, the Allocation of Land and Social Welfare*

In this section we examine the effects of a monetary expansion on welfare. Following a purely utilitaristic approach we define total welfare as the sum of the utilities of the farmer and the gatherer. In the steady state:

$$U = \gamma \ln c^F + (1-\gamma)\ln m^F + \gamma \ln c^G + (1-\gamma)\ln m^G \tag{35}$$

It is easy to show that

$$\begin{aligned}
c^F &= \gamma(\alpha-\mu)K^F = \gamma AY \\
m^F &= \frac{1-\gamma}{\pi}(\alpha-\mu)K^F = \frac{1-\gamma}{\pi}AY \\
c^G &= \gamma\left[G\left(\bar{K}-K^F\right)+\mu K^F\right] = \gamma(1-A)Y \\
m^G &= \frac{1-\gamma}{\pi+\varphi^G}\left[G\left(\bar{K}-K^F\right)+\mu K^F\right] = \frac{1-\gamma}{\pi+\varphi^G}(1-A)Y
\end{aligned}$$

Substituting the expressions above in (35) we get:

$$U = 2\left[\gamma\ln\gamma + (1-\gamma)\ln(1-\gamma)\right] + 2\ln Y + \ln A(1-A) - (1-\gamma)\ln\pi\left(\pi+\varphi^G\right) \tag{36}$$

Total welfare is increasing in total output and decreasing in inflation. It is also affected (in an ambiguous way) by the distribution of total income, incorporated in the term $A(1-A)$.

We know that $A = A(\pi), A' > 0$. In the steady state, moreover, $K_1^F = K(A) = K(A(\pi))$. Total output in the steady state therefore will be: $Y = Y(\pi) = G\left(\bar{K} - K(A(\pi))\right) + \alpha K(A(\pi))$.

The condition for welfare maximization (with respect to the inflation rate) therefore is:

$$\frac{2}{Y}Y'(\pi) = (1-\gamma)\frac{2\pi+\varphi^G}{\pi\left(\pi+\varphi^G\right)} - \frac{1-2A}{A(1-A)}A'(\pi) \tag{37}$$

whose solution we indicate with π_f.

If the RHS of the expression above is positive, i.e. if

$$(1-\gamma)\frac{2\pi+\varphi^G}{\pi\left(\pi+\varphi^G\right)} > \frac{1-2A}{A(1-A)}A'(\pi) \tag{38}$$

then the output maximizing inflation rate π_m is greater than the welfare maximising inflation rate π_f.

In Fig. 8 we report the relationship between total welfare and π in case (38) holds.

Of course, also the opposite scenario can occur.

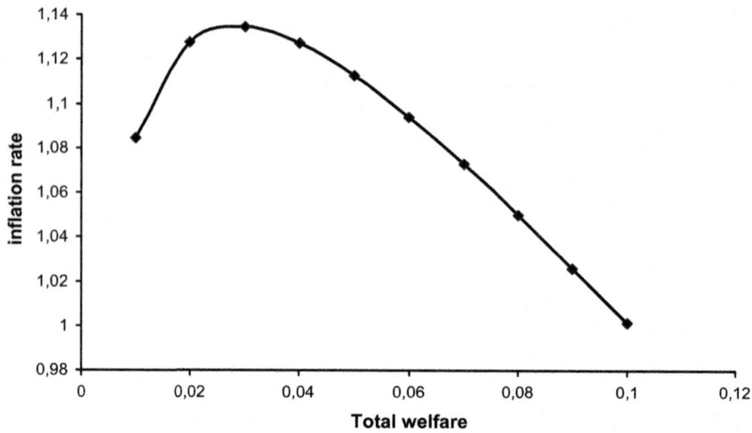

Fig. 8 Total welfare and inflation rate

5 The Effects of an Unexpected Productivity Shock

In this section, following the original KM approach we analyze the effects of a small unexpected and temporary shock to technology on output and asset prices by means of a linear approximation around the saddle point.

Suppose that at time 0 the economy is in the saddle point and an unexpected technological shock occurs so that the productivity of the farmer increases from α_0 to α_1. As in KM we assume that the farmer decides whether to supply labour or not before the shock. If the farmer chooses to cultivate land, when the shock occurs it is too late to change his mind.Moreover the shock is temporary, i.e. the parameter α goes back to α_0 immediately after the shock.

In order to study the effects of a shock to productivity, we start from the definition of net worth, i.e. the sum of tradable output $\left(\alpha K_{t-1}^F\right)$ and the *current* value of real assets $\left(q_t K_{t-1}^F\right)$ net of interest payments $\left(Rb_{t-1}^F\right)$ and of (a multiple of) the increase in real money balances $(\delta g_M m_t^F)$:

$$n_t^F = (\alpha + q_t) K_{t-1}^F - Rb_{t-1} - \delta g_M m_t^F$$

Net worth is employed as downpayment, i.e. $n_t^F = \mu_t K_t^F$. Moreover $R\mu_t = G'\left(\bar{K} - K_t^F\right)$ and $m_t^F = \dfrac{1-\gamma}{g_M(1+\sigma)+\varphi\sigma}\left[\alpha K_{t-1}^F + G\left(K_{t-1}^G\right)\right]$. Therefore:

$$\frac{G'\left(\bar{K} - K_t^F\right)}{R} K_t^F = \left[\alpha(1-A) + q_t\right] K_{t-1}^F - AG\left(\bar{K} - K_{t-1}^F\right) - Rb_{t-1} \qquad (39)$$

At time 0, before the shock, $n_0^F = \alpha_0(1-A) K_0^F + q_0 K_0^F - Rb_0 - AG\left(\bar{K} - K_0^F\right)$ and $q_0 K_0^F = Rb_0$—i.e. the current value of the farmer's land is equal to interest payments on debt inherited from the past—so that (39) boils down to:

$$\frac{G'\left(\bar{K}-K_0^F\right)}{R}K_0^F = (1-A)\,\alpha_0 K_0^F - AG\left(\bar{K}-K_0^F\right) \tag{40}$$

Suppose now that in the same period the productivity parameter goes up by $\Delta\alpha = \alpha_1 - \alpha_0$. By assumption, the first round effect of the shock on net worth concerns tradable output and the price of land, *given the (steady state) landholding of the farmer* K_0^F. Immediately after the shock, net worth becomes

$$n_{AS}^F = \left[(\alpha_0 + \Delta\alpha)(1-A) + (q_t - q_0)\right] K_0^F - AG\left(\bar{K}-K_0^F\right) \tag{41}$$

since interest payments have been predetermined on the basis of the steady state price of land: $q_0 K_0^F = Rb_0$.The first round effect of the shock creates a wedge between the current (after shock) value of land $q_t K_0^F$ and interest payments $q_0 K_0^F$. In particular, as will be clear in the following, the current price of land *jumps* from q_0 to q_t creating an (unexpected) *capital gain.*

Substituting (41) into the RHS of (39) we obtain

$$\frac{G'\left(\bar{K}-K_t^F\right)}{R}K_t^F = \left[(\alpha_0 + \Delta\alpha)(1-A) + (q_t - q_0)\right] K_0^F - AG\left(\bar{K}-K_0^F\right) \tag{42}$$

which describes the impact of the shock on K_t^F.

Assuming that the shock is temporary, in period 1 and all the following periods the situation goes "back to normal", i.e.

$$n_{t+s}^F = \alpha_0 (1-A) K_{t+s-1}^F - AG\left(\bar{K}-K_{t+s-1}^F\right) = \frac{G'\left(\bar{K}-K_{t+s}^F\right)}{R}K_{t+s}^F \qquad s = 0, 1, \ldots \tag{43}$$

Consider now (42). Let's take a first order approximation of the LHS in K_0^F:

$$\frac{G'\left(\bar{K}-K_t^F\right)}{R}K_t^F \approx \frac{G'\left(\bar{K}-K_0^F\right)}{R}K_0^F + \frac{G'\left(\bar{K}-K_0^F\right)}{R}(\varepsilon_0 + 1)\left(K_t^F - K_0^F\right) \tag{44}$$

where:

$$\varepsilon_0 = \frac{-G''\left(\bar{K}-K_0^F\right)K_0^F}{G'\left(\bar{K}-K_0^F\right)}K_0^F$$

is the elasticity of the marginal productivity of the land of the gatherer with respect to the land of the farmer *evaluated in the steady state* K_0^F.

The rate of change of total downpayment, i.e. of the LHS of (44), relative to the steady state is $\left[\frac{G'\left(\bar{K}-K_t^F\right)}{R}K_t^F / \frac{G'\left(\bar{K}-K_0^F\right)}{R}K_0^F\right] - 1$. Denoting the rate of change of a variable with respect to the steady state with a hat, the rate of change of the LHS becomes:

$$\widehat{\mu_t K_t^F} = (\varepsilon_0 + 1)\,\hat{K}_t^F \tag{45}$$

where $\hat{K}_t^F = \dfrac{K_t^F - K_0^F}{K_0^F}$ is the rate of change of the farmer's landholding.

On the other hand, the rate of change of net worth, i.e. of the RHS of (42), is $\dfrac{[(\alpha_0 + \Delta\alpha)(1-A) + (q_t - q_0)]\,K_0^F - AG\left(\bar{K} - K_0^F\right)}{\alpha_0(1-A)\,K_0^F - AG\left(\bar{K} - K_0^F\right)} - 1$. But $q_0\left(1 - \dfrac{1}{R}\right)K_0 = \mu_0 K_0 = \dfrac{G'\left(\bar{K} - K_0^F\right)}{R}K_0 = \alpha_0(1-A)\,K_0^F - AG\left(\bar{K} - K_0^F\right)$ so that the rate of change of the RHS becomes:

$$\hat{n}_{AS}^F = \hat{\alpha}\,(1-A)\,\frac{\alpha_0}{q_0}\frac{R}{R-1} + \hat{q}_t\frac{R}{R-1} \tag{46}$$

where $\hat{\alpha} = \dfrac{\Delta\alpha}{a_0}$ and $\hat{q}_t = \dfrac{q_t - q_0}{q_0}$ are the rates of change of the farmer's productivity and of the price of land.

After the productivity shock, the farmer's net worth goes up for two reasons: the direct effect $\hat{\alpha}\,(1-A)\,\dfrac{\alpha_0}{q_0}\dfrac{R}{R-1}$ and the *indirect effect through asset prices* $\hat{q}_t\dfrac{R}{R-1}$. Notice that $0 < A < 1$. Moreover in our context $\mu_0 = q_0\dfrac{R-1}{R} < \alpha_0$. Hence the ratio of the productivity to the downpayment $\dfrac{\alpha_0}{q_0}\dfrac{R}{R-1}$,which we will denote with θ_0 in the following, is greater than one. In symbols $\theta_0 := \dfrac{\alpha_0}{q_0}\dfrac{R}{R-1} > 1$.

In the original KM framework the rate of change of the farmer's net worth is $\hat{n}_{AS}^F = \hat{\alpha} + \hat{q}_t\dfrac{R}{R-1}$. From a comparison with (46) it is clear that in the present model the indirect effect is specified exactly as in KM while the direct effect $\hat{\alpha}\,(1-A)\,\theta_0$ is greater than $\hat{\alpha}$ (as in KM) if $A < 1 - \dfrac{1}{\theta_0}$. The smaller the policy parameter A, the higher the direct effect of the productivity shock on the rate of change of net worth after the shock.

Equating (45) and (46) we get

$$(\varepsilon_0 + 1)\,\hat{K}_t^F = \hat{\alpha}\,(1-A)\,\theta_0 + \hat{q}_t\frac{R}{R-1} \tag{47}$$

We have to determine now how the asset price q_t changes over time. Following KM, we note that from the definition of the downpayment and (61) follows $q_t = \dfrac{G'\left(\bar{K} - K_t^F\right)}{R} + \dfrac{q_{t+1}}{R}$, $q_{t+1} = \dfrac{G'\left(\bar{K} - K_{t+1}^F\right)}{R} + \dfrac{q_{t+2}}{R}$ and so on. Substituting the second expression in the first one and iterating the procedure we end up with:

$$q_t = \frac{G'\left(\bar{K} - K_t^F\right)}{R} + \frac{G'\left(\bar{K} - K_{t+1}^F\right)}{R^2} + \ldots = \sum_{s=0}^{\infty} R^{-s} \frac{G'\left(\bar{K} - K_{t+s}^F\right)}{R} \tag{48}$$

i.e. the current price of land is equal to the present value of the stream of future downpayments over an infinite horizon.

Taking a first order approximation of $G'\left(\bar{K} - K_{t+s}^F\right)$ in K_0^F, we can write $G'\left(\bar{K} - K_{t+s}^F\right) \approx G'\left(\bar{K} - K_0^F\right) - G''\left(\bar{K} - K_0^F\right)\left(K_{t+s}^F - K_0^F\right)$. Noting that this approximation holds true for any time period s, Eq. (48) boils down to:

$$q_t = \sum_{s=0}^{\infty} R^{-s} \frac{G'\left(\bar{K} - K_0^F\right) - G''\left(\bar{K} - K_0^F\right)\left(K_{t+s}^F - K_0^F\right)}{R}$$

Recalling that $\sum_{s=0}^{\infty} R^{-s} = \frac{R}{R-1}$ and that $\frac{G'\left(\bar{K} - K_0^F\right)}{R} = q_0 \frac{R-1}{R}$, from the expression above we obtain:

$$\hat{q}_t = \frac{1}{q_0} \frac{-G''\left(\bar{K} - K_0^F\right)}{R} \sum_{s=0}^{\infty} R^{-s}\left(K_{t+s}^F - K_0^F\right)$$

Notice now that $\frac{1}{q_0} \frac{-G''\left(\bar{K} - K_0^F\right)}{R} = \frac{-G''\left(\bar{K} - K_0^F\right)}{G'\left(\bar{K} - K_0^F\right)} \frac{R-1}{R} = \frac{\varepsilon_0}{K_0^F} \frac{R-1}{R}$. Moreover $\sum_{s=0}^{\infty} R^{-s} K_0^F = K_0^F \frac{R}{R-1}$. Therefore $\hat{q}_t = \varepsilon_0 \frac{R-1}{R} \sum_{s=0}^{\infty} R^{-s} \hat{K}_{t+s}^F$. In the end we obtain:

$$\hat{q}_t = \varepsilon_0 \frac{R-1}{R} \sum_{s=0}^{\infty} R^{-s} \left(\frac{1}{\varepsilon_0 + 1}\right)^s \hat{K}_t^F$$

Moreover $\sum_{s=0}^{\infty} R^{-s} \left(\frac{1}{\varepsilon_0 + 1}\right)^s = \sum_{s=0}^{\infty} \left(\frac{1}{R\left(\varepsilon_0 + 1\right)}\right)^s = \frac{R\left(\varepsilon_0 + 1\right)}{R\left(\varepsilon_0 + 1\right) - 1}$. Therefore

$$\hat{q}_t = \varepsilon_0 \frac{(R-1)\left(\varepsilon_0 + 1\right)}{R\left(\varepsilon_0 + 1\right) - 1} \hat{K}_t^F \tag{49}$$

Solving (47) and (49) for $\hat{q}_t$ and $\hat{K}_t^F$ yields

$$\hat{q}_t = \varepsilon_0 (1 - A)\, \theta_0 \hat{\alpha} \tag{50}$$

$$\hat{K}_t^F = \frac{R\left(\varepsilon_0 + 1\right) - 1}{(R-1)\left(\varepsilon_0 + 1\right)} (1 - A)\, \theta_0 \hat{\alpha} \tag{51}$$

The rate of change of net worth $\hat{n}^F_{AS} = (1 - A)\,\theta_0 \left(1 + \frac{R}{R-1}\varepsilon_0\right)\hat{\alpha}$ is larger than the rate of growth of productivity thanks to the indirect effect of the shock on the price of land. The rate of change of total downpayment $(\varepsilon_0 + 1)\,\hat{K}^F_t$ should match to keep equilibrium on the market for land. Therefore the rate of change of landholding is a multiple of the rate of change of productivity. In fact the multiplier $\frac{1}{\varepsilon_0 + 1}\left(1 + \frac{R}{R-1}\varepsilon_0\right)$ is greater than one.

What happens to aggregate output? Recalling Eq. (30) aggregate output Y_t is defined as:

$$Y_t = \frac{\alpha K^F_{t-1}}{A} - \frac{G'\left(\bar{K} - K^F_t\right)}{RA} K^F_t \tag{52}$$

At time 0, before the shock, $Y_0 = \frac{\alpha_0 K^F_0}{A} - \frac{G'\left(\bar{K}-K^F_0\right)}{RA} K^F_0$. Suppose now that in the same period the productivity parameter goes up by $\Delta\alpha = \alpha_1 - \alpha_0$. By assumption, the first round effect of the shock on aggregate output concerns tradable output, *given the (steady state) landholding of the farmer* K^F_0. Immediately after the shock, aggregate output becomes:

$$Y_{AS} = \frac{(\alpha_0 + \Delta\alpha)\, K^F_0}{A} - \frac{G'\left(\bar{K} - K^F_0\right)}{RA} K^F_0 \tag{53}$$

Assuming that the shock is temporary, in period 1 and all the following periods the situation goes "back to normal". The rate of change of aggregate output is $\frac{\frac{(\alpha_0+\Delta\alpha)K^F_0}{A} - \frac{G'\left(\bar{K}-K^F_0\right)}{RA}K^F_0}{\frac{\alpha_0 K^F_0}{A} - \frac{G'\left(\bar{K}-K^F_0\right)}{RA}K^F_0} - 1$ i.e., the RHS of (52) Therefore the rate of change of aggregate output will be:

$$\hat{Y}_t = \hat{\alpha}\frac{R\alpha_0}{R\alpha_0 - G'\left(\bar{K} - K^F_0\right)}$$

notice that $\hat{Y}_t > 0$ iff $R\alpha_0 - G'\left(\bar{K} - K^F_0\right) > 0$ i.e., $K^F_0 < \bar{K} - G'^{-1}\left(R\alpha_0\right)$.

6 The Effects of an Unexpected Monetary Shock

In this section we analyze the effects of a small shock to the rate of growth of money supply on output and asset prices by means of a linear approximation around the saddle point. Suppose that at time 0 the economy is in a steady state and an unexpected increase of the rate of growth of money supply makes A increase from A_0 to A_1. Assume moreover that the shock is temporary, i.e. the parameter A goes back to A_0 immediately after the shock.

By assumption, the first round effect of the shock on net worth concerns the price of land, *given the (steady state) landholding of the farmer* K_0^F. Immediately after the shock, denoting with $\Delta A = A_1 - A_0$, net worth becomes

$$n_{AS}^F = \alpha \left(1 - A_0 - \Delta A\right) K_0^F - A_0 G\left(\bar{K} - K_0^F\right) - \Delta A G\left(\bar{K} - K_0^F\right) + q_t K_0^F - R b_0.$$

Since interest payments have been predetermined on the basis of the steady state price of land ($q_0 K_0^F = R b_0^F$) we can write:

$$\frac{G'\left(\bar{K} - K_t^F\right)}{R} K_t^F = \alpha \left(1 - A_0 - \Delta A\right) K_0^F - A_0 G\left(\bar{K} - K_0^F\right) + \\ -\Delta A G\left(\bar{K} - K_0^F\right) + \left(q_t - q_0\right) K_0^F$$

The rate of change of net worth, i.e. of the RHS, is:

$$\hat{n}_{AS}^F = -\left(\theta_0 - 1\right) \hat{A} + \hat{q}_t \frac{R}{R-1} \tag{54}$$

where $\hat{A} = \dfrac{\Delta A}{A_0}$ and $\theta_0 := \dfrac{\alpha_0}{q_0} \dfrac{R}{R-1} > 1$.

After the shock, the farmer's net worth changes for two reasons: the direct effect of the rate of growth of money supply $-\left(\theta_0 - 1\right) \hat{A}$ and the *indirect effect* of the rate of growth of money supply on net worth *through asset prices* $\hat{q}_t \dfrac{R}{R-1}$. This indirect effect is a multiple $\dfrac{R}{R-1}$ of the asset price increase. The direct effect of the rate of growth of money supply is negative.

Equating (45) and (54) we get:

$$\left(\varepsilon_0 + 1\right) \hat{K}_t^F = -\left(\theta_0 - 1\right) \hat{A} + \hat{q}_t \frac{R}{R-1} \tag{55}$$

Solving (55) and (49) for $\hat{q}_t$ and $\hat{K}_t^F$ yields:

$$\hat{q}_t = -\left(\theta_0 - 1\right) \hat{A}$$

$$\hat{K}_t^F = -\left(\theta_0 - 1\right) \frac{R\left(\varepsilon_0 + 1\right) - 1}{\left(R - 1\right)\left(\varepsilon_0 + 1\right)} \hat{A}$$

An increase in the rate of change of money supply, has negative effects both on the rate of change of the farmer's landholding and the rate of change of the asset price. The rate of change of net worth in the end is $\hat{n}_{AS}^F = -\left(\theta_0 - 1\right) \hat{A} \dfrac{2R-1}{R-1}$.

What happens to aggregate output? Recalling Eq. (30) aggregate output Y_t is defined as:

$$Y_t = \frac{1}{A}\left[\alpha K_{t-1}^F - G'\left(\bar{K} - K_t^F\right) K_t^F\right] \tag{56}$$

At time 0, before the shock, $Y_0 = \frac{1}{A_0}\left[\alpha K_0^F - G'\left(\bar{K} - K_0^F\right) K_0^F\right]$. Suppose now that in the same period the monetary policy parameter goes up by $\Delta A = A_1 - A_0$. Immediately after the shock, aggregate output becomes:

$$Y_{AS} = \frac{1}{A_0 + \Delta A}\left[\alpha K_0^F - G'\left(\bar{K} - K_0^F\right) K_0^F\right] \tag{57}$$

Assuming that the shock is temporary, in period 1 and all the following periods the situation goes "back to normal". The rate of change of aggregate output is $\frac{\frac{1}{A_0+\Delta A}\left[\alpha K_0^F - G'\left(\bar{K} - K_0^F\right) K_0^F\right]}{\frac{1}{A_0}\left[\alpha K_0^F - G'\left(\bar{K} - K_0^F\right) K_0^F\right]} - 1$ i.e., the RHS of (56) Therefore the rate of change of aggregate output will be:

$$\hat{Y}_t = -\frac{1}{\frac{1}{\hat{A}} + 1}$$

notice that $\hat{Y}_t$ decreasing in $\hat{A}$. In a linearized system, in fact, we cannot capture the positive effects of the shock on output discussed in Sect. 4.2.

7 Conclusions

In this paper we have developed a model of a monetary economy with financing constraints. We borrow some of the basic ingredients of Kiyotaki and Moore's financial accelerator framework in order to keep the appealing feature of the intertwined dynamics of asset price changes and borrowing constraints. In order to evaluate the impact of monetary policy, however, we model a monetary economy with financing constraints adopting the Money In the Utility function (MIU) approach.

The basic difference with respect to the original framework is the likely occurrence of multiple equilibria. A change in the growth rate of money supply can affect real output through *the impact of inflation on the borrower's net worth, landholding and output*. In a sense the monetary transmission mechanism we are focusing on consists of a combination of the *inflation tax* effect and the *net worth channel*. Contrary to the traditional view, at least for some parameter configurations, an increase of the inflation tax can bring about an increase of aggregate output.

In this context, it is possible to determine an optimal inflation rate. For instance we can determine an output maximizing inflation rate, i.e. a rate of inflation which equals the marginal productivity of land of the farmer and of the gatherer. In our context, we can also determine an inflation rate which maximizes social welfare.

Appendix 1: The Farmer's Maximization Problem

The farmer maximizes (1) subject to the FF constraint (3) and the financing constraint (4). From the Lagrangian:

$$\mathcal{L} = \sum_{s=0}^{\infty} (\beta^F)^s \left[\gamma \ln c^F_{t+s} + (1-\gamma)\ln m^F_{t+s}\right] +$$

$$+\sum_{s=0}^{\infty} (\beta^F)^s \lambda^F_{t+s} \left[\alpha K^F_{t-1+s} - Rb_{t-1+s} + \right.$$

$$\left. -q_{t+s}\left(K^F_{t+s} - K^F_{t-1+s}\right) + b_{t+s} - g_M m^F_{t+s} - c^F_{t+s}\right] +$$

$$+\sum_{s=0}^{\infty} (\beta^F)^s \phi_{t+s} \left(\frac{q_{t+1+s}}{R} K^F_{t+s} - b^F_{t+s}\right)$$

we obtain the following *FOCs*:

(*iF*) $\dfrac{\partial \mathcal{L}}{\partial c^F_t} = 0 \Longrightarrow \dfrac{\gamma}{c^F_t} = \lambda^F_t$

(*iiF*) $\dfrac{\partial \mathcal{L}}{\partial c^F_{t+1}} = 0 \Longrightarrow \dfrac{\gamma}{c^F_{t+1}} = \lambda^F_{t+1}$

(*iiiF*) $\dfrac{\partial \mathcal{L}}{\partial m^F_t} = 0 \Longrightarrow \dfrac{1-\gamma}{m^F_t} = \lambda^F_t g_M$

(*ivF*) $\dfrac{\partial \mathcal{L}}{\partial m^F_{t+1}} = 0 \Longrightarrow \dfrac{1-\gamma}{m^F_{t+1}} = \lambda^F_{t+1} g_M$

(*vF*) $\dfrac{\partial \mathcal{L}}{\partial b^F_t} = 0 \Longrightarrow \lambda^F_t - \phi_t - R\lambda^F_{t+1}\beta^F = 0$

(*viF*) $\dfrac{\partial \mathcal{L}}{\partial b^F_{t+1}} = 0 \Longrightarrow \lambda^F_{t+1} - \phi_{t+1} - R\lambda^F_{t+2}\beta^F = 0$

From (*vF*) and (*viF*) we conclude that the financing constraint is binding if:

$$\frac{\lambda^F_t}{\lambda^F_{t+1}} > \beta^F R \tag{58}$$

Moreover after trivial substitutions we get:

$$\frac{\lambda^F_t}{\lambda^F_{t+1}} = \frac{c^F_{t+1}}{c^F_t} = \frac{m^F_{t+1}}{m^F_t} \tag{59}$$

i.e. consumption and real money balances should grow at the same rate. This condition ensures that the ratio of consumption to real money balances remains constant as implicitly stated in (5).

From (58) and (59) it follows:

$$\frac{c_{t+1}^F}{c_t^F} = \frac{m_{t+1}^F}{m_t^F} > \beta^F R$$

Notice that $\frac{m_{t+1}^F}{m_t^F} = \frac{1+g_M}{1+\pi_{t+1}}$ by construction. Therefore from (59) we conclude that consumption grows at the rate $\frac{1+g_M}{1+\pi_{t+1}}$.

Appendix 2: The Gatherer's Maximization Problem

The maximization problem of the gatherer is:

$$\max \sum_{s=0}^{\infty} \left(\beta^G\right)^s \left[\gamma \ln c_{t+s}^G + (1-\gamma) \ln m_{t+s}^G\right]$$

$$s.t.\ c_t^G + q_t \left(K_t^G - K_{t-1}^G\right) + b_t + (g_M + \varphi^G) m_t^G \leq G\left(K_{t-1}^G\right) + Rb_{t-1}$$

From the Lagrangian:

$$\begin{aligned}
\mathcal{L} = & \sum_{s=0}^{\infty} \left(\beta^G\right)^s \left[\gamma \ln c_{t+s}^G + (1-\gamma) \ln m_{t+s}^G\right] + \\
& + \sum_{s=0}^{\infty} \left(\beta^G\right)^s \lambda_{t+s}^G \left[G\left(K_{t-1+s}^G\right) + Rb_{t-1+s} + \right. \\
& \left. - q_{t+s}\left(K_{t+s}^G - K_{t-1+s}^G\right) - b_{t+s} - (g_M + \varphi^G) m_{t+s}^G - c_{t+s}^G\right]
\end{aligned}$$

we obtain the following *FOCs*:

$$(iG) \quad \frac{\partial \mathcal{L}}{\partial c_t^G} = 0 \Longrightarrow \frac{\gamma}{c_t^G} = \lambda_t^G$$

$$(iiG) \quad \frac{\partial \mathcal{L}}{\partial c_{t+1}^G} = 0 \Longrightarrow \frac{\gamma}{c_{t+1}^G} = \lambda_{t+1}^G$$

$$(iiiG) \quad \frac{\partial \mathcal{L}}{\partial m_t^G} = 0 \Longrightarrow \frac{1-\gamma}{m_t^G} = (g_M + \varphi^G)\lambda_t^G$$

$$(ivG) \quad \frac{\partial \mathcal{L}}{\partial m_{t+1}^G} = 0 \Longrightarrow \frac{1-\gamma}{m_{t+1}^G} = (g_M + \varphi^G)\lambda_{t+1}^G$$

$$(vG) \quad \frac{\partial \mathcal{L}}{\partial b_t^G} = 0 \Longrightarrow -\lambda_t^G + \beta^G R \lambda_{t+1}^G = 0$$

$$(viG) \quad \frac{\partial \mathcal{L}}{\partial b_{t+1}^G} = 0 \Longrightarrow -\lambda_{t+1}^G + R\beta^G \lambda_{t+2}^F = 0$$

$$(viiG) \quad \frac{\partial \mathcal{L}}{\partial K_t^G} = 0 \Longrightarrow -\lambda_t^G q_t + \beta^G \lambda_{t+1}^G \left[G'\left(K_t^G\right) + q_{t+1}\right] = 0$$

From (*iG*)–(*ivG*) follows that the flow of funds constraint of the gatherer is binding in each period. Moreover:

$$\left(g_M + \varphi^G\right) m_t^G = \frac{1-\gamma}{\gamma} c_t^G \qquad \forall t \tag{60}$$

i.e. the ratio of the increase in money holdings $\left(g_M + \varphi^G\right) m_t^G$ to consumption c_t^G is constant and equal to $\frac{1-\gamma}{\gamma}$. Considering (*vG*) and (*viG*) we reach the following conclusion:

$$\frac{\lambda_t^G}{\lambda_{t+1}^G} = \frac{c_{t+1}^G}{c_t^G} = \frac{m_{t+1}^G}{m_t^G} = \beta^G R$$

In words: real money balances and consumption must grow at the same rate $\beta^G R$.

Substituting (*vG*) into (*viiG*) we get:

$$-\lambda_t^G q_t + \frac{\lambda_t^G}{R}\left[G'\left(K_t^G\right) + q_{t+1}\right] = 0$$

from which, using (*iG*), we obtain:

$$G'\left(K_t^G\right) = R\mu_t \tag{61}$$

References

Bernanke, B., & Gertler, M. (1989). Agency costs, net worth and business fluctuations. *The American Economic Review*, 79(1), 14–31.

Bernanke, B., & Gertler, M. (1990). Financial fragility and economic performance. *Quarterly Journal of Economics*, 105(1), 87–114.

Bernanke, B., Gertler, M., & Gilchrist, S. (1999). The financial accelerator in quantitative business cycle framework. In J. Taylor & M. Woodford (Eds.), *Handbook of macroeconomics* (Vol. 1C, pp. 1341–1393). Amsterdam: North Holland.

Cordoba, J. C., & Ripoll, M. (2004a). Collateral constraints in a monetary economy. *Journal of the European Economic Association, 2*(6), 1172–1205.

Cordoba, J. C., & Ripoll, M. (2004b). Credit cycles redux. *International Economic Review, 45*(4), 1011–1046.

Greenwald, B. C., & Stiglitz, J. E. (1988). Imperfect information, finance constraints and business fluctuations. In M. Kohn & S. Tsiang (Eds.), *Finance constraints, expectations and macroeconomics* (pp. 141–165). Oxford: Clarendon Press.

Greenwald, B. C., & Stiglitz, J. E. (1993). Financial market imperfections and business cycles. *Quarterly Journal of Economics, 108*(1), 77–114.

Iacoviello, M. (2005). House prices, borrowing constraints and monetary policy in the business cycles. *The American Economic Review, 95*(3), 739–764.

Kiyotaki, N. (1998). Credit and Business Cycles. *The Japanese Economic Review*, 49(1), 18–35.

Kiyotaki, N., & Moore, J. (1997). Credit cycles. *Journal of Political Economy, 105*(2), 211–248.

Kiyotaki, N., & Moore, J. (2002a). Balance-sheet contagion. *The American Economic Review, 92*(2), 46–50.

Kiyotaki, N., & Moore, J. (2002b). Evil is the root of all money. *The American Economic Review, 92*(2), 62–66.

Ljungqvist, L., & Sargent, T. (2000). *Recursive macroeconomic theory*. Cambridge, MA: MIT Press.

Modeling Climate Change Effects on Renewable and Non-Renewable Resources

Aleksandr V. Gevorkyan and Arkady Gevorkyan

1 Introduction

The phenomenon of climate change shifting the demand and the supply of renewable and non-renewable resources has shed light on developments in the food and energy industries. It is now an established fact among scholars of various disciplines that climate change, if not addressed urgently, will pose a major threat to global stability and development in the future (e.g. Bernard and Semmler 2015). In one way or another climate change is most likely to eventually affect all areas of economic activity in every country. This chapter focuses on the potential impact climate change may exert on production and consumption of agricultural products (renewable or replenished resources) and non-agricultural products (non-renewable or exhaustible).

It is common to view climate change as a constantly increasing mean temperature. Globally rising sea levels, increasingly prolonged droughts, and severe storms are just some impacts of intensifying climate change. At the same time, weather changes (e.g. levels of precipitation) represent the stochastic factor moving around the mean. As a result, there is a clear, often negative, effect on the prevailing agricultural and industrial production processes. For example, Schlenker and Roberts (2009) argue that an increase in the average temperatures by 2 °C would devastate agricultural production in the Northern hemisphere.

Climate change not only affects the agricultural sector, but also significantly impacts the production (supply side) of non-renewable resources (e.g. oil, gas and

A.V. Gevorkyan (✉)
The Peter J. Tobin College of Business, St. John's University, New York, NY, USA
e-mail: gevorkya@stjohns.edu

A. Gevorkyan
New School for Social Research, The New School, New York, NY, USA
e-mail: gevoa020@newschool.edu

L. Bernard, U. Nyambuu (eds.), *Dynamic Modeling, Empirical Macroeconomics, and Finance*, DOI 10.1007/978-3-319-39887-7_5

copper). Notably, production of oil and copper each account for approximately 40 % of global warming emissions from fossil fuels worldwide (WDI 2015). Therefore, active policy involvement at the pan-national level may soon become inevitable as governments collectively strive to diminish hazardous energy use consistent with a sustainable development model of rising urban population dependency and structural economic growth. The dynamic is further reinforced by the growing effort to seek alternative energy, with active development of substitutes for fossil fuels (e.g. biofuels, wind and solar energy, etc.).

This chapter focuses on the main aspects of the production of renewable and non-renewable resources affected by climate change. We also develop a theoretical framework of analysis. The goal here is to present a dynamic path for renewable, non-renewable resources production in an environment of increasing CO_2 emissions.

The rest of the chapter is structured as follows. After a review of relevant literature in Sect. 2 we proceed to discuss stylized facts of climate change and commodity markets in Sect. 3. The theoretical model is developed in Sect. 4, where some key results of the model calibration are also discussed. Final remarks follow in the concluding section of the chapter.

2 Literature Review

Expanding significantly over the last decade, the literature on the problem of climate change is quite substantive, rich, and multifaceted. For purposes of our analysis, we direct the attention to some representative studies of the impact ongoing climate change may have on renewable and non-renewable resources. Recall that we define renewable resources as those that can be reproduced (some with application of new technology). Normally the commodities that fall into this category are agricultural products harvested on land. Recently other types of renewable energies (i.e. solar, biofuels, wind, etc.) have been added to the group as well. The non-renewable group consists of exhaustible primary commodities extracted from the ground and with no possibility of natural replenishment (i.e. crude oil, coal, metals, etc.).

Based on the definitions above it becomes clear that the production process for the renewable group is disproportionately affected by climate change. For example, variations in temperature make production of staple commodities such as corn or rice infeasible as soil and weather patterns change. At the same time, demand and consumption factors for non-renewable commodities are most impacted by climate change. First is the notion of exhaustibility and second is the ongoing search for new energy alternatives, as mentioned above, pushing into the renewable sector. When analyzing this phenomenon it is critical to maintain a broader perspective of the global commodities trade, which interacts with the capital markets and impacts individual countries' development.

In a recent study, Roache (2010) analyzes volatilities in food prices. Differentiating between high and low volatility, Roache focuses on low volatility, as it is a higher concern to the economy's stability. Low volatility implies gradually increasing destabilizing trends in the resource's supply-chain. Low frequency volatility is positively correlated among different commodities, perhaps pointing to significant common factors.

The economic impact of climate change is one of the factors contributing to the volatility of commodity markets. Schlenker and Roberts (2009) provide empirical evidence of the effects of climate change on agricultural production. They find that the yield growth of crops production increases with temperatures up to 29–31 °C (depending on the particular crop). However, eventually the growth in yields decreases sharply for corn, cotton, and soybeans as temperature continues to increase. At the same time, the increase of the average temperature to 28°C can be beneficial for agricultural production.

Related is the problem of accumulating greenhouse gas emissions and the effect on agricultural productivity. Jones and Sands (2013) introduce a dynamic global computable general equilibrium model, covering 15 regions. The model simulates the impacts of changes in projected crop and livestock productivity on the scale and composition of agricultural output and its effect on the green-house gases (GHGs) emissions. Based on the analysis of three dynamic scenarios running through 2034, the model predicts that the output shares are set to decline for economies in transition (Eastern Europe and the former Soviet Union) and industrial economies, and will increase for the developing-country regions with average or higher than average emissions intensities.

Elsewhere, Havlík et al. (2013) look at the effects of land use change and greenhouse gas emissions on crop productivity and the livestock sector. The authors develop a partial equilibrium model for the agricultural and forestry sectors using several alternative scenarios. The study finds crop yield production growth to be a cost effective way of abating GHG. By 2030 GHG emissions are projected to be lower by 2 GtCO_2—per year if crop yields grow according to past trends. A similar simulation points to the problem of sustaining agricultural production consistent with population growth while continuing to reduce GHG emissions from farming, reducing transaction costs, and improving risk management. As a possible solution, Herrero et al. (2010), advocates to develop policies for public investment in infrastructure and market strengthening.

As stated earlier, production and consumption of exhaustible resources also comes under pressure with climate change. At the moment, the pressure has been clearly visible with the high volatility of futures prices, albeit a known phenomenon for economists, but also for other reasons. Hotelling (1931) was one of first to offer a theoretically fundamental analysis of non-renewable resources dynamics. Hotelling's rule applied to exhaustible product pricing predicts a gradual increase in the price of the non-renewable resources. The underlying assumption is that volatility for these futures will increase substantially over time.

On the other hand, the Prebisch-Singer hypothesis argues that primary commodity prices (renewables and non-renewables) have a tendency to decline over a longer run with negative implications for economic development in the low-income countries. Ruta and Venables (2012) emphasize the fact that volatility in oil is predominantly caused by low price elasticities. Volatility in the renewables futures prices has also been a great concern for countries experiencing export led growth.

More concretely, fluctuations—declining or soaring—in prices of crude oil or corn, for example, affect all commodity exporter nations. The greater the proportion of the commodity in the overall country's exports, the stronger the impact of price volatility. For example, Gevorkyan and Gevorkyan (2012) attempt to link trading in the commodity derivatives market with social costs to macroeconomic development in emerging markets. As a transmission channel, they rely on the exchange rate index of countries whose primary commodity exports are traded in the U.S. futures market. Because of high dollarization and foreign currency debt pile up, spillover effects of high volatility in the commodity markets may lead to devastating social costs that pass through across commodity exporters in emerging markets. Potentially significant social and economic costs justify the study of commodities volatility.

A peculiar and important strand in commodity markets literature that has expanded since the latest crisis deals with the effects of crude oil price fluctuations on energy and non-energy commodity futures. Employing conventional econometric tools, Baffes (2007) and Baffes and Haniotis (2010), show the potential effects of high oil prices (at the time of the research) on energy and non-energy commodities. The first paper finds that during the period of 1960–2005 a 10 % increase in the price of crude oil led to a 1.6 % increase in the price of non-energy commodities. Subsequently, Baffes and Haniotis (2010) show that the recent financial crisis has strengthened the crude oil to non-energy commodity association, where a 10 % increase in the price of oil leads to a 2.6 % increase in the non-energy index. Manera et al. (2012) also test the spillover effects from the energy commodities trade to agricultural commodity futures. The main findings of their multivariate GARCH modeling suggest that agricultural commodities are dependent on the trading of energy futures.

Analyzing energy and grain price behavior, Enders and Holt (2013) applied econometric modeling (VAR estimations) and determined that energy prices, interest rates, and exchange rates directly influence grain prices and their subsequent run. In the second part of the paper, Enders and Holt (2013) identify the mean shifting pattern across the entire researched sample group of commodities. The trend for mean switching is evident among real energy prices, but unclear among agricultural products.

Knittel and Pindyck (2013) steer their analysis of oil prices as a leading indicator for other energy and non-energy resources. Relying on a supply and demand model, the authors determine that speculation is not the key futures price driver, consistent with data on production, consumption, inventory change, and changes in the yields. The empirical analysis carried out by Hamilton and Wu (2014) supports this finding.

A number of research papers have analyzed the issue of new competition for land that becomes apparent with intensifying climate change. Harvey and Pilgrim (2011), Tilman et al. (2011) analyzed this problem in detail.

As can be inferred from the preceding review, literature on commodities market determinants is vast and growing. The focus is now increasingly being shifted towards analyzing the dynamic of renewable and non-renewable resources production and consumption in the context of intensifying climate change. In this paper we bring the two together. This is a rare effort, as much of the literature tends to develop a theoretical framework for just one type of commodity group. However, before we elaborate on the model, we first address some stylized facts on the commodity markets patterns and climate change. We focus on a range of country groups, as per the World Bank's classification, to assess the overall global situation.

3 Stylized Facts

The analytical and theoretical evidence presented in the preceding section can be summarized in the following four points. First, there is a clear concern about the price volatility (specifically in the futures markets) of primary commodities across the world. Second, aside from financial market impacts, once the relative production and consumption shares are accounted for, this volatility has immediate macroeconomic ramifications for commodities exporter and importer nations. Third, much of the recent volatility has acquired a somewhat new qualitative aspect that looms as a greater impeding concern requiring collective sub-national attention. Specifically, climate change—roughly defined as a change in the global and regional climate patterns—induced by the increased usage of fossil fuels (the non-renewables above) since approximately the 1950s, is now having a reverse effect on the agricultural and industrial capacity across a wide range of countries.

Finally, the fourth point yields a realization of the confluence of the related concerns. Namely, merging the first three observations with the evident multifaceted pressures of the global rising population poses a problem of sustaining the prevalent living standards. Notably, in economic development the growing population question has long been debated in the academic and policy fields (e.g. for recent trends refer to the UN's Population Division, UN 2015).

Figure 1 offers an intriguing illustration of the population aspect of the global economy. The focus here is on the proportion of urban residents to the total population and correlating that with average urban population growth between 1961 and 2014. According to the available evidence, many of the industrialized nations (e.g. HIC, EUU, OED—see Appendix Table 1 for an explanation of country group codes) report high proportions of urban population (between 73 and 82 % of total population). At the same time many developing and some middle-income countries (mainly in Africa and Asia) are predominantly rural, with less than 50 % of urban population (UN 2015).

Importantly, it is the lower income developing, nations that have seen the fastest rates of urbanization (e.g. LDC, SSA, LIC, etc.) ahead of the advanced economies. This is partly due to the ongoing structural change in these countries and rise in economic growth during the post-World War II era. Elsewhere, e.g. China and Southeast Asia, the rising urbanization is a reflection of industrial strengthening, as much of the economic activity and employment shifts from the agricultural sector to industrial and economically diverse urban areas.

Cities remain the primary drivers of breakthrough industry, mass employment and innovation. As the agricultural sector gains efficiency, a substantial number of workers are pushed into urban economies (for the global track record see, e.g. Quigley 1998; WB 2000; or Henderson 2002). As urbanization rises, the urban population's dependency on energy sources and other commodities increases, pushing consumption levels high above historical trends.

Putting the two data points in Fig. 1 together brings up the concern of sustainability of urban infrastructure. According to UN data, global urban population is set to reach close to 7.4 billion people by 2050, with a significant rise in urban shares (between 48 and 57 %) in the low and lower-middle-income economies respectively.

In the immediate future, assuming no significant breakthroughs in prevailing technology and in the absence of cost-effective substitutes for fossil fuels, all of the above translates into the world's growing demand for primary commodities across agricultural and extracting industries. However, such increase in demand, as Fig. 2 suggests, would also cause higher CO_2 emissions, thus exacerbating the climate change problem. Moreover, and concerning, the assumptions are not likely to hold as new technologies emerge at a faster rate.

Figure 2 also reveals another story. Decomposing the global CO_2 emissions per capita by country groups makes it evident that high-income economies (i.e. countries in North America and Western Europe) are also among the highest contributors to the global CO_2 emissions pool. Low income and least-developed economies, while significantly lower on the scale, have been seeing their CO_2 levels rapidly increasing in the last two decades. The trend is partially a reflection of

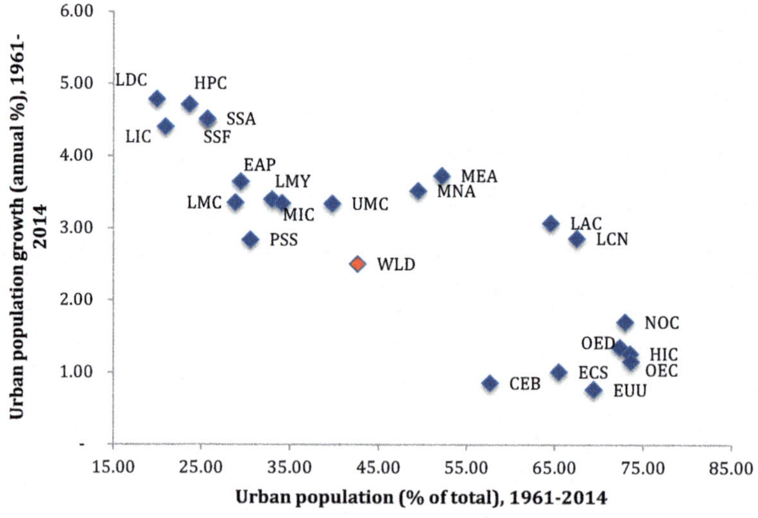

Fig. 1 Urban population shares and growth rate, 1961–2014. *Source*: authors' calculation based on data from WDI (2015). *Note*: see Appendix Table 1 for country group codes explanation. The world average is WLD

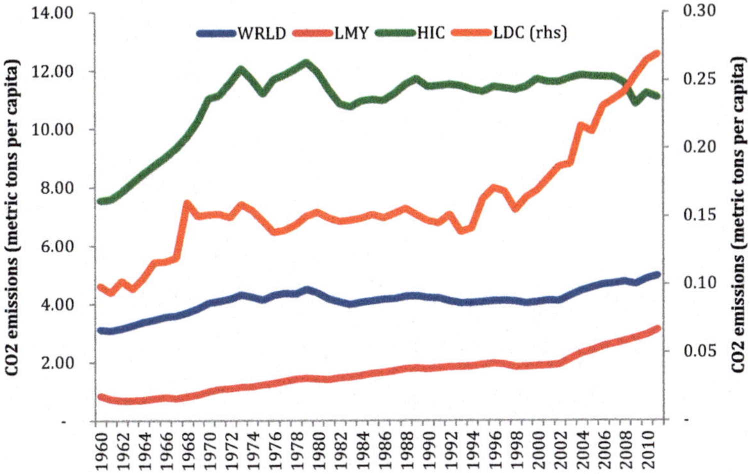

Fig. 2 CO_2 emissions (metric tons per capita), annual. *Source*: authors' calculation based on data from WDI (2015)

the industrialization in the LMY or growth of extractive industries in the LDCs as economies opt for cheaper energy sources (e.g. coal). At the same time the HIC economies are showing modest declines in the CO_2 emissions as efforts for renewable energy are gaining traction.

If CO_2 emissions are seen as the primary contributors to the intensifying climate change, variations in land-ocean temperatures (Fig. 3) reflect the extent to which the change has been occurring. Figure 3 tracks annual mean temperature from 1880 through mid-2015. The data is obtained from the NASA/GISS database taking the 1951–1980 period as a base with an average temperature of 14 °C.

Irrespective of the scale, the end result from Fig. 3 is clear—global temperature index has been rising. Consistent with the evidence available for the CO_2 emissions, the most dramatic rise in temperature has been occurring since the second half of the twentieth century. The last two decades have been the warmest with year 2014 claiming the highest mark across the 135-year record.

Turning our discussion back to global primary commodities, Fig. 4 relates the factor of price volatility in the market. There is a clearly identifiable trend, dubbed in the financial industry as a "super-cycle". The trend roughly corresponds to the general dynamic in prices across all commodity groups as gradually increasing from 1990s up until early 2014, when prices started to decline. There is a visible dive down during the 2008 crisis, followed by a rapid recovery in the immediate years. Whether the trend is to be seen as permanent or temporary in a larger historical perspective is yet to be determined (e.g. Canuto 2014).

However, the most recent downward trend in commodity prices highlights the volatility that has increased in the commodity markets and continues to rise, perhaps likely due to the spread of commodity based financial instruments (e.g. Arezki

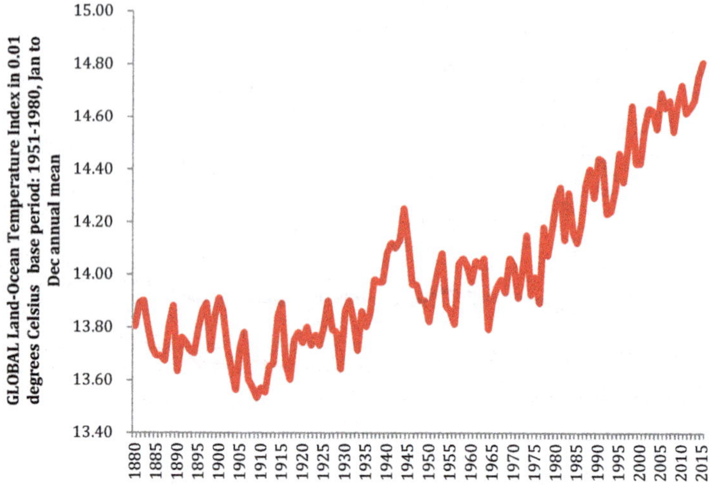

Fig. 3 Land-Ocean Temperature Index, Jan-Dec annual mean for 1880–2015. *Source*: based on data from NASA/GISS (2015). *Note*: Figure shows absolute degrees Celsius with the base of absolute global mean for 1951–1980 of 14 °C

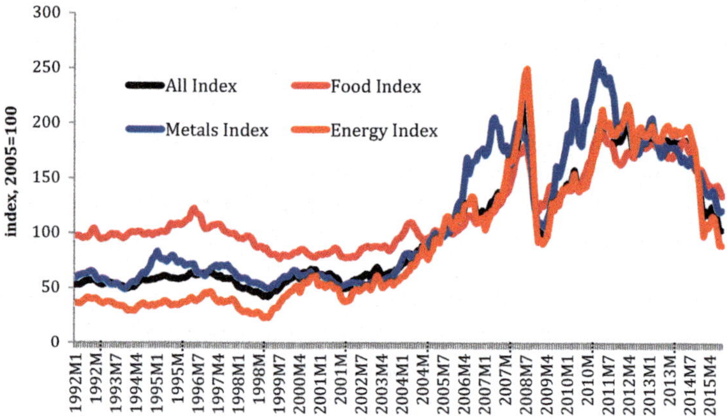

Fig. 4 Primary commodities price indexes, 2005 = 100. *Source*: based on the IMF (2015). *Note*: All Commodity Price Index, 2005 = 100, includes both Fuel and Non-Fuel Price Indices

et al. 2013; Gevorkyan and Gevorkyan 2012). The downward trend also poses a difficult dilemma for the exporter nations among the low and lower-middle income economies that rely primarily on commodity exports. Gevorkyan and Kvangraven (2016) show how instability spills across the real sector, as in the case of Sub-Saharan African exporters, and affects the countries' ability to tap into international capital markets and sustain competitive rates on sovereign debt.

Over the latter half of the twentieth century, crude oil has evolved as a benchmark commodity that sets the stage for the behavior of prices across renewable and non-renewable resources markets. The higher correlation between primary commodities comes with ongoing technological evolution (including the search for renewables) and stronger price shock transmissions as substitutability between non-renewables and renewables rises (e.g. Canuto 2014 or Dobbs et al. 2013). Such positioning also explains the close movement of the energy index with all commodity indices in Fig. 4, as other resources follow the price of oil. Relative changes in oil consumption are also reflective of the world's demand for fossil (exhaustible) resources.

On the latter point, Fig. 5 summarizes the available evidence from the oil industry since 1965. Here, the emerging nations of the Middle East, Asia Pacific, and Africa comprise the top three largest growing consumers of oil. However, in terms of absolute volumes of consumption, with the exception of Asia Pacific region, their numbers pale in comparison with the high-income economies.

For instance, according to the BP (2015) report, Middle East region oil consumption in 2014 accounted for 9.3 % of the total oil demand in the world. In Africa that share was at 4.3 %, with South and Central America making up 7.8 %. In contrast, Europe and Eurasia accounted for 20.4 %, North America's share was 24.3 %, and Asia Pacific totaled 34 % (largely due to China's consumption of 12.4 % of the global total and second only to the US with a 20 % oil consumption share). Yet, persistent demand for fossil fuels energy sources, as stated earlier, is making a significant contribution to the rise in the CO_2 emissions, exacerbating the factors causing climate change and adding to the inherent volatility of the commodities market.

Finally, Fig. 6 puts all aspects of climate change and the dichotomy of renewable and non-renewable resources in a comprehensive framework. Taking the levels of 1960 as the base point, it is evident that growth in CO_2 emissions has outpaced

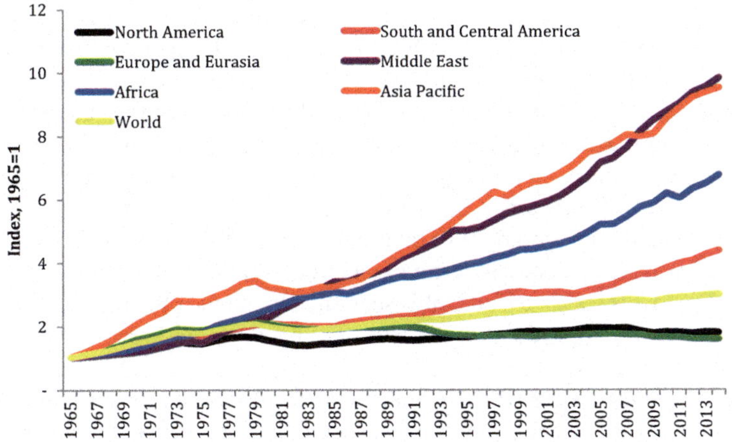

Fig. 5 Crude oil consumption across the regions (1965 = 1). *Source*: BP Statistical Review of World Energy (BP 2015). *Note*: oil consumption based on actual thousands of barrels daily data, 1965 = 1

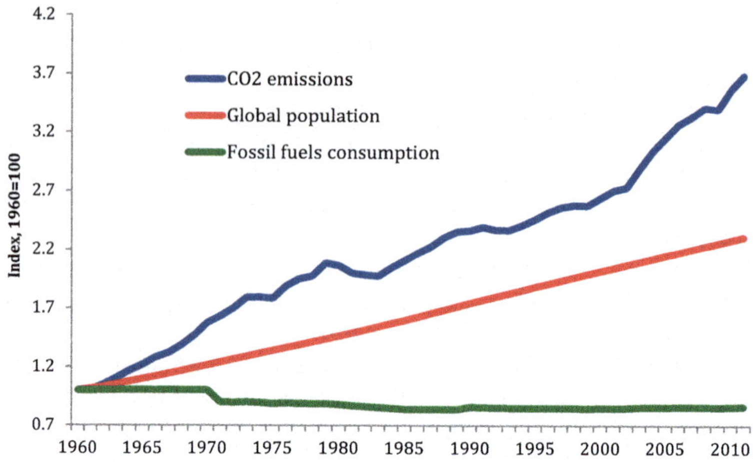

Fig. 6 CO_2 emissions, population growth, and fossil fuels consumption. *Source*: authors' calculation based on data from WDI (2015)

population growth on the global scale by many folds. Importantly, the gap has grown wider through the 2000s as population continued a steady increase and CO_2 emissions intensified. An intuitive explanation may refer to the above-mentioned industrialization, coupled with urbanization and more intensive use of fossil fuels across the world.

The latter observation may initially appear to contradict the third trend clearly visible in Fig. 6. That is, the decline in the share of the fossil fuels in the overall global energy consumption (including biofuels, solar, wind, and geothermal resources). However, once the data in Fig. 5 is correlated with evidence on rising shares of renewable energy consumption (in particular in the developed world, as per BP, 2015) the picture becomes somewhat more consistent.

The decline in fossil fuels consumption, while noticeable, is clearly not yet significant enough as the high-income economies struggle to introduce renewable sources of energy on a major scale. Until the renewable energy trend picks up in the two main oil consumer markets in the high-income and emerging Asia economies, the global economy will continue through the commodities cycles and price volatility, with subsequent detrimental effects on the climate.

The stylized facts summarized in this section help put the phenomenon of climate change along with the production and use of exhaustible and renewable resources in a globally dynamic perspective. There is a clear pressure emanating from rising urban population across the world, primarily in developing countries that remain at the core of commodities (non-renewables, specifically) production, while advanced economies with already high urban population shares act as the major commodities consumers.

The population rise is checked by innovation, technological efficiency, and yet ongoing industrialization and greater CO_2 emissions even as fossil fuels are substituted for renewables. Importantly, data reported in this section is consistent with much of the abovementioned theoretical and empirical analysis. These observations offer additional support for the purposes of sketching the theoretical model of climate change and renewable and non-renewable resources use in the next section.

4 Theoretical Model

In the theoretical part of the chapter, we work with a general concept of a resource (renewable and non-renewable) and its use in ongoing socio-economic processes. The model we develop here is similar to and based on the work in Greiner and Semmler (2008) and Bernard et al. (2012).

Non-renewable resources (exhaustible) $x(t)$, where t is a time argument, represent the part of resources that have already been discovered. There are resources that are known but have not yet been extracted—$x^k(t)$. The portion of resources that has not yet been discovered is denoted by $x^n(t)$. Following this logic, the total amount of the non-renewable resources will be equal to:

$$x(t) = x^k(t) + x^n(t)$$

The model applies a certain rate at which resources are to be discovered, noted as: $f \geq 0$. We treat f, the discovery rate, as a function of the resources that have not yet been discovered. Intuitively, if there are more resources that have not been discovered, then it is easier to discover new ones. In the model, u stands for the amount of resources supplied at each point in time.

With the discount rate being constant, $r > 0$, the optimization problem is formulated the following way:

$$\max_{u,c} \int_0^\infty U(C_t)\, e^{-rt} dt \tag{1}$$

Subject to:

$$\dot{x}^k = -u + f\left(x_0 - y - x^k\right), \quad x^k(0) > 0 \tag{2}$$

$$\dot{y} = u \tag{3}$$

$$\dot{g} = f(c, z, u) - am = (g - c)^\alpha u^\beta b z^{1-\alpha-\beta} - am \tag{4}$$

$$\dot{m} = u - \delta m \tag{5}$$

$$\lim_{t \to \infty} x^k \geq 0$$

Cumulative past exploitation is presented by y in Eqs. (2) and (3). C_t represents resources consumption. Equation (2) represents the demand for the non-renewable resources. In the optimization problem u and c are the decision variables.

In Eq. (4), $g(t)$ stands for the stock of agricultural products. The change in the stock of agricultural products positively depends on land, z, which for modeling purposes is taken as constant. Agricultural products that are invested in are presented by $g-c$, with c denoting agricultural products that are consumed. Further, growth of agricultural products is negatively affected by climate change, which is impacted by the m, levels of CO_2 emissions accumulation. In the model, $a, b, \delta, \alpha, \beta > 0$ are constant parameters. In Eq. (5), δm is the decay of the CO_2 emissions. The amount of resources supplied is u at time t.

We follow convention, assuming that the process runs over the infinite horizon time period. To provide estimations of the model we use the receding finite horizon by a relatively new numerical procedure—non-linear model predictive control (NMPC)—introduced and explored at length by Grune and Pannek (2011). This method allows for a multi-period model, which in a longer horizon approaches the usual infinite horizon solution. NMPC computes single trajectories instead of calculating an optimal value function.

To solve the theoretical model we use the algorithm where we choose the following parameter values: $\alpha = 0.2$, $\beta = 0.05$, $a = 0.08$, $\delta = 0.5$, $b = 0.75$ and $z = 2$. In Fig. 7, the red line (+) shows the path of non-renewable resources and their diminishing availability over time. The solid black line (0) represents renewable/agricultural resources, and the green line (x) shows the accumulation of CO_2 emissions. Finally, the blue line (*) is the accumulation of the past exploitation of non-renewable resources y.

Calculation of the model through the NMPC algorithm and its graphical presentation confirm the common assumption that with continuous use of exhaustible resources, the CO_2 emissions will continue to grow, impacting the changing climate. In a feedback effect, a changing climate affects production of the renewable, i.e., agricultural, resources.

Active development of biofuels and use of agricultural products for the purposes of energy production eventually displaces the use of the traditional energy fossil fuels (i.e. oil, copper). Such substitution is expected to gradually spread, encouraged and promoted by policy makers. While the private sector's activist campaigns are gaining popularity, successful substitution on a significant scale implies government's active involvement in considerably reducing CO_2 emissions. The presented dynamic path, modeled in Fig. 7, represents this optimistic scenario.

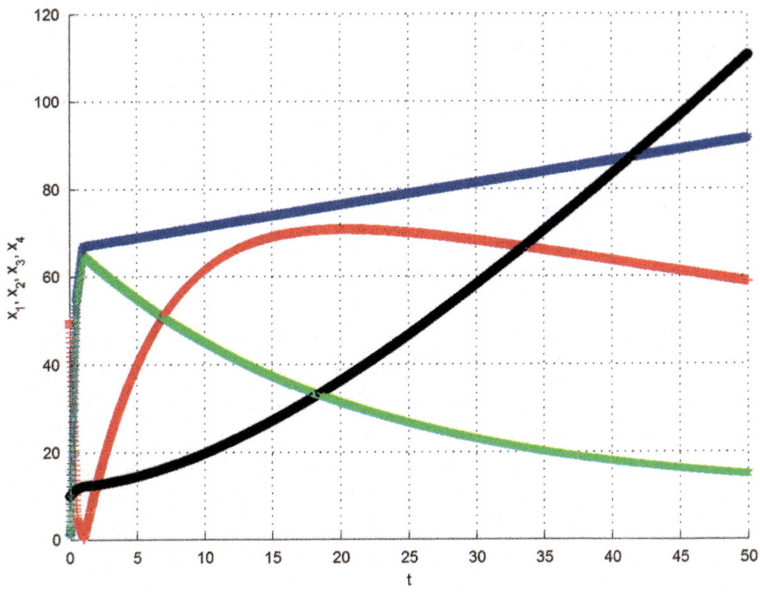

Fig. 7 Dynamic path of renewable resources, non-renewable resources, past exploitation of the non-renewable resources and accumulation of CO_2 emissions. *Note*: Initial conditions for $x(0) = 50$, $y(0) = 1$, $g(0) = 10$, $m(0) = 1$

It is difficult to define specific timeframes when the gradual shift to replace oil and copper with agricultural products for energy purposes could potentially occur. And if the shift occurs later, then it would be much harder to continuously develop biofuel energy as the average temperatures would be rising with the increase in occurrences of severe cataclysms.

Overall, this model strives to provide a dynamic path for the ongoing processes of renewable and nonrenewable resources connecting with climate change. The theoretical model is unique in a sense that it is solved using NMPC modeling, which shows that the points of stability would only be reached once nonrenewable energy use is diminished and replaced with renewable products.

5 Final Remarks

This paper models the dynamic of renewable and non-renewable resources in the context of climate change. Based on the available literature and empirical evidence, climate change—a gradual change in the climate across the world intensified by the use of fossil fuels since late 1950s—poses a challenge to the sustainability of the global economy faced with the needs of the growing urban-dependence population and exhaustibility of some of the primary resources, particularly in the energy sector.

The model developed suggests that with active development of biofuel and green energy, the use of exhaustible resources for energy purposes will be diminished over time. Contraction in the use of non-renewable resources would have a positive effect on CO_2 emissions. Thus, according to this chapter's theoretical model, the shift from non-renewable to renewable energy may be occurring gradually over time.

Looming climate change and its potentially devastating weather effects have already started to adjust global consumption and production of renewable resources. According to our analysis, mitigating the negative impact of climate change on economic activity requires two critical steps. First, sustained spread of renewable resources across known and evolving industries with continuous integration of the renewables within the prevailing production processes. Second, and in parallel, for mitigation to gain any serious traction, it must be accompanied by lower consumption of exhaustible resources, complemented by the implementation of relevant economic and environmental development policies.

In this tug of economic rationale and sustainable development, the advanced economies may be held to a higher accountability as the major consumers of exhaustible resources. Yet, rising urbanization and sporadic industrialization holds developing economies equally accountable for production and consumption of exhaustible resources. Exactly how things will work out is yet to be discovered. Clearly, a global consensus on climate policy action is urgently needed.

Appendix

See Table 1.

Table 1 Country group code and name

Code	Country group name
CEB	Central Europe and the Baltics
EAP	East Asia & Pacific (developing only)
ECS	Europe & Central Asia (all income levels)
EUU	European Union
HIC	High income
HPC	Heavily indebted poor countries (HIPC)
LAC	Latin America & Caribbean (developing only)
LCN	Latin America & Caribbean (all income levels)
LDC	Least developed countries: UN classification
LIC	Low income
LMC	Lower middle income
LMY	Low & middle income
MEA	Middle East & North Africa (all income levels)
MIC	Middle income
MNA	Middle East & North Africa (developing only)
NOC	High income: non OECD
OEC	High income: OECD
OED	OECD members
PSS	Pacific island small states
SSA	Sub-Saharan Africa (developing only)
SSF	Sub-Saharan Africa (all income levels)
UMC	Upper middle income
WLD	World

Source: World Bank definitions (WDI 2015)

References

Arezki, R., Hadri, K., Loungani, P., & Rao, Y. (2013). Testing the Prebisch-Singer hypothesis since 1650: Evidence from panel techniques that allow for multiple breaks. IMF Working Paper No. 13/180.

Baffes, J. (2007). Oil spills on other commodities. *Policy Research Working Paper Series, 32*(3), 126–134.

Baffes, J., & Haniotis, T. (2010). Placing the 2006/08 commodity price boom into perspective. Policy Research Working Paper 5371.

Bernard, L., & Semmler W. (Eds.). (2015). *The Oxford handbook of the macroeconomics of global warming*. New York, NY: Oxford University Press.

Bernard, L., Greiner, A., & Semmler, W. (2012). Agricultural commodities and their financialization. *AESTIMATIO The IEB International Journal of Finance, 5*, 2–25.

BP. (2015). 64th edition of the BP Statistical Review of World Energy – Data workbook. Available online: http://www.bp.com/en/global/corporate/energy-economics/statistical-review-of-world-energy.html. (Accessed October 31, 2015)

Canuto, O. (2014). Commodity Super Cycle to Stick Around a Bit Longer. Available online. http://cfi.co/africa/2014/08/world-bank-group-commodity-super-cycle-to-stick-around-a-bit-longer/. (Accessed October 31, 2015)

Dobbs, R., Oppenheim, J., Thompson, F., Mareels, S., Nyquist, S., & Sanghvi, S. (2013). *Resource revolution: Tracking global commodity markets.* McKinsey Global Institute. Available online: http://www.mckinsey.com/insights/energy_resources_materials/resource_revolution_tracking_global_commodity_markets. (Accessed October 1, 2015)

Enders, W., & Holt, M. (2013). The evolving relationships between agricultural and energy commodity prices: A shifting-mean vector autoregressive analysis. NBER 12810.

Gevorkyan, A. V., & Kvangraven, I. H. (2016, forthcoming). Assessing recent determinants of borrowing costs in Sub-Saharan Africa. *Review of Development Economics.* doi:10.1111/rode.12195.

Gevorkyan, A., & Gevorkyan, A. (2012). Derivatives, commodities, and social costs: Exploring correlation in economic uncertainty. *Contemporary Studies in Economic and Financial Analysis, 94*, 47–70.

Greiner, A., & Semmler, W. (2008). *The global environment, natural resources, and economic growth.* Oxford: University Press.

Grune, L., & Pannek, J. (2011). *Nonlinear model predictive control.* New York, NY: Springer.

Hamilton, J., & Wu, J. (2014). Effect of index-fund investing on commodity futures prices. NBER Working Paper doi:10.3386/w19892

Harvey, M., & Pilgrim, S. (2011). The new competition for land: Food, energy, and climate change. *Food Policy, 36*(1), S40–S51.

Havlík, P., et al. (2013). Crop productivity and the global livestock sector: Implications for land use change and greenhouse gas emissions. *American Journal of Agricultural Economics, 95*(2), 442–448.

Henderson, V. (2002). Urbanization in developing countries. *The World Bank Research Observer, 17*(1), 89–112.

Herrero, M., et al. (2010). Smart investments in sustainable food production: Revisiting mixed crop-livestock systems. *Science, 327*, 822–825.

International Monetary Fund. (2015). IMF Primary Commodity Prices. Available online: http://www.imf.org/external/np/res/commod/index.aspx. (Accessed October 25, 2015)

Jones, C., & Sands, R. (2013). Impact of agricultural productivity gains on greenhouse gas emissions: A global analysis. *American journal of agricultural economics, 95*(5), 1309–1316.

Knittel, C., & Pindyck, R. (2013). The simple economics of commodity price speculation. NBER Working Paper doi:10.3386/w18951

Manera, M., Nicolini, M., & Vignati, I. (2012). Returns in commodities futures markets and financial speculation: A multivariate GARCH approach. FEEM Working Paper No. 23.2012. Available at SSRN: http://ssrn.com/abstract=2041177

NASA/GISS. (2015). Global Temperature. Available online: http://climate.nasa.gov/vital-signs/global-temperature/. (Accessed October 25, 2015)

Quigley, J. M. (1998). Urban diversity and economic growth. *Journal of Economic Perspectives, 12*, 127–138.

Roache, K. (2010). What explains the rise in food price volatility? IMF Working Paper, WP/10/129

Ruta, M., & Venables, A. (2012). International trade in natural resources: Practice and policy. World Trade Organization, Working Paper ERSD-2012-07.

Schlenker, W., & Roberts, M. (2009). Nonlinear temperature effects indicate severe damages to U.S. crop yields under climate change. *PNAS, 106*(37), 15594–15598.

Tilman, D., Balzerb, C., Hillc, J., & Beforta, B. (2011). Global food demand and the sustainable intensification of agriculture. *Proceedings of the National Academy of Sciences of the United States of America, 108*(50), 20260–20264.

United Nations. (2015). *World urbanization prospects the 2014 revision.* New York, NY: United Nations.

World Bank. (2000). Dynamic cities as engines of economic growth, *Entering the 21st Century: World Development Reporter:* 125–138.

World Bank Development Indicators. (2015). World Development Indicators – online database. Available online: http://data.worldbank.org/products/wdi. (Accessed October 1, 2015)

A North-South Model with Technological Spillovers, Environmental Degradation and Structural Change

Anton Bondarev and Alfred Greiner

1 Introduction

Climate change mitigation is widely discussed in the recent economics literature (for a survey of the current state of knowledge see Bernard and Semmler 2015). However, the question of how structural change in the world economy can mitigate the effects of climate changes has been given only few attention, despite the fact that the question of how to foster green innovations as a medium to reduce global warming is frequently analyzed. Among the available options are: international carbon emissions trading, as in Eichner and Pethig (2014), subsidizing the R&D to fight off the technology lock-in, as in Krysiak (2011) and cooperative R&D schemes. We argue that stimulating innovations in order to promote endogenous structural change in the economy is another possibility to replace older and dirtier technologies by newer and cleaner ones.

This paper has been written in honor of Willi Semmler who contributed tremendously to the field of economics. Besides his achievements in dynamic macroeconomics and finance, his articles on environmental and resource economics have considerably added to our understanding of the mechanisms at work in this field of research. Therefore, we decided to prepare a paper dealing with climate change in an integrated world and analyze how structural change can help to mitigate that problem.

A. Bondarev (✉)
Department of Economics, University of Basel, Peter Merian-Weg, 6, 4002 Basel, Switzerland
e-mail: anton.bondarev@unibas.ch

A. Greiner
Department of Business Administration and Economics, Bielefeld University, Universitätstraße 25, 33615 Bielefeld, Germany
e-mail: agreiner@wiwi.uni-bielefeld.de

L. Bernard, U. Nyambuu (eds.), *Dynamic Modeling, Empirical Macroeconomics, and Finance*, DOI 10.1007/978-3-319-39887-7_6

The goal of this paper is to present a stylized North-South endogenous growth model that allows for structural change and environmental degradation, where the South benefits form R&D investment in the North through spillover effects of knowledge. Structural change occurs as an endogenous phenomenon resulting from the introduction of new technologies that are developed with the help of R&D investment, with new technologies replacing old ones. Simultaneously, the existing technologies are continuously improved through vertical innovations. Both the North and the South invest in horizontal and vertical innovations but the North represents the more developed economy that disposes of a higher stock of physical capital and of a higher level of knowledge. Since we consider a world economy in which both economies invest in new technologies and their improvement, our framework describes the situation between industrialized countries and emerging market economies, or the case of two industrialized economies with one lagging behind, rather than the one between industrialized economies and developing economies. In addition, production goes along with environmental degradation that negatively affects output in both economies.

An early paper that deals with technology spillovers in an endogenous growth framework is the contribution by Rivera-Batiz and Romer (1991). These authors assume that the spillovers go in both directions such that the increase in productivity in each country positively depends on the level of technology in the other country. The analysis of this paper shows that allowing for spillover effects implies a higher growth rate compared to the autarky case since the productivity grows at a higher rate leading to a larger spectrum of intermediate goods. The effects of technology diffusion and trade within endogenous growth models allowing for heterogeneous firms has been analyzed by Baldwin and Robert-Nicoud (2008) and Unel (2010). These authors find that the exposure to trade has an ambiguous effect on economic growth although it raises the average productivity. The answer to the question of whether economic growth rises or declines depends on the exact nature of the innovation technology and its connection to international trade.

As concerns the effects of environmental pollution in a multi-country work, a seminal paper is the contribution by Chichilnisky (1994). There, the focus is put on property rights and it is shown that the latter create a motive for trade among otherwise identical regions. Two identical regions will trade if the South has badly defined property rights on environmental resources. Trade with a region with well defined property rights, the North, leads to an overconsumption of resource intensive goods imported from the South. Imposing a tax on the use of resources in the South can lead to even more overextraction and a property rights policies may be more effective. The latter contribution resorts to a static framework to derive its results but does not take into account dynamic aspects as it is done in the later paper by Copeland and Taylor (2009).

Dynamic North-South models that study the interrelation between economic activities and the environment often resort to dynamic game theory. For example, Alemdar and Özyildirim (1998) and Alemdar and Özyildirim (2002) present North-South models, where the North imports raw materials from the South at a monopoly price to produce manufactured goods that are consumed in both regions. There exists a technology diffusion process from the North to the South and the extraction of resources causes environmental degradation. The second contribution, in contrast to the first, assumes that waste material is dumped in the South and it allows for multiple resource owners in the South and damages from resource extraction are only local. The analysis demonstrates that an uncoordinated resource extraction can cause a significant reduction of welfare in the South and cooperation between resource producers in the South raises global welfare, with the South gaining to a larger degree. Further, even without cooperation, both regions are better off when productive activities are less polluting and when knowledge spillovers are larger.

With this paper we intend to contribute to the literature that analyzes the effects of international technology spillovers as concerns economic growth and environmental degradation, where we pay special attention to structural change. Our main findings are as follows:

- Structural change is boosted in both economies as a result of technological spillover;
- This raises economic growth in both the North and the South, but to a greater extent in the South;
- Since new cleaner technologies are adopted faster in the South, emissions of the world economy decrease despite higher economic growth.

The rest of the paper is organized as follows. The next section briefly presents the structure of the North-South growth model. Section 3 gives the solution of the model and Sect. 4 derives the impacts of technology spillovers and of environmental pollution. Section 5, finally, concludes.

2 The Basic Model

The baseline model represents two decentralised economies which interact with each other only through R&D channels and do not compete on product markets. First, we present the model neglecting environmental degradation. Later on this assumption is relaxed to account for the influence of the environment on the overall dynamics.

There are two countries, marked N and S for North and South, respectively. Every country $\mathrm{k} \in \{\mathrm{N}, \mathrm{S}\}$ is described by the framework with endogenous structural change as in Bondarev and Greiner (2014a) with both vertical and horizontal R&D. We assume symmetric economies with respect to the labour force that is constant over time. In every economy labour equals total population and is distributed across the existing range of sectors at every point in time t:

$$\forall \mathrm{k} \in \{\mathrm{N}, \mathrm{S}\} : L^{\mathrm{k}} = \int_{N_{min}(t)}^{N_{max}(t)} L^{\mathrm{k}}(i, t)di; N_{min}(t) < N_{max}(t) < N(t) \tag{1}$$

Here L^{k} is the total labour in each of the economies, $L(i,t)$ is the employment in sector i at time t (changing in time), $N(t)$ is the number of technologies (range) being invented up to time t, $N_{max}(t)$ is the range of manufacturing sectors with positive operating profit and $N_{min}(t)$ is the range of technologies which are no longer profitable and are not used in production. Note that both limits in the integral above are dynamic denoting ongoing structural change in the economy.

The range of invented technologies $N(t)$ is common for both countries and represents the state of fundamental knowledge in the world. Provided symmetry in exogenous parameters of the model, the ranges of operating sectors are also similar across countries and given by the number $N_{max}(t) - N_{min}(t)$ at any point in time.

2.1 Households

In each country households are maximizing their utility from consumption of the available range of products (the same as the range of operating sectors). The objective functional of the household is

$$\forall \mathrm{k} \in \{\mathrm{N}, \mathrm{S}\} : J^{\mathrm{k},H} = \int_0^\infty \mathbf{e}^{-\rho t} U(C^{\mathrm{k}}) dt \,, \tag{2}$$

with $U(C) = \ln C$ being the utility function.

The representative household in each country is maximizing utility from consumption C^k over a continuum of differentiated products from existing sectors

$$C^{\mathrm{k}} = \left[\int_{N_{min}}^{N_{max}} \left(C_i^{\mathrm{k}}\right)^{\frac{\varepsilon-1}{\varepsilon}} di \right]^{\frac{\varepsilon}{\varepsilon-1}} , \tag{3}$$

where ε is the elasticity of substitution between goods.

The flow budget constraint of the household is for both countries

$$\dot{K}^{\mathrm{k}}(t) = rK^{\mathrm{k}}(t) + wL^{\mathrm{k}} - E^{\mathrm{k}}(t) \,, \tag{4}$$

where $K^{\mathrm{k}}(t)$ is country-specific capital, $E^{\mathrm{k}}(t)$ denotes expenditures and r is the interest rate, being assumed equal across countries. Since the total labor is constant and similar in size in each of the economies, the wage rate w is taken as a numeraire and normalized to one further on.

Demand for each product follows standard derivations as in the benchmark model:

$$C_i^{\mathrm{k}}(t) = E^{\mathrm{k}}(t) \frac{\left(P_i^{\mathrm{k}}(t)\right)^{-\varepsilon}}{\int_{N_{min}}^{N_{max}} \left(P_j^{\mathrm{k}}(t)\right)^{1-\varepsilon} dj} \,, \tag{5}$$

where P_i^{k} denotes the price of good i.

The standard Euler equation implies that the optimal growth rate for expenditure is given by

$$\frac{\dot{E}^{\mathrm{k}}}{E^{\mathrm{k}}} = r - \rho \,. \tag{6}$$

2.2 The Manufacturing Sector

For both countries manufacturing sectors are isolated and do not compete with foreign producers. Goods producers employ labor and buy technology from the R&D sector. With these inputs they produce the goods which they sell to the consumer. Output of good i is given by:

$$Y_i^{\mathrm{k}}(t) = \left(A_i^{\mathrm{k}}(t)\right)^{\alpha} L_i^{\mathrm{k}}(t) \,, \tag{7}$$

with A_i^{k} giving the productivity. The profit of firm i is

$$\Pi_i^{\mathrm{k}}(t) = P_i^{\mathrm{k}}(t) Y_i^{\mathrm{k}}(t) - L_i^{\mathrm{k}}(t) - \Psi \,, \tag{8}$$

where Ψ is a fixed operating cost assumed to be equal for both countries.

The only use for output is consumption, so that $C_i = Y_i$. Firm i, therefore, sets its price to

$$P_i^{\mathrm{k}}(t) = \frac{\varepsilon}{\varepsilon - 1} \left(A_i^{\mathrm{k}}(t)\right)^{-\alpha} \,. \tag{9}$$

Inserting (5) and (9) into (7) yields (piecewise defined) labour demand (for each country)

$$L^{\mathrm{k}}(i) = \begin{cases} 0, \; t < \tau_{max}(i), \tau_{max}(i) : \Pi_i^{\mathrm{k}} = 0, \dot{\Pi}_i^{\mathrm{k}} > 0; \\ \frac{\epsilon - 1}{\epsilon} E^{\mathrm{k}} \frac{(A_i^{\mathrm{k}})^{-\alpha(1-\epsilon)}}{\int_{N_{min}}^{N_{max}} (A_j^{\mathrm{k}})^{-\alpha(1-\epsilon)} dj}, \; \tau_{max}(i) < t \leq \tau_{min}(i), \tau_{min}(i) : \Pi_i^{\mathrm{k}} = 0, \dot{\Pi}_i^{\mathrm{k}} < 0; \\ 0, t > \tau_{min}(i). \end{cases}$$

Here and further throughout the paper denote

- $\tau_{min} = N_{min}^{-1}(i)$, time when product (technology) i becomes outdated and the profit of the manufacturing sector decreases below zero;
- $\tau_{max} = N_{max}^{-1}(i)$, time when product (technology) i becomes profitable and the manufacturing sector starts producing positive amounts of the consumption good;
- $\tau_0 = N^{-1}(i)$, time when technology i is invented through horizontal innovations.

Technology is acquired by the manufacturing sector in the form of a patent of finite duration. Pricing for this patent therefore follows Nordhaus (1967), Romer (1990) and Grimaud and Rouge (2004): the price for the patent equals the total value of profits which can be derived from it. Since positive profits may be extracted by the manufacturing firm only during a limited period of time, the price of the patent is also defined over a limited duration. After the patent expires because the technology does not yield positive profits any longer, the technology is freely available for production to everyone. However, due to the process of out-dating of technologies, older technologies are not used in production despite their zero price. The price of the patent is,

$$p_A^{\mathrm{k}}(i) \stackrel{def}{=} \int_{\tau_{max}}^{\tau_{min}} \mathrm{e}^{-r(t-\tau_0)} \Pi_i^{\mathrm{k}} dt = \int_{\tau_{max}}^{\tau_{min}} \mathrm{e}^{-r(t-\tau_0(i))} \left(\frac{1}{\epsilon} E^{\mathrm{k}} \frac{\left(A_i^{\mathrm{k}}\right)^{-\alpha(1-\epsilon)}}{\int_{N_{min}}^{N_{max}} \left(A_j^{\mathrm{k}}\right)^{-\alpha(1-\epsilon)} dj} - \Psi \right) dt. \tag{10}$$

The point in time at which patent i starts, τ_{max}, is endogenously determined through the process of horizontal innovations while the effective time of the expiration of the patent, τ_{min}, is endogenously determined from the demand for the manufactured patented product i.

2.3 The R&D Sector

For each country the process of vertical innovations is described by the same laws as in the stand-alone model except for the possibility of technological spillover. We limit ourselves to the case where one of the countries benefits from the technology spillover (constant leadership case). For the advanced country (North) the process of development of new products is fully similar to the baseline model:

$$V^{\mathrm{N}} = \max_g \int_0^{\infty} e^{-rt} \int_{N_{min}(t)}^{N(t)} p_A^{\mathrm{N}}(i) - \frac{1}{2} \left(g^{\mathrm{N}}(i,t)\right)^2 dt; \tag{11}$$

$$s.t.\ \forall i \in [N_{min}, N] : \dot{A}^{\mathrm{N}}(i,t) = \gamma^{\mathrm{N}} g^{\mathrm{N}}(i,t) - \beta A^{\mathrm{N}}(i,t) \tag{12}$$

$$\int_{N_{min}(t)}^{N(t)} g^{\mathrm{N}}(i,t) di = K^{\mathrm{N}}(t) - u^{\mathrm{N}}(t). \tag{13}$$

At the same time the less developed country benefits from the technological spillover proportional to the technology gap between itself and the developed economy:

$$V^{\mathrm{S}} = \max_{g} \int_{0}^{\infty} e^{-rt} \int_{N_{min}(t)}^{N(t)} p_{A}^{\mathrm{S}}(i) - \frac{1}{2}\left(g^{\mathrm{S}}(i,t)\right)^{2} didt; \tag{14}$$

$$s.t.\, \forall i \in [N_{min}, N] : \dot{A}^{\mathrm{S}}(i,t) = \gamma^{\mathrm{S}} g^{\mathrm{S}}(i,t) - \beta A^{\mathrm{S}}(t) + \theta\left(A^{\mathrm{N}}(t) - A^{\mathrm{S}}(t)\right) \tag{15}$$

$$\int_{N_{min}(t)}^{N(t)} g^{\mathrm{S}}(i,t)di = K^{\mathrm{S}}(t) - u^{\mathrm{S}}(t). \tag{16}$$

where:

- $g^{\mathrm{k}}(i,t)$ are investments into the increase of productivity of technology i at time t by country k;
- $A^{\mathrm{k}}(i,t)$ is the state of productivity of technology i at time t in country k;
- γ^{k} is the efficiency of investments into productivity in country k;
- θ is the speed of technological spillover from the North to the South;
- β is the decay rate of technology in the absence of investments;
- $K^{\mathrm{k}}(t) - u^{\mathrm{k}}(t)$ are resources available for vertical innovations given by the accumulated capital minus horizontal innovations investments.

As concerns the source for those spillovers, one can think of two sources. First, it is possible that more developed economies foster technical progress in less developed countries as a means of development aid. This may occur in form of a direct knowledge transfer or by training students of the less developed country in the more developed one, for example. In that case, one can speak of a cooperation between these two economies. Second, knowledge can never be completely kept secret so that it may be transferred from one country to another even if the developed country does not actively contribute to its dissemination. This holds all the more when the less developed economy is trying to acquire the knowledge of the developed economy.

The only incentive for horizontal innovations is the potential profit from selling the new technology to manufacturing firms. Since horizontal innovations have zero productivity at the time when they are invented,[1] the value of horizontal R&D consists in expected future profits that arise when an innovation becomes profitable as a result of vertical innovations (analogous to Peretto and Connolly 2007):

$$V_{N}^{\mathrm{k}} = \max_{u(\cdot)} \int_{0}^{\infty} \mathrm{e}^{-rt} \left(\left(\delta_{\mathrm{N}} u^{\mathrm{N}}(t) + \delta_{\mathrm{S}} u^{\mathrm{S}}(t)\right) \pi_{\mathrm{k}}^{R}(i,t)|_{i=N} - \frac{1}{2}\left(u^{\mathrm{N,S}}(t)\right)^{2} \right) dt \tag{17}$$

[1]In this sense, we differ between the invention of a new technology and its economic use.

where $\pi_{\mathrm{k}}^{R}(i,t)|_{i=N}$ denotes the profit from the subsequent development of the new technology $i = N$ for country k. It is defined as:

$$\pi_{\mathrm{k}}^{R}(i,t) = p_{A}^{\mathrm{k}}(i) - \frac{1}{2} \int_{\tau_0(i)}^{\tau_{min}(i)} \mathrm{e}^{-r(t-\tau_0(i))} \left(g^{\mathrm{k}}(i,t)\right)^2 di, \tag{18}$$

stating that the profit equals the difference between the price of the patent and the accumulated investments into the technology development during the life of the technology. The change in the range of technologies in the world, then, is the result of the R&D investments in these two regions:

$$\dot{N} = \delta_{\mathrm{N}} u^{\mathrm{N}}(t) + \delta_{\mathrm{S}} u^{\mathrm{S}}(t). \tag{19}$$

3 Solution and Basic Results

3.1 *Vertical Innovations*

We limit ourselves to the open-loop solution, since it is difficult to formulate a HJB pair for the resource-constrained differential game, see e.g. Dockner et al. (2000). With homogeneous efficiency of investments across technologies within the country, the optimal investments for every technology are just proportional to the total available research capital minus variety expansion investments as long as the derivative of the patent price (10) w.r.t. productivity A_i does not depend on i. That this is indeed the case is formally proved in the benchmark model. It is sufficient to note that the patent price equation is the same for the technological spillover model as in the benchmark case, thus giving the same result:

$$g^{\mathrm{k}}(i,t) = \frac{K^{\mathrm{k}} - u^{\mathrm{k}}}{N - N_{min}}. \tag{20}$$

At the same time the **evolution** of productivity is different for the North and the South:

$$\dot{A}^{\mathrm{N}} = \frac{K^{\mathrm{N}} - u^{\mathrm{N}}}{N - N_{min}} - \beta A^{\mathrm{N}}, \dot{A}^{\mathrm{S}} = \frac{K^{\mathrm{S}} - u^{\mathrm{S}}}{N - N_{min}} - \beta A^{\mathrm{S}} + \theta \left(A^{\mathrm{N}} - A^{\mathrm{S}}\right). \tag{21}$$

as long as $A^{\mathrm{N}}(t) > A^{\mathrm{S}}(t)$.

Since horizontal investments are constant, the dynamics solely depends on the capital evolution. The latter is analogous to the stand-alone baseline model and given

Fig. 1 Evolution of productivities of both economies

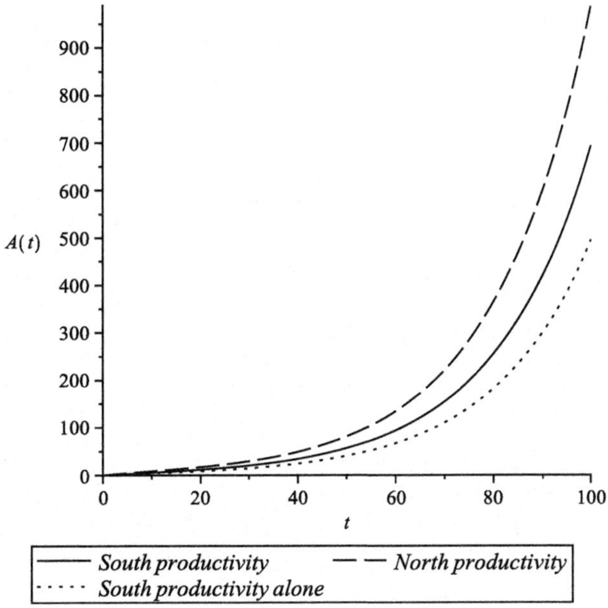

by:

$$K^{\mathrm{k}}(t) = \mathrm{e}^{rt}\left(K_0^{\mathrm{k}} - \frac{1}{(\epsilon-1)r}L\right) + \frac{1}{r(\epsilon-1)}L. \tag{22}$$

With similar labor in both countries, the difference in evolution of capital is fully defined by the difference in initial asset holdings of households which is a natural measure of the state of development of the economy. With similar initial asset holdings no technological spillover is possible as further discussions below show.

Denote the capital available for vertical innovations as $G^{\mathrm{N}}, G^{\mathrm{S}}$ for both countries. For leadership to be constant it is sufficient to have $G^{\mathrm{N}} > G^{\mathrm{S}},\ \forall t$. Given linear variety investments (because of homogeneous technologies) and monotonic capital accumulation (because of constant expenditures E) it amounts to the condition on initial capital endowments in both countries. With $K_0^{\mathrm{N}} > K_0^{\mathrm{S}}$ the follower will never catch-up with the leader in productivity, but the South productivity will be still higher than in the autarky case as the illustration in Fig. 1 shows. This gives rise to the following Proposition.

Proposition 1 (Evolution of Productivities) *For constant technological spillovers from the North to the South, it is sufficient to have $K_0^{\mathrm{N}} > K_0^{\mathrm{S}}$ in the symmetric case. In this case, productivity in the South $A^{\mathrm{S}}(i,t)$ grows faster for each sector than in the absence of spillovers while the North productivity $A^{\mathrm{N}}(i,t)$ is unaffected by it.*

Proof As long as $K_0^{\mathrm{N}} > K_0^{\mathrm{S}}$ we have $\forall t,\ K^{\mathrm{N}} > K^{\mathrm{S}}$. As long as horizontal innovations investments u are independent of time (which is indeed the case for homogeneous technologies), it follows that $\forall t$, $K^{\mathrm{N}} > K^{\mathrm{S}} \rightarrow G^{\mathrm{N}} > G^{\mathrm{S}}$. The initial productivity for every new technology is zero, thus in (21) it is always the case that $A^{\mathrm{N}}(t) > A^{\mathrm{S}}(t)$. ■

3.2 *Horizontal Innovations*

The solution of a pair of HJB equations derived from (17) under the assumption of constant profits for every next technology yields horizontal innovations investments proportional to expected profit for both countries:

$$u^{\mathrm{N}}(t) = \delta_{\mathrm{N}}\pi_{\mathrm{N}}^{R}(N,t);\, u^{\mathrm{S}}(t) = \delta_{\mathrm{S}}\pi_{\mathrm{S}}^{R}(N,t). \tag{23}$$

Now, we show that the South, while benefiting from the technological spillover in the development of productivities, invests more than the North in the creation of new technologies. This creates an endogenous specialization effect similar to the one obtained for the dynamic regional monopolies setup in Bondarev (2014).

The expected profit from each technology is defined by two components: accumulated investments and price of the patent. The specialization of innovative activities comes from the fact that investments for the follower are smaller than for the leader, while the price of the patent is the same in both countries.

We state the first part of this result as a Lemma:

Lemma 2 (Patent Prices) *Patent prices for all technologies are the same across countries,*

$$p_A^{\mathrm{N}}(i) = p_A^{\mathrm{S}}(i). \tag{24}$$

Proof To see this recall that the patent price is defined as the profit stream of the manufacturing sector and thus amounts to the time integral over the relative productivity of the technology i within the operational time, (10).

Subtracting the patent price of the South from that of the North we have an expression constant in time but growing in i:

$$p_A^{\mathrm{N}}(i) - p_A^{\mathrm{S}}(i) = \int_{\tau_{max}}^{\tau_{min}} e^{-r(t-\tau_0)}\Pi_i^{\mathrm{N}}dt - \int_{\tau_{max}}^{\tau_{min}} e^{-r(t-\tau_0)}\Pi_i^{\mathrm{S}}dt = \int_{\tau_{max}}^{\tau_{min}} e^{-r(t-\tau_0)}\left(\Pi_i^{\mathrm{N}} - \Pi_i^{\mathrm{S}}\right)dt =$$

$$\left(\frac{1}{\epsilon}\right)\int_{\tau_{max}}^{\tau_{min}} e^{-r(t-\tau_0(i))}\left(E^{\mathrm{N}}\frac{\left(A_i^{\mathrm{N}}\right)^{-\alpha(1-\epsilon)}}{\int_{N_{min}}^{N_{max}}\left(A_j^{\mathrm{N}}\right)^{-\alpha(1-\epsilon)}dj} - E^{\mathrm{S}}\frac{\left(A_i^{\mathrm{S}}\right)^{-\alpha(1-\epsilon)}}{\int_{N_{min}}^{N_{max}}\left(A_j^{\mathrm{S}}\right)^{-\alpha(1-\epsilon)}dj}\right)dt \tag{25}$$

At the same time the difference between productivities for both countries is the same for all the technologies:

$$A_i^{\mathrm{N}}(t) - A_i^{\mathrm{S}}(t) = \frac{(\mathrm{e}^{rt} - \mathrm{e}^{-(\theta+\beta)t})\left((u^{\mathrm{N}} - u^{\mathrm{S}})(K_0^{\mathrm{N}} - K_0^{\mathrm{S}})\right)}{(\theta + \beta + r)(N - N_{min})} \quad (26)$$

The relative productivity of each technology is then the same in both countries (but total productivity is different) and thus the price for the patent is the same, provided expenditures are the same. This is indeed the case since expenditures are constant and proportional to the labor force which is assumed to be equal across countries:

$$E = \int_{N_{min}}^{N_{max}} p^{\mathrm{k}}(i,t)C^{\mathrm{k}}(i,t)di = \int_{N_{min}}^{N_{max}} p^{\mathrm{k}}(i,t)Y^{\mathrm{k}}(i,t)di = \frac{\epsilon}{\epsilon - 1}\int_{N_{min}}^{N_{max}} L(i,t)di = \frac{\epsilon}{\epsilon - 1}L. \quad (27)$$

Then, it follows

$$p_A^{\mathrm{N}}(i) - p_A^{\mathrm{S}}(i) = 0. \quad (28)$$

■

At the same time the accumulated investments for every technology i are higher for the developed country, since the capital accumulation is faster.

Lemma 3 (Accumulated Investments) *For every technology i accumulated along the total life-cycle, investments into productivity are lower for the follower. In case $K_0^{\mathrm{S}} < K_0^{\mathrm{N}}$ this turns out to be the South:*

$$\frac{1}{2}\int_{\tau_0(i)}^{\tau_{min}(i)} \mathrm{e}^{-r(t-\tau_0(i))}\left(g^{\mathrm{S}}(i,t)\right)^2 di < \frac{1}{2}\int_{\tau_0(i)}^{\tau_{min}(i)} \mathrm{e}^{-r(t-\tau_0(i))}\left(g^{\mathrm{N}}(i,t)\right)^2 di \quad (29)$$

Proof As long as horizontal investments are constant and initial capital endowment is as in the condition of the Lemma it follows,

$$G^{\mathrm{N}}(t) = K^{\mathrm{N}}(t) - u^{\mathrm{N}} > G^{\mathrm{S}}(t) = K^{\mathrm{S}}(t) - u^{\mathrm{S}}. \quad (30)$$

Since the investments into each technology are given by (20) and $N - N_{min}$ is constant we have,

$$\forall t, \in [\tau_0(i); \tau_{min}(i)],\ \forall i \in [0; N] : g^{\mathrm{S}}(i,t) < g^{\mathrm{N}}(i,t). \quad (31)$$

Since investments are always nonnegative, the integration and power operations are monotonic w.r.t. the order of relations and the result follows. ■

Thus, the profit for every new technology (including the boundary one) is higher for the less developed country:

$$\pi_{\mathrm{N}}^{R}(N) - \pi_{\mathrm{S}}^{R}(N) < 0 \tag{32}$$

and investments of the follower into variety expansion are higher than that of the leader.

Proposition 4 (Endogenous Specialization of Innovations) *With equal labor force and fixed operating costs across countries, the less developed country invests more into new products creation since the expected benefit from a new technology for this country is higher:*

$$\pi_{\mathrm{N}}^{R}(N,t) < \pi_{\mathrm{S}}^{R}(N,t) \to u^{\mathrm{N}} < u^{\mathrm{S}}. \tag{33}$$

Proof The profit from each new technology is given by (18) with $i = N$. The price of the patent is the same by Lemma 2. The accumulated investments are in the relation given by Lemma 3. The result thus follows. ∎

4 Effects of International Spillovers and Environmental Degradation

4.1 Comparison with the Benchmark Model

Now, we study the effects of international technological spillovers as concerns the growth rates in both economies, where we first neglect environmental degradation.

First, it should be noted that the productivity of individual technologies in the North is the same as without spillovers in the benchmark model. At the same time, the speed of variety expansion is higher since the process of discoveries now benefits from the investments of the other country. Assume the initial range of available technologies is the same for both countries,

$$N_0^{\mathrm{N}} = N_0^{\mathrm{S}} = N_0. \tag{34}$$

Recall that under autarky the variety expansion for both countries is a linear process, yielding a constant range of existing sectors,

$$\forall \mathrm{k} \in \{\mathrm{N},\mathrm{S}\} : \dot{N}^{\mathrm{k}} = \dot{N}^{\mathrm{k}}_{min} = \dot{N}^{\mathrm{k}}_{max}. \tag{35}$$

However, it follows that the higher is the speed of structural change, the higher is the existing diversity of technologies. Thus, for the leading country the effect of technological spillovers would be a faster turnover of sectors, that is the life-cycle of each technology would be shorter. This boosts structural change in the economy and increases the overall productivity. To see this, compare the range of existing sectors under technological spillovers and without them for both countries:

$$\forall \mathrm{k} \in \{\mathrm{N}, \mathrm{S}\} : N^{\mathrm{k}}(t) = \delta^2 \pi_{\mathrm{k}}^R t + N_0,$$
$$N_{min}^{\mathrm{k}}(t) = \delta^2 \pi_{\mathrm{k}}^R t + N_0 - \tau_{min}(N_0), \qquad N_{max}^{\mathrm{k}}(t) = \delta^2 \pi_{\mathrm{k}}^R t + N_0 - \tau_{max}(N_0), \quad (36)$$

where $\tau_{max}(N_0), \tau_{min}(N_0)$ are the times when the technology $i = N_0$ becomes operational and out-dated respectively. The range of existing sectors of the economy is thus defined by,

$$\forall \mathrm{k} \in \{\mathrm{N}, \mathrm{S}\} : N_{max}^{\mathrm{k}}(t) - N_{min}^{\mathrm{k}}(t) = \mathcal{O}^{\mathrm{k}} = \delta^2 \pi_{\mathrm{k}}^R (\tau_{min}(N_0) - \tau_{max}(N_0)) \quad (37)$$

At the same time, with technological spillovers the variety expansion is faster and, therefore, the range of the operational sectors (core) is wider as long as the profit from each individual technology under autarky is not lower than under spillovers. This is indeed the case since we neglect product market competition and profits from patents in the North are unchanged while in the South they are greater (hence greater productivity than under autarky):

$$N^{\mathrm{T}} = \left(\delta_{\mathrm{N}}^2 \pi_{\mathrm{N}}^R + \delta_{\mathrm{S}}^2 \pi_{\mathrm{S}}^R\right) t + N_0,$$
$$N_{max}^{\mathrm{T}}(t) - N_{min}^{\mathrm{T}}(t) = \mathcal{O}^{\mathrm{T}} = \left(\delta_{\mathrm{N}}^2 \pi_{\mathrm{N}}^R + \delta_{\mathrm{S}}^2 \pi_{\mathrm{S}}^R\right) (\tau_{min}(N_0) - \tau_{max}(N_0)), \quad (38)$$

where the superscript T denotes quantities for the world economy with technological spillovers and the superscript A denotes quantities without spillovers (the autarky case). Then, it is straightforward to see that,

$$\pi_{\mathrm{N}}^R \geq \pi_{\mathrm{A}}^R, \pi_{\mathrm{S}}^R \geq \pi_{\mathrm{A}}^R \rightarrow \mathcal{O}^{\mathrm{T}} > \mathcal{O}_{\mathrm{N}}^{\mathrm{A}} \geq \mathcal{O}_{\mathrm{S}}^{\mathrm{A}}. \quad (39)$$

It is important to note that the effect is strictly positive for the North with any level of the technology gap between countries, since the patent profits in the North are at least the same and in the South they are at least non-zero. The effect of faster structural change and sectoral turnover is observed for both economies, but to a stronger degree for the South, since the patent profits in the South rise to a larger extent. The leading North economy will also benefit from the wider diversity of technologies being operational. The higher speed of structural change thus generates a larger variety of technologies. Figure 2 illustrates the result.

Proposition 5 (Effects of Technological Spillovers on Structural Change) *When technological spillover effects occur,* $\theta > 0, K_0^{\mathrm{N}} > K_0^{\mathrm{S}}$, *the speed of expansion of variety of technologies,* $\dot{N}(t)$, *as well as of the out-dating of technologies,* $\dot{N}_{min}(t)$, *is faster for both countries:*

$$\forall \mathrm{k} \in \{\mathrm{N}, \mathrm{S}\} : \dot{N}^{\mathrm{T}} = \dot{N}_{min}^{\mathrm{T}} = \dot{N}_{max}^{\mathrm{T}} > \dot{N}^{\mathrm{k}} = \dot{N}_{min}^{\mathrm{k}} = \dot{N}_{max}^{\mathrm{k}}. \quad (40)$$

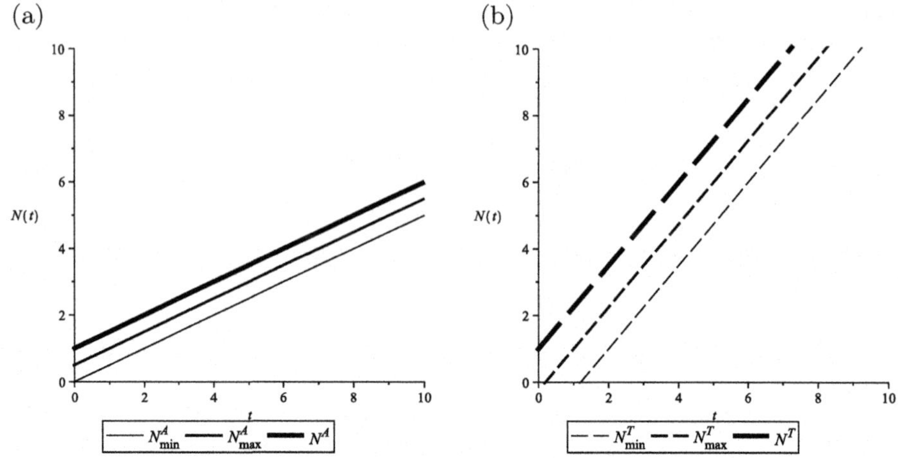

Fig. 2 Horizontal innovations and expansion of the core. (**a**) Variety expansion under autarky, (**b**) variety expansion under spillovers

Therefore, the structural change in the economy with technological spillovers is faster,

$$\mathcal{O}^{\mathrm{T}} > \mathcal{O}^{\mathrm{A}}_{\mathrm{N}} \geq \mathcal{O}^{\mathrm{A}}_{\mathrm{S}}. \tag{41}$$

Proof In the absence of spillovers, the profit from each technology in the North is given by,

$$\pi^{R}_{\mathrm{N}}(i,t) = p^{\mathrm{N}}_{A}(i) - \frac{1}{2}\int_{\tau_0(i)}^{\tau_{min}(i)} \mathrm{e}^{-r(t-\tau_0(i))}\left(\frac{K^{\mathrm{N}}-u^{\mathrm{N}}}{N^{\mathrm{N}}-N^{\mathrm{N}}_{min}}\right)^2 di. \tag{42}$$

With spillovers, the productivity of each technology in the North is the same, and the difference in profits may come only from the changes in integration limits in patent price and accumulated investments. However, these limits change in the same direction and by the same amount given a linear variety expansion process. Thus, the profits from vertical innovations in the North would remain unchanged under spillovers.

In the South productivity of technologies is higher under the spillover and thus profits are higher than without the spillover. Thus we have (39).

Because of this, the variety expansion is boosted in comparison to the autarky case and we have (40). Since processes of N_{max} and N_{min} are just shifts of a linear N process, we have (41). ■

Now, consider that the rate of output growth for such an economy is,

$$\frac{\dot{Y}^{\mathrm{N,S}}}{Y^{\mathrm{N,S}}} = \alpha \frac{\dot{\bar{A}}^{\mathrm{N,S}}}{\bar{A}^{\mathrm{N,S}}} (N_{max} - N_{min}) > 0, \tag{43}$$

where $\bar{A}$ denotes the average productivity. For the North this means that the average productivity growth rate is unchanged but the core of the economy is larger and, thus, economic growth is higher:

$$\frac{\dot{Y}^{\mathrm{N}}_{\mathrm{A}}}{Y^{\mathrm{N}}_{\mathrm{A}}} = \alpha \frac{\dot{\bar{A}}}{\bar{A}} \mathcal{O}^{\mathrm{A}} < \frac{\dot{Y}^{\mathrm{N}}_{\mathrm{T}}}{Y^{\mathrm{N}}_{\mathrm{T}}} = \alpha \frac{\dot{\bar{A}}}{\bar{A}} \mathcal{O}^{\mathrm{T}} \tag{44}$$

where superscript N indicates North and subscripts A, T denote quantities in autarky and under technological spillovers.

Now, turn to the South. For the less developed country the technological spillover is even more beneficial, since both the range of technologies being used is increased compared to the autarky regime and the average productivity of each technology is higher because of technological spillovers from the leader:

$$\frac{\dot{\bar{A}}^{\mathrm{S}}_{\mathrm{A}}}{\bar{A}^{\mathrm{S}}_{\mathrm{A}}} < \frac{\dot{\bar{A}}^{\mathrm{S}}_{\mathrm{T}}}{\bar{A}^{\mathrm{S}}_{\mathrm{T}}}; \mathcal{O}^{\mathrm{A}} < \mathcal{O}^{\mathrm{T}}; \frac{\dot{Y}^{\mathrm{S}}_{\mathrm{A}}}{Y^{\mathrm{S}}_{\mathrm{A}}} << \frac{\dot{Y}^{\mathrm{S}}_{\mathrm{T}}}{Y^{\mathrm{S}}_{\mathrm{T}}} \tag{45}$$

Thus, the technology sharing would be beneficial for both economies without any drawbacks. The rate of growth of the world economy is the sum of the growth rates of North and South and is higher with technological spillovers than without them.

Proposition 6 (World Economy with Technological Spillovers) *When the technological spillover from the North to the South takes place, the following effects are observed:*

1. *The range of operating sectors in both economies is higher than without the spillover,* $\mathcal{O}^{\mathrm{T}} > \mathcal{O}^{\mathrm{A}}$;
2. *The productivity of each technology is the same in the North as if no spillover occurred,* $A^{\mathrm{N}}_{\mathrm{T}}(i,t) = A^{\mathrm{N}}_{\mathrm{A}}(i,t)$;
3. *Productivity for each technology in the South is higher due to the technology spillover,* $A^{\mathrm{S}}_{\mathrm{T}}(i,t) > A^{\mathrm{S}}_{\mathrm{A}}(i,t)$;
4. *Output growth in both economies is higher as well as the growth rate of the world economy,* $\frac{\dot{Y}^{\mathrm{N}}_{\mathrm{T}}}{Y^{\mathrm{N}}_{\mathrm{T}}} > \frac{\dot{Y}^{\mathrm{N}}_{\mathrm{A}}}{Y^{\mathrm{N}}_{\mathrm{A}}}, \frac{\dot{Y}^{\mathrm{S}}_{\mathrm{T}}}{Y^{\mathrm{S}}_{\mathrm{T}}} > \frac{\dot{Y}^{\mathrm{S}}_{\mathrm{A}}}{Y^{\mathrm{S}}_{\mathrm{A}}}, \frac{\dot{Y}^{\mathrm{W}}_{\mathrm{T}}}{Y^{\mathrm{W}}_{\mathrm{T}}} > \frac{\dot{Y}^{\mathrm{W}}_{\mathrm{A}}}{Y^{\mathrm{W}}_{\mathrm{A}}}$.

Proof 1. amounts to Proposition 5; 2. and 3. follow from (21); 4. is obtained by direct computations performed above. ■

4.2 The Extended Model with Environmental Degradation

Now, we extend the basic model analyzed above to take into account the interrelation between the economy and the environment. To do so we assume that the change of the environment is described by the following differential equation:

$$\dot{T} = -\mu T + eY^{\mathrm{W}}; \tag{46}$$

where:

- T is some aggregate measure of the environment (temperature increase above pre-industrial level);
- μ is the regeneration rate in the absence of industrial activity;
- e is the intensity of emissions, defined by the state of technology;
- Y^{W} is the aggregate output of the world economy.

The intensity of emissions is defined as a mix of technologies currently in use in both countries, weighted by the share of output produced with these technologies:

$$e = e_0 \frac{\int_{N_{min}}^{N_{max}} (1/i) y_i di}{\mathcal{O}}, \tag{47}$$

with y_i denoting the share of world output for technology i. Note that, due to the assumption of a common range of technologies for both countries, the emissions intensity is defined over the common range and no separate functions are necessary. The parameter e_0 is the estimate of the initial intensity of emissions. In the numerical example presented below, we assume that 0.0475 of the total output is transferred into the temperature increase along the lines of Nordhaus (2007).

The influence of the environment on economic activity is modelled through a damage function. Environmental degradation reduces production capabilities in the economies but, since it is an externality, it is not taken into account by the manufacturing sector. The aggregate output now is given by,

$$Y^{\mathrm{W}}(t) = \frac{1}{1+T} \left(\int_{N_{min}}^{N_{max}} Y^{\mathrm{N}}(i,t) di + \int_{N_{min}}^{N_{max}} Y^{\mathrm{S}}(i,t) di \right), \tag{48}$$

with $1/(1+T)$ reflecting environmental damages. Since all of the output of the manufacturing sector is consumed, such a specification is equivalent to the reduction in consumption of every product within the operational range and to a proportional increase in prices:

$$\forall \mathrm{k} \in \{\mathrm{N}, \mathrm{S}\} : C_i^{\mathrm{k},T} = \frac{1}{1+T} Y_i^{\mathrm{k},O}; P_i^{\mathrm{k},T} = (1+T) P_i^{\mathrm{k},O}; L_i^{\mathrm{k},T} = \frac{1}{1+T} L_i^{\mathrm{k},O}, \tag{49}$$

with superscripts T, O denoting the world economy with and without environmental impact, respectively.

Thus, total expenditures in both economies are still constant and are unaffected by the state of environment:

$$\forall \mathrm{k} \in \{\mathrm{N}, \mathrm{S}\} : E^{\mathrm{k},T} = \int_{N_{min}}^{N_{max}} P_i C_i di = E^{\mathrm{k},O}, \tag{50}$$

since the impact on the prices and on labour employed by different sectors cancels out. Following the same lines as for the stand-alone model without technological spillovers in Bondarev and Greiner (2014b), it can be demonstrated that environmental degradation leads to a decrease in labour income and in capital accumulation, making it harder to raise the productivity of the economy:

$$\forall \mathrm{k} \in \{\mathrm{N}, \mathrm{S}\} : \dot{K}^{\mathrm{k},T} = rK^{\mathrm{k},T} - E^O + \frac{1}{1+T} < \dot{K}^{\mathrm{k},O}. \tag{51}$$

The capital accumulation is decreased in the same way for both the North and the South, decreasing productivity growth symmetrically in both countries:

$$\dot{A}^{\mathrm{N},T}(i,t) = \frac{K^{\mathrm{N},T} - u^{\mathrm{N}}}{N - N_{min}} - \beta A^{\mathrm{N},T}(i,t) < \dot{A}^{\mathrm{N},O}; \tag{52}$$

$$\dot{A}^{\mathrm{S},T}(i,t) = \frac{K^{\mathrm{S},T} - u^{\mathrm{S}}}{N - N_{min}} - \beta A^{\mathrm{S},T}(i,t) + \theta \left(A^{\mathrm{N},T}(i,t) - A^{\mathrm{S},T}(i,t)\right) < \dot{A}^{\mathrm{S},O}. \tag{53}$$

Nevertheless, the patent prices are higher than in the model without environmental damages, since the profits of the manufacturing sector are unaffected. The decrease in output is balanced by the increase in prices and labor costs are lower:

$$\forall \mathrm{k} \in \{\mathrm{N}, \mathrm{S}\} : \Pi_i^{\mathrm{k},T} = P_i^{\mathrm{k},T} Y_i^{\mathrm{k},T} - L_i^{\mathrm{k},T} - \Psi = \left(\frac{\epsilon}{\epsilon - 1} - \frac{1}{1+T}\right) L_i^{\mathrm{k},O} - \Psi; \tag{54}$$

$$\Pi_i^{\mathrm{k},O} = P_i^{\mathrm{k},O} Y_i^{\mathrm{k},O} - L_i^{\mathrm{k},O} - \Psi = \left(\frac{\epsilon}{\epsilon - 1} - 1\right) L_i^{\mathrm{k},O} - \Psi; \tag{55}$$

$$T \geq 0 : \Pi_i^{\mathrm{k},T} \geq \Pi_i^{\mathrm{k},O}. \tag{56}$$

Due to the magistrale property of the productivity dynamics (during the operational phase all of the technologies follow the same evolution path) and due to the definition of the time integration limits of the patent price, it follows that patent prices are unchanged by the slowdown of capital accumulation. This results from the fact that productivities of all technologies within the operational phase are lowered by exactly the same amount. But, this does not mean that the productivity of new technologies is the same.

At the same time, the accumulated investments into productivity growth are lower for both countries because of lower capital stocks. As a result, the expected profit for each new technology is increasing despite the decrease in productivity itself. Indeed, the decrease of productivity does not lead to a decrease in the patent price so that the development of new technologies becomes more attractive: one may get the same price with lower investments.

Since the profit from patents for both countries is higher under environmental degradation, the process of variety expansion is accelerated and the operational range of technologies rises:

$$\pi^T_{\mathrm{N,S}} > \pi^O_{\mathrm{N,S}} \rightarrow \dot{N}^T_{\mathrm{N,S}} > \dot{N}^T_{\mathrm{N,S}} \rightarrow \mathcal{O}^T_{\mathrm{N,S}} > \mathcal{O}^O_{\mathrm{N,S}}. \tag{57}$$

We summarize these results in the following Proposition:

Proposition 7 (Influence of Environmental Degradation) *In the world economy with technological spillovers, the presence of environmental degradation described by* (46) *and* (48) *leads to the following effects:*

1. *Decrease in labour demand compared to the basic model,* $L_i^{\mathrm{k},T} = \frac{1}{1+T} L_i^{\mathrm{k},O}$*;*
2. *Decrease in productivity of all the technologies, given by* (52), (53)*;*
3. *Increase in the speed of variety expansion and of structural change,* (57).

Proof 1. Follows from (48) and (49); 2. follows from depressed capital accumulation, (51); 3. follows from increased final producers profits, (56). ■

Recall that the growth rate of the economy is given by (43). It is straightforward to see that the growth rate of the average productivity is lower under environmental pollution while the range of operational sectors is wider. Hence, the exact difference between the economic growth rates of the world economy with and without environmental degradation cannot be determined in general.

It should be noted that the range of operational sectors does not change over time and depends only on the initial range of technologies, as in (38). At the same time, if the world temperature rises, $\dot{T} > 0$, the growth rate of the average productivity declines and it is not constant as in the model without the environment, since the capital growth rate decreases:

$$\frac{\dot{\bar{A}}}{\bar{A}} \sim \frac{\dot{K}}{K}. \tag{58}$$

This creates additional stimuli for international technological spillovers as defined above. That holds because boosting structural change would not only increase overall economic growth but also speed up the introduction of cleaner technologies in both economies, thus, decreasing the emissions intensity and slowing down output degradation.

4.3 Effects of Environmental Degradation with and Without Technological Spillovers

In this subsection, we compare the evolution of the environment in the economy without technological spillovers to the case of an economy featuring such spillovers. We already established the stimulating effect of technological spillovers on the economic and R&D development of both countries and discussed the changes being brought about by the presence of environmental damages in that model. It is then straightforward to expect that the environmental impact of the world economy should be lower with technological spillovers than without them. This result would be of limited interest since it has already been discussed in the environmental economics literature, but our model enables us to highlight the role of endogenous structural change in this process.

Consider the total environmental impact of the world economy:

$$\dot{T} = -\mu T + e_0 \frac{\int_{N_{min}}^{N_{max}} (1/i) y_i^{\mathrm{W}} di}{\mathcal{O}} Y^{\mathrm{W}}. \tag{59}$$

In what follows we denote with the superscript T quantities for the world economy with technological spillovers and environmental pollution and by the superscript A quantities for the autarky regime with environmental degradation (but no technological spillovers).

The world output is equivalent to the sum of sectoral outputs in both economies. With technological spillovers, both countries share the same technological space:

$$Y^{\mathrm{W,T}} = \int_{N_{min}}^{N_{max}} \left(Y_i^{\mathrm{N}} + Y_i^{\mathrm{S}}\right) di, \tag{60}$$

while under autarky the range of technologies can be different:

$$Y^{\mathrm{W,A}} = \int_{N_{min}^{\mathrm{N}}}^{N_{max}^{\mathrm{N}}} Y_i^{\mathrm{N}} di + \int_{N_{min}^{\mathrm{S}}}^{N_{max}^{\mathrm{S}}} Y_i^{\mathrm{S}} di, \tag{61}$$

yielding a different denominator $\mathcal{O}$ in the evolution of the environment (59).

Sectoral outputs (with environmental impact) are functions of relative productivities in both scenarios:

$$Y_i^{\mathrm{k}} = \frac{1}{1+T} \left(\frac{(A^{\mathrm{k}}(i,t))^{\alpha\epsilon}}{\int_{N_{min}^{\mathrm{k}}}^{N_{max}^{\mathrm{k}}} (A^{\mathrm{k}}(j,t))^{\alpha(\epsilon-1)} dj} \right), \tag{62}$$

so that the world output can be expressed as,

$$Y^{\mathrm{W}} = \frac{1}{1+T}\left(\frac{\int_{N_{min}^{\mathrm{N}}}^{N_{max}^{\mathrm{N}}}(A^{\mathrm{N}}(i,t))^{\alpha\epsilon}di}{\int_{N_{min}^{\mathrm{N}}}^{N_{max}^{\mathrm{N}}}(A^{\mathrm{N}}(j,t))^{\alpha(\epsilon-1)}dj} + \frac{\int_{N_{min}^{\mathrm{S}}}^{N_{max}^{\mathrm{S}}}(A^{\mathrm{S}}(i,t))^{\alpha\epsilon}di}{\int_{N_{min}^{\mathrm{S}}}^{N_{max}^{\mathrm{S}}}(A^{\mathrm{S}}(j,t))^{\alpha(\epsilon-1)}dj}\right). \tag{63}$$

The world share of technology i is

$$y_i^{\mathrm{W}} = \frac{Y_i^{\mathrm{N}} + Y_i^{\mathrm{S}}}{Y^{\mathrm{W}}} = \frac{\frac{(A^{\mathrm{N}}(i,t))^{\alpha\epsilon}}{\int_{N_{min}^{\mathrm{N}}}^{N_{max}^{\mathrm{N}}}(A^{\mathrm{N}}(j,t))^{\alpha(\epsilon-1)}dj} + \frac{(A^{\mathrm{S}}(i,t))^{\alpha\epsilon}}{\int_{N_{min}^{\mathrm{S}}}^{N_{max}^{\mathrm{S}}}(A^{\mathrm{S}}(j,t))^{\alpha(\epsilon-1)}dj}}{\left(\frac{\int_{N_{min}^{\mathrm{N}}}^{N_{max}^{\mathrm{N}}}(A^{\mathrm{N}}(i,t))^{\alpha\epsilon}di}{\int_{N_{min}^{\mathrm{N}}}^{N_{max}^{\mathrm{N}}}(A^{\mathrm{N}}(j,t))^{\alpha(\epsilon-1)}dj} + \frac{\int_{N_{min}^{\mathrm{S}}}^{N_{max}^{\mathrm{S}}}(A^{\mathrm{S}}(i,t))^{\alpha\epsilon}di}{\int_{N_{min}^{\mathrm{S}}}^{N_{max}^{\mathrm{S}}}(A^{\mathrm{S}}(j,t))^{\alpha(\epsilon-1)}dj}\right)}. \tag{64}$$

Depending on whether the variety of technologies coincides for both countries (under the spillover scenario) or not, one obtains two different expressions for the evolution of the environment. For the economy with technological spillover one gets:

$$\dot{T}^{\mathrm{T}} = \frac{e_0}{1+T^{\mathrm{T}}}\frac{1}{\mathcal{O}^{\mathrm{T}}}\left(\frac{\int_{N_{min}^{\mathrm{T}}}^{N_{max}^{\mathrm{T}}}(1/i)(A^{\mathrm{N,T}}(j,t))^{\alpha\epsilon}di}{\int_{N_{min}^{\mathrm{T}}}^{N_{max}^{\mathrm{T}}}(A^{\mathrm{N,T}}(j,t))^{\alpha(\epsilon-1)}dj} + \frac{\int_{N_{min}^{\mathrm{T}}}^{N_{max}^{\mathrm{T}}}(1/i)(A^{\mathrm{S,T}}(j,t))^{\alpha\epsilon}di}{\int_{N_{min}^{\mathrm{T}}}^{N_{max}^{\mathrm{T}}}(A^{\mathrm{S,T}}(j,t))^{\alpha(\epsilon-1)}dj}\right) - \mu T^{\mathrm{T}}, \tag{65}$$

and for the economy without technological spillover (autarky):

$$\dot{T}^{\mathrm{A}} = \frac{e_0}{1+T^{\mathrm{A}}}\left(\frac{1}{\mathcal{O}^{\mathrm{N,A}}}\frac{\int_{N_{min}^{\mathrm{N}}}^{N_{max}^{\mathrm{N,A}}}(1/i)(A^{\mathrm{N,A}}(j,t))^{\alpha\epsilon}di}{\int_{N_{min}^{\mathrm{N,A}}}^{N_{max}^{\mathrm{N,A}}}(A^{\mathrm{N,A}}(j,t))^{\alpha(\epsilon-1)}dj} + \frac{1}{\mathcal{O}^{\mathrm{S,A}}}\frac{\int_{N_{min}^{\mathrm{S,A}}}^{N_{max}^{\mathrm{S,A}}}(1/i)(A^{\mathrm{S,A}}(j,t))^{\alpha\epsilon}di}{\int_{N_{min}^{\mathrm{S,A}}}^{N_{max}^{\mathrm{S,A}}}(A^{\mathrm{S,A}}(j,t))^{\alpha(\epsilon-1)}dj}\right) - \mu T^{\mathrm{A}}. \tag{66}$$

The terms in brackets are the environmental impacts of the individual economies of the North and the South. As long as the world economy is in steady state, all productivities grow at the same average speed (country-specific), $\dot{A}_i^{\mathrm{k}} = \dot{\bar{A}}^{\mathrm{k}}$, and one can get rid of integration terms, yielding for the economy under spillovers,

$$\dot{T}^{\mathrm{T}} = \frac{e_0}{1+T^{\mathrm{T}}}\left(\frac{1}{(\mathcal{O}^{\mathrm{T}})^2}\ln(N^{\mathrm{T}}_{max}/N^{\mathrm{T}}_{min})\left((\bar{A}^{\mathrm{N,T}})^{\alpha}+(\bar{A}^{\mathrm{S,T}})^{\alpha}\right)\right)-\mu T^{\mathrm{T}} \tag{67}$$

and for the economy without it,

$$\dot{T}^{\mathrm{A}} = \frac{e_0}{1+T^{\mathrm{A}}}\left(\frac{1}{(\mathcal{O}^{\mathrm{N,A}})^2}\ln(N^{\mathrm{N,A}}_{max}/N^{\mathrm{N,A}}_{min})(\bar{A}^{\mathrm{N,A}})^{\alpha}+\frac{1}{(\mathcal{O}^{\mathrm{S,A}})^2}\ln(N^{\mathrm{S,A}}_{max}/N^{\mathrm{S,A}}_{min})(\bar{A}^{\mathrm{S,A}})^{\alpha}\right)-\mu T^{\mathrm{A}}. \tag{68}$$

From the previous analysis it follows that the intensity of emissions is lower under spillovers leading to,

$$\frac{1}{(\mathcal{O}^{\mathrm{T}})^2}\ln(N^{\mathrm{T}}_{max}/N^{\mathrm{T}}_{min}) < \frac{1}{(\mathcal{O}^{\mathrm{N,A}})^2}\ln(N^{\mathrm{N,A}}_{max}/N^{\mathrm{N,A}}_{min})+\frac{1}{(\mathcal{O}^{\mathrm{S,A}})^2}\ln(N^{\mathrm{S,A}}_{max}/N^{\mathrm{S,A}}_{min}) \tag{69}$$

since the presence of spillovers boosts structural change and the turnover of sectors in the economy, yielding a wider diversity of sectors, $\mathcal{O}$, and a lower intensity $\ln(N_{max}/N_{min})$.

However, the average productivity of the world economy is higher in the scenario with technological spillover than without it. Thus, the actual dynamics of the environment depends on the relative size of these two effects: reduction in intensity of emissions and increase in productivity. We established in Proposition 1 above that the North productivity is unaffected by the spillover itself, $\bar{A}^{\mathrm{N,T}} = \bar{A}^{\mathrm{N,A}}$. Hence, the increase of the overall influence on the environment only comes from the increase of the productivity in the South. At the same time, the intensity of emissions in the South is also affected by the technological spillover, given a lower productivity in this country. Therefore, overall the environmental degradation is expected to be lower in the case with spillovers than under autarky.

We resort to a numerical example with the following parameters to illustrate this discussion: $N^{\mathrm{T}}_0 = N^{\mathrm{N,A}}_0 = N^{\mathrm{S,A}}_0 = 1$, $r = 0.05$, $\beta = 0.1$, $\mu = 0.4$, $\theta = 0.4$, $e_0 = 0.0475$, $\pi^R_{\mathrm{N,T}} = \pi^R_{\mathrm{N,A}} = 0.25$, $\pi^R_{\mathrm{S,T}} = 0.75$, $\pi^R_{\mathrm{S,A}} = 0.5$, $\tau_{min}(N_0) = 3$, $\tau_{max}(N_0) = 1$, $\delta_{\mathrm{N}} = \delta_{\mathrm{S}} = 0.5$.

In Fig. 3 it is shown that the productivity grows only in the South, while the overall intensity of emissions of both countries, given in (69), is lower in the spillover case. Since the contributions to the environmental degradation of both the South and the North are dynamic and are given as products of productivities and intensities, one can conclude that the effect of cleaner technologies should dominate the effect of the output increase.

Proposition 8 summarizes the results of our discussions.

Proposition 8 (Influence of Technological Spillover on Environment) *In the world economy with environmental degradation given by* (49)*, the technological spillover is expected to reduce environmental pollution in the world economy and leads to a boost of output both in the North and in the South.*

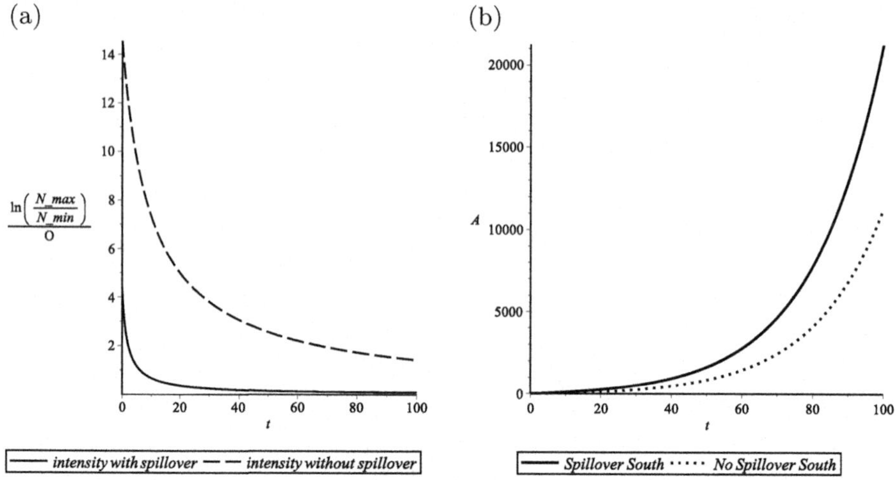

Fig. 3 Relative dynamics of intensities and productivity. (**a**) Intensity of emissions, (**b**) productivity growth

Proof The boost in output growth follows from the Proposition 6. A comparison of the growth rates of the intensity of emissions, (69), and of the productivity increase yields the result of a slowdown in environmental degradation. ■

Figure 4 illustrates the environmental dynamics where we use the same parameter values as for Fig. 3. It can be seen that the change of the environment (measured as the increase in the average surface temperature) is much less drastic in the presence of technological spillovers: the temperature decreases to pre-industrial levels in 100 years before it begins to rise, while without spillovers we get the typical result of an increasing temperature over 3.5 °C, which is in line with other more detailed and empirical estimations.

It should also be pointed out that environmental degradation continues in the long-run as output grows unless resource are used for abatement. The simplest way to achieve a constant level of the environment would be to levy a lump-sum tax and to use the tax revenue for abatement, for example. The question of how environmental pollution can be stabilized in growing economies has been the subject of a great many studies (see e.g. the models in Greiner and Semmler 2008). Therefore, we do not treat this problem but, rather, focus on the relation between structural change, economic growth and environmental pollution in a North-South context, with the environment determined by the decisions of private agents alone.

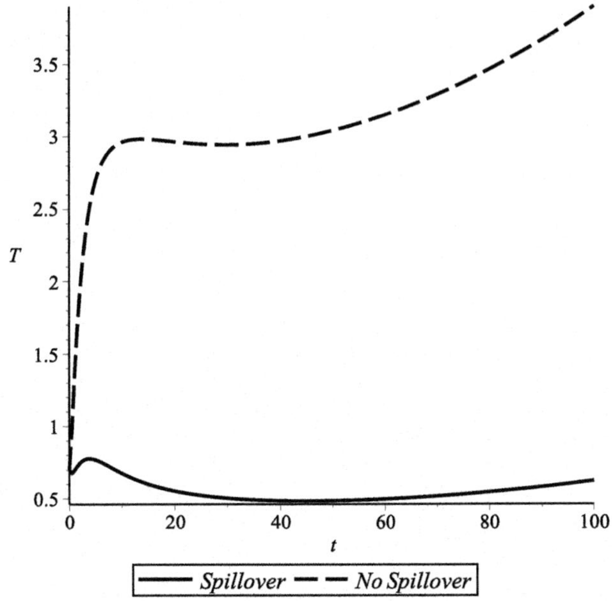

Fig. 4 Evolution of environment with and without technological spillover

5 Conclusion

Technological spillovers between countries lead to an acceleration of structural change and to an increase of output growth rates. At the same time, if one assumes newer technologies to be cleaner and less harmful for the environment than older ones, this technological spillover also slows down environmental degradation, thus, reducing the temperature increase. Hence, the goal of fostering structural change through technological spillovers is worth pursuing both from economic and environmental perspectives.

The main results of the paper are twofold: First, structural change is boosted both in the leading economy and in the lagging one, even without resorting to the concept of cooperative agreements. Rather, it is the multidimensionality of R&D and spillovers that raise the amount of fundamental research and, thus, structural change. But, the lagging country must be advanced enough to be able to adopt new technologies. Thus, our analysis applies to spillovers of technology from a developed country to some emerging market economy, rather than to a developing one. In terms of the model, this means that both countries should be comparable in their technology levels, e.g. they should share the common technological space.

Second, the existence of ongoing technological spillover slows down the environmental degradation significantly while boosting economic growth in both economies. The environmental benefit results from the fact that productivity increases only in the South, while both economies use progressively cleaner technologies because of their faster turnover in the economy. For this result to hold it is important that technologies become obsolete in both economies simultaneously. Indeed, if some of the technologies were outdated faster in the North than in the South, the production of these outdated products might generate an additional source of emissions which would harm the environment. Thus, the result of the beneficial nature of technological R&D for the environment is again limited to economies that do not differ too much with respect to their technological levels.

Acknowledgements Financial support from the Bundesministerium für Bildung und Forschung (BMBF) is gratefully acknowledged (grant 01LA1105C).

References

Alemdar, N. M., & Özyildirim, S. (1998). A genetic game of trade growth and externalities. *Journal of Economic Dynamics and Control, 22*(6), 811–832.

Alemdar, N. M., & Özyildirim, S. (2002). Knowledge spillover, transboundary pollution and growth. *Oxford Economic Papers, 54*, 597–616.

Baldwin, R. E., & Robert-Nicoud, F. (2008). Trade and growth with hetergeneous firms. *Journal of International Economics, 74*, 21–34.

Bernard, L., & Semmler, W. (Eds.). (2015). *The oxford handbook of the macroeconomics of global warming*. Oxford: Oxford University Press.

Bondarev, A. (2014). Endogenous specialization of heterogeneous innovative activities of firms under the technological spillovers. *Journal of Economic Dynamics and Control, 38*, 235–249.

Bondarev, A., & Greiner, A. (2014a). Endogenous growth and structural change through vertical and horizontal innovations. *Working papers in economics and management 03–2014*. Bielefeld: Bielefeld University.

Bondarev, A., & Greiner A. (2014b). Environmental pollution in a growing economy with endogenous structural change. *Working papers in economics and management 07–2014*. Bielefeld: Bielefeld University.

Chichilnisky, G. (1994). North-south trade and the global environment. *American Economic Review, 84*, 851–874.

Copeland, B. R., & Taylor, M. S. (2009). Trade, tragedy, and the commons. *American Economic Review, 99*, 725–749.

Dockner, E., Jorgensen, S., Long, N., & Sorger, G. (2000). *Differential games in economics and management sciences*. Cambridge: Cambridge University Press.

Eichner, T., & Pethig, R. (2014). International carbon emissions trading and strategic incentives to subsidize green energy. *Resource and Energy Economics, 36*, 469–486.

Greiner, A., & Semmler, W. (2008). *The global environment, natural resources, and economic growth*. Oxford: Oxford University Press.

Grimaud, A., & Rouge, L. (2004). Polluting non-renewable resources, tradeable permits and endogenous growth. *International Journal of Global Environmental Issues, 4*, 38–57.

Krysiak, F. C. (2011). Environmental regulation, technological diversity, and the dynamics of technological change. *Journal of Economic Dynamics and Control, 35*, 528–544.

Nordhaus, W. D. (1967). The optimal life of a patent. *Cowles foundation discussion paper* (Vol. 241). Yale: Yale University.

Nordhaus, W. (2007). *The challenge of global warming: economic models and environmental policy* (Vol. 4). Yale: Yale University.

Peretto, P. F., & Connolly, M. (2007). The manhattan metaphor. *Journal of Economic Growth, 12*, 329–350.

Rivera-Batiz, L. A., & Romer, P. M. (1991). International trade with endogenous technological change. *European Economic Review, 35*, 971–1001.

Romer, P. M. (1990). Endogenous technological change. *Journal of Political Economy, 98*, S71-S102.

Unel, B. (2010). Technology diffusion through trade with heterogeneous firms. *Review of International Economics, 18*, 465–481.

Foreign Exchange Volatility and its Implications for Macroeconomic Stability: An Empirical Study of Developing Economies

Unurjargal Nyambuu

1 Introduction

Opening of the economy to the world through increased trade and investment by reducing trade barriers and liberalizing capital mobility is generally thought to benefit developing and emerging economies. Foreign trade policies often promote exports of raw materials and commodities to fast-growing countries, e.g., China, as well as already developed economies, e.g., the United Kingdom. Foreign direct investment (FDI) builds new factories, creates jobs, improves the skills of workers, and encourages use of technology. Moreover, government borrowing from international capital markets results in investment in domestic industries and improved infrastructure. Cross-country data show an increased economic growth in such countries and in their contribution to the world economy.

However, the above advantages also come with costs as these countries become more dependent on external demand and international flows of the capital, and more vulnerable to external shocks especially foreign exchange (FX) rate volatility. When the local currency loses its value (or depreciates) against the major importers' currencies or against the major lenders' currencies, the imported goods become more expensive and can cause inflation pressure. For the external debt, it becomes more costly with the increase in borrowing costs. If the local currency depreciation persists for a long time, default risk increases and the debt can become unsustainable, possibly leading to a debt crisis. For developing countries, international capital inflows increased sharply in the 1990s, with private capital dominating compared to public capital, but when the crisis hit, there were reversals in the direction of capital flow, with a massive capital flight. These problems appear to

U. Nyambuu (✉)
Department of Social Science, The New York City College of Technology,
The City University of New York, Brooklyn, NY, USA
e-mail: UNyambuu@citytech.cuny.edu

L. Bernard, U. Nyambuu (eds.), *Dynamic Modeling, Empirical Macroeconomics, and Finance*, DOI 10.1007/978-3-319-39887-7_7

be more visible when the world economy faces changes and challenges, e.g., oil prices crumbling or a slowdown in the Chinese economy. For example, almost all emerging and developing countries' exports have been affected by China's slow growth in 2015. China is the largest trade partner for Brazil thus the recent decline in demand for Brazil's raw material from China caused the Brazilian economy contract significantly.

The value of capital is affected by the FX rate as foreign currency-denominated capital or as debt from international investors or lenders when it is converted into the local currency to be used in the domestic economy. But, at the same time, local capital needs to be converted back to foreign currency when debt or premiums are repaid and also when the capital flows out the economy. Thus, as high fluctuations in FX rates can cause large capital volatility, we should study the currency composition of the capital. Some countries take economic measures, e.g., tightening of capital controls to reduce FX rate fluctuations and the effects of currency mismatches. Countries such as Chile, Colombia, Malaysia, and Thailand have imposed capital controls in the past.

FX rates in the developing countries under study show significant volatility, as measured by different statistical indicators. Indonesia's FX rate, in particular, against the US Dollar shows extremely large volatility during the period between 2000 and 2015. In this paper, FX rates of developing countries against the US Dollar are considered. There is also a high correlation between the FX rates of these countries. In addition, nominal effective exchange rate (NEER) as well as real effective exchange rate (REER) are analyzed in this paper to better assess the macroeconomic effects. NEER shows changes of the value of the local currency against the currencies of the trading partners. REER is derived from the NEER by adjusting for inflation.

Historical studies show that developing countries borrowed heavily from abroad, especially during economic boom periods. However, during times characterized by external and/or domestic shocks, in some cases, debt became unsustainable and these countries became unable to repay the debt (principal) or its interest. In some regions, these trends were contagious[1] and affected other countries with similar problems, as international investors lost confidence in developing economies. When this occurs, countries either default or restructure their debt with reductions on par or debt rescheduling, sometimes with reduced interest rate and extended maturity, sometimes for decades. External debt default and rescheduling for developing countries date back to the 1800s, with defaults in Latin America (e.g., Colombia, Mexico, Peru, Venezuela), Africa (e.g., Algeria, Egypt), and Turkey. Then, in the 1900s, there were again a number of defaults and related rescheduling, e.g., in Algeria, Brazil, Chile, Ecuador, India, Indonesia, Mexico, Nigeria, Peru, South Africa, and Turkey. External debt to GDP ratios peaked for many Latin American and Asian countries during the debt crisis in the 1980s and 1990s: Peru (120 %), Ecuador (80 %), Mexico (80 %), Brazil (50 %), and Indonesia (150 %).

[1]Debt crisis spreading into other regions.

In this paper, we study how the economic growth of developing countries are affected by the opening of their economy to the world, especially through foreign trade and external borrowing. From the accounting of the balance of payments, we can show how the current account (consisting mostly of trade) deficits are financed by the financial account or an increase in capital inflows into the economy, e.g., in the form of external debt. This is taken into account in our growth model where external debt dynamics affect the GDP growth of the economy in addition to other constraints, e.g., capital. The model considers the interest rate paid on external debt as well as FX rates, which are used to convert them. Furthermore, the model is solved and optimal total external debt to GDP ratios are calculated following Stein (2006, 2012) and stressing the importance of the FX volatility. Empirical findings suggest an increase in excess external debt during the debt crises, e.g., 1980s, and the emerging market's financial crises, e.g., 1997, and an individual country's debt default or debt restructuring period, associated with high volatility of relevant FX rates. Covered countries are from different regions: Africa (South Africa), Asia (India, Indonesia), Eastern Europe (Bulgaria, Hungary), Latin America (Brazil, Ecuador, Peru). Five of these developing countries, Brazil, India, Indonesia, South Africa and Turkey, with high inflation, large current account deficits and large depreciation of the currency were named as 'fragile five' by Morgan Stanley in 2013 and considerable attention was paid by articles in many newspapers, journals, and media.

The paper is organized as follows: Section 1 is the introduction. Section 2 reviews the existing literature. Section 3 presents relevant statistics on economic growth, foreign trade, foreign exchange, external debt, and credit default. Section 4 sketches the theoretical model and its solutions. Section 5 presents the empirical results for the selected countries. Finally, Sect. 6 concludes the paper.

2 Literature Review

Stiglitz (2000) has discussed the weakness of the liberalization of capital markets and how it affects economic growth and leads to instability. He (Stiglitz 2000, p. 1085) raises the issue of the volatility of short-term capital flows and states that "The risks are recognized to be greater, the benefits lower, the circumstances in which countries should engage in full liberalization more restrictive than was the case before the crisis." Stallings (2007) studied financial globalization and its effect on capital flows and how volatility can have a negative impact on the economy by studying benefit and costs, especially for developing countries. Stallings (2007, p. 212) argues that "Volatility undermines the value of financial flows for almost all recipients. The tendency for these flows to come in surges and withdraw just as quickly makes it much more difficult to manage economies in a sensible way and, in especially complicated situations, can even provoke financial crises. These retard economic growth for years, in part because the financial sector itself is severely damaged."

Volatility of FX rates are associated with capital volatility. In order to reduce sudden capital movements, some countries impose capital controls and this measure, in turn, can also ease fluctuations of FX rates. There are a number of studies on the effectiveness of these measures and how it affects FX rates and their volatility in different countries. Some of the measures include increased limitations on capital mobility which impacts capital inflows. For example, Stiglitz (2000) suggested different interventions in capital flows (short term) while discussing commonly taken measures, e.g., restrictions imposed on capital inflows (referring to Chile's case), capital outflows (referring to Malaysia's case), and the banking system (East Asian case in particular Malaysia). Stallings (2007) discussed capital volatility in terms of both short and long term capital and measures used on capital inflows in Latin American countries. Edwards and Rigobon (2009) study the case of Chile's external shock vulnerability in the 1990s and find that capital control makes the FX rate more volatile.

Earlier studies in favor of capital controls' effect on FX rate volatility include Stiglitz (2002), Eichengreen (2000), Eichengreen and Hausmann (1999). A connection between financial fragility and FX rates is investigated in emerging economies in Eichengreen and Hausmann (1999). They analyzed the policy implications and the FX rate regimes with case studies of Argentina, Australia and Panama (as dollarization) (also see Krugman 2014). Calvo et al. (1993) studied the external shocks, such as recession and balance of payments in the US and low world interest rates, and their effects on capital inflows into Latin America and how these factors caused the appreciation of real FX rates using monthly data between 1988 and 1991.

As countries accumulate large external debt, they become more vulnerable to external shocks such as reductions in exports due to a sharp fall in external demand for their goods, or a collapse of commodity prices. Also, external debt service can become very costly if the unfavorable FX rates persist. Other factors can be also associated with the increased borrowing cost, weak macroeconomic conditions, contagion from other countries, social and political tension, or geographical conflicts and challenges. When a country fails to repay its debt, it may default or restructure their debt. These and other contributors to a default are discussed in the literature, e.g., Catão and Kapur (2006), Sturzenegger and Zettelmeyer (2006), Reinhart and Rogoff (2008), Manasse and Roubini (2009), Kolb (2011), Mellios and Paget-Blanc (2011), Das et al. (2012), Reinhart and Kenneth (2014) and Davis (2015). Davis (2015) studies the short-run effect of debt (e.g., bank loans and bonds) versus equity (e.g., FDI) based capital inflows on the macroeconomic performance of the economy. Davis (2015, p. 95) concludes that "debt-based capital inflows, not equity-based inflows, provide the real threat to stability."

There is a vast literature on the differences between debt default and debt restructurings, and their effects on the economy. For restructuring, while Standard & Poor (2006) considers distressed debt exchanges, IMF working papers such as by Das et al. (2012, p. 4) argue the importance of distinguishing "distressed debt exchanges from routine liability management operations (LMOs) aimed at improving the profile of public debt, such as debt swaps, which could occur in normal times."

Reinhart and Rogoff (2008) present a historical database on external debt defaults and banking crises, and other variables such as inflation, and currency crashes. Their study covers 66 countries from different regions. Moreover, they study sovereign default, and how factors like commodity prices affect the risk of default. There were multiple external debt defaults since the 1300s in different regions. According to Reinhart and Rogoff (2008), defaults and rescheduling in the 1800s and 1900s for selected countries are listed below with the dates:

- **Africa:** Algeria (1991), Egypt (1831, 1984), Nigeria (1982, 1986, 1992, 2001, 2004), South Africa (1985, 1989, 1993)
- **Asia**: India (1958, 1969, 1972), Indonesia (1998, 2000, 2002)
- **Europe:** Spain (1867, 1872, 1882), Turkey (1876, 1978, 1982)
- **Latin America:** Brazil (1898, 1983), Chile (1826, 1880, 1972, 1974, 1983), Colombia (1850, 1873, 1880), Ecuador (1894, 1982, 1999), Mexico (1866, 1898, 1982), Peru (1976, 1978, 1980, 1984), Venezuela (1892, 1898).

3 Descriptive Statistics

3.1 Economic Growth

Typically, developing countries experience high economic growth during boom years. For example, between 1970 and 1980, Brazil's average GDP growth rate was 9 %, with a peak of 14 % in 1973. During this same period, Mexico and Ecuador grew, on average, at an annual rate of 7 %, respectively. However, their growth contracted in 2009, due to the Global Financial crisis; this impacted other countries as well, e.g., with GDP growth rates in Hungary at −6.6 %, in Turkey at −4.8 %, and in Mexico at −4.8 %. India has been growing with positive growth rate since the early 2000s and has shown an average growth of 8 % between 2003 and 2014. Similarly, Indonesia has shown a positive growth rate since the 1970s—the only exception was a sharp contraction of 13 % during the crisis period of 1998. These data are obtained from the World Bank's Database (Fig. 1).

3.2 Foreign Trade

As the world economy has opened up, developed countries have become more involved in foreign trade. Merchandise trade (exports plus imports) as percentage of GDP shows an increasing trend for most countries, except during crisis or recession periods (see Fig. 2), thus indicating the importance of trade. In 2014, this ratio reached 62 % in Mexico compared to 11 % in 1970, Ecuador rose to 53 % from 16 %, and Turkey 50 % from 9 % in 1970.

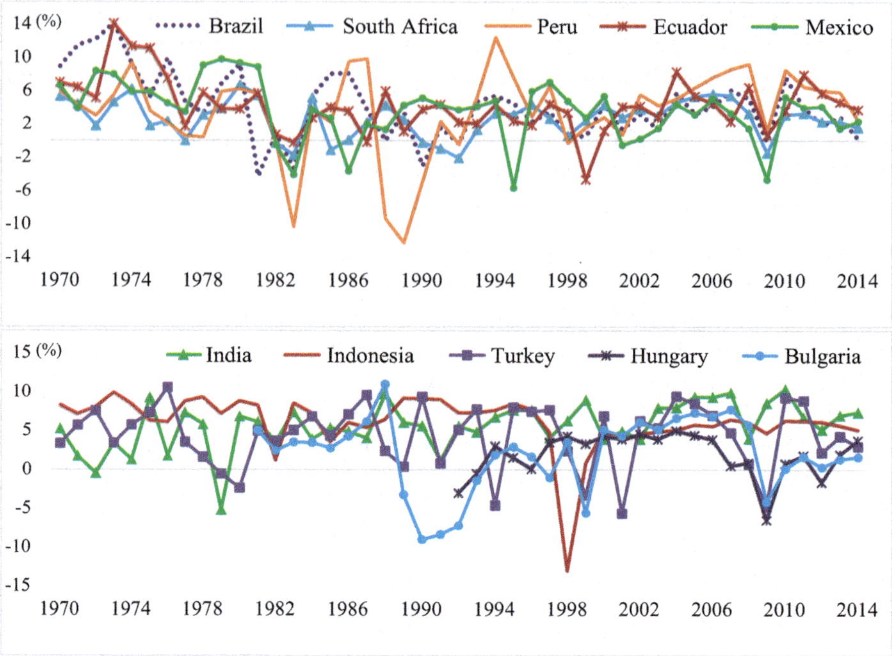

Fig. 1 GDP annual growth. Data source: The World Bank

Above mentioned countries benefit immensely from the expansion of exports. Positive net exports or trade surplus, defined as a positive difference between exports and imports, contribute to the aggregate national output or GDP of the economy. Merchandise trade balance (net exports = exports-imports), as a percentage of GDP, is shown in Fig. 3. Indonesia has had a trade surplus since 1970, but its surplus declined during the past few years and has even turned into a deficit of 0.2 % of GDP. Other developing countries have experienced both surpluses and deficits for different periods. Deficits dominated during the challenging and crisis periods such as in 2007–2009.

Export revenues generally depend on the demand coming from major trade partners. Thus, the economic performance of the trade partner plays an important role in the macroeconomic stability of the domestic economy. Top 5 export partners of selected countries are shown in Fig. 4. These data are obtained from the International Monetary Fund (IMF) Database (IMF 2016). Almost all developing countries' exports are highly dependent on China; especially for Brazil and South Africa, China is the top importer of their raw material. Thus, when China's economy started slowing down between 2014 and 2016, emerging economies started slowing down as well. More recently, the collapse of commodity prices since 2014 has contributed to a further decline in resource exports. These countries also rely on exports to the US and Japan.

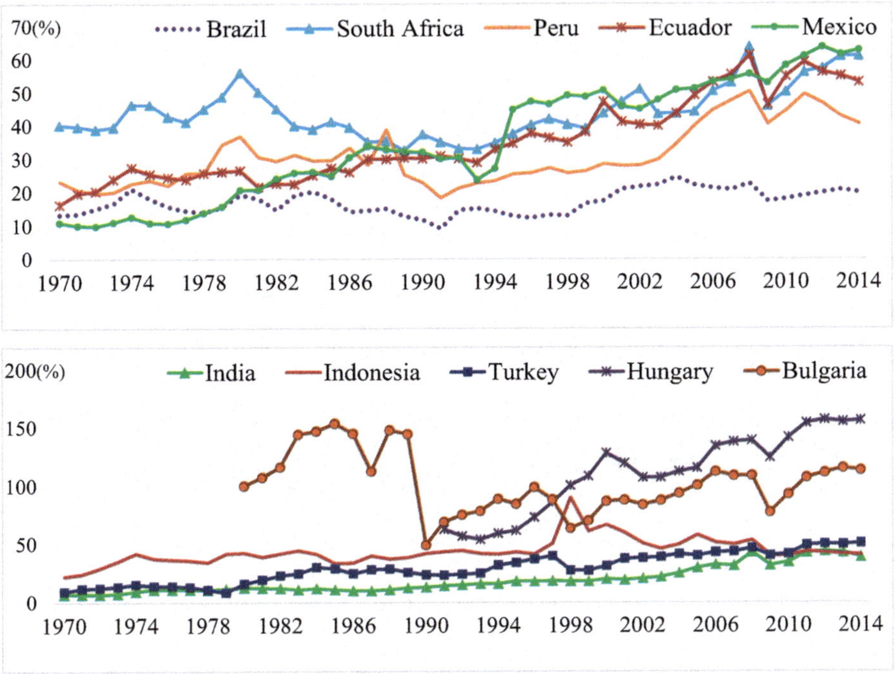

Fig. 2 Merchandise trade (% of GDP). Data source: The World Bank

3.3 *Foreign Exchange*

Historical, monthly exchange rates as local currency per US Dollar are shown in Fig. 5 for different developing and developed countries, between 1990 and 2015. Monthly exchange rate data were obtained from the IMF (2016) for Brazil, Bulgaria, Hungary, India, Indonesia, Mexico, Peru, South Africa, and Turkey. We observe that exchange rates are very volatile in emerging and developing economies, as compared to developed economies. The US Dollar appreciated to its peak, especially against the currencies of Brazil and South Africa, until early 2000s. Then the next round of US Dollar appreciation occurred right after the financial crisis of 2008; currencies of almost all countries depreciated sharply against the US Dollar. However, the latest US Dollar appreciation (or local currency depreciation) has exceeded previous peaks, especially for Mexico, South Africa, Brazil, Turkey, and India. For Indonesia, its currency has been depreciating since 2011 and almost exceeded the historically significant deprecation that occurred during the Asian Financial Crisis of 1997. Such a large depreciation of a local currency can help boost the exports. However, it can also make external debt more expensive to take on, especially if the debt is issued in US Dollars. These fluctuations are illustrated below using statistical indicators of volatility and risk in Table 1.

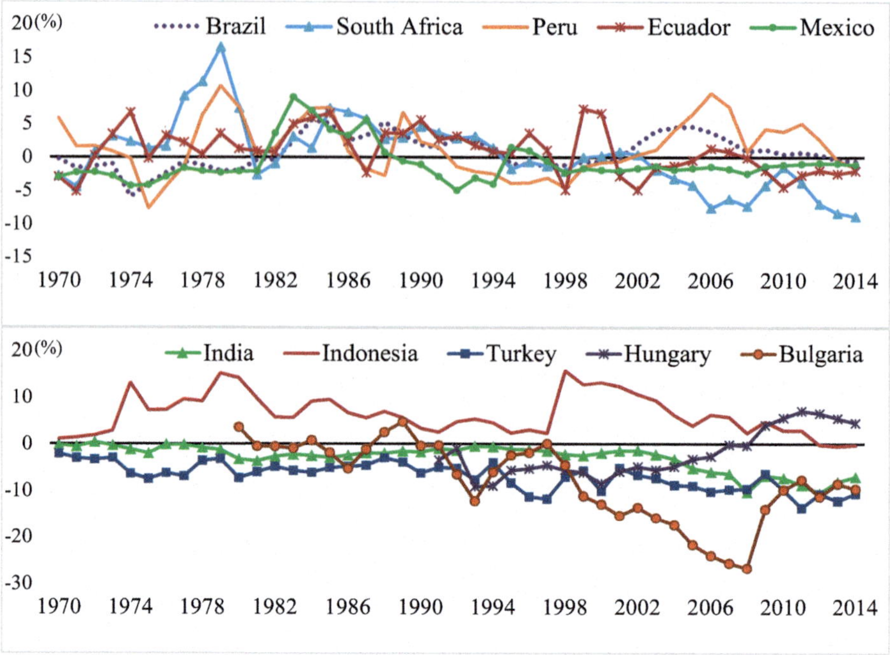

Fig. 3 Merchandise net exports (% of GDP). Data source: The World Bank

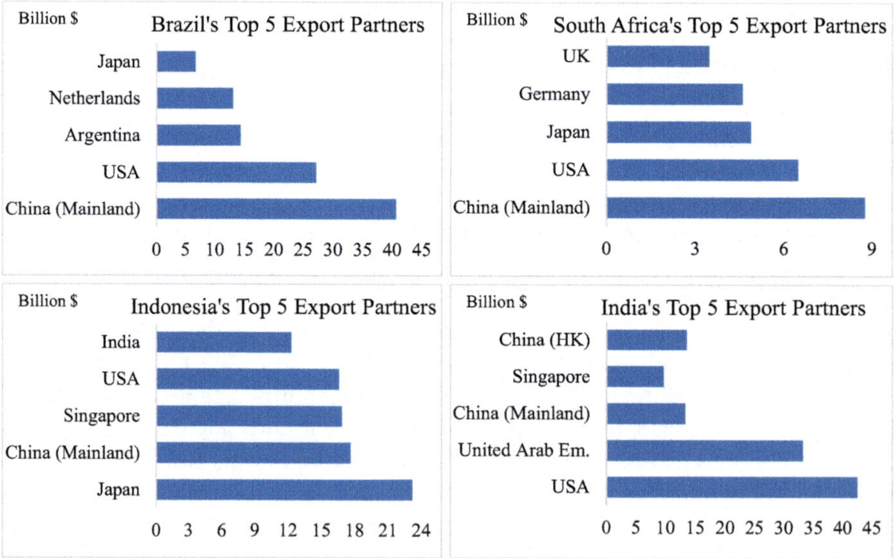

Fig. 4 Exports by major trading partner (2014). Data source: IMF

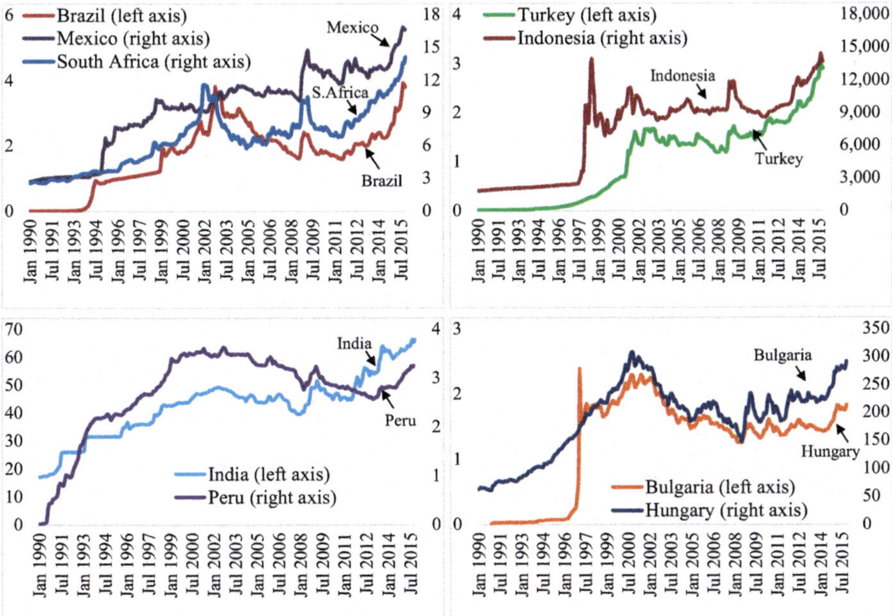

Fig. 5 Exchange rate (local currency value per USD) (1990. Jan-2015. Nov). Data source: IMF

Table 1 Volatility of FX rates (2000 Jan-2015 Nov). Data source: IMF

	Brazil	Bulgaria	Hungary	India	Indonesia	Mexico	Peru	South Africa	Turkey
Mean	2.3	1.6	224.1	49.0	9764.1	11.7	3.1	8.3	1.6
St. Deviation	0.5	0.3	35.7	6.6	1332.9	1.7	0.3	1.8	0.5
Variance	0.3	0.1	1276.8	43.8	1,776,561.7	3.0	0.1	3.3	0.2
Kurtosis	0.4	0.0	−0.5	0.3	1.5	0.0	−1.4	0.2	1.6
Skewness	1.0	1.1	0.5	1.1	1.4	0.6	−0.1	1.0	0.6
Range	2.3	1.1	161.2	26.8	7117.3	7.8	1.1	8.4	2.5
Minimum	1.6	1.2	147.1	39.4	7278.8	9.1	2.6	5.7	0.5
Maximum	3.9	2.3	308.3	66.2	14,396.1	16.8	3.6	14.1	3.0

In Table 1, descriptive statistics of monthly FX rates are shown for the period of 2000–2015. These results reflect the high volatility of FX rates for almost all countries. In particular, we see extremely high variance or standard deviations, especially for Indonesia and Hungary, as compared with other countries under study.

There is a high and mostly positive correlation between emerging markets currencies. Using the same monthly data as above, the correlation matrix for the exchange rate against the US Dollar are shown in Table 2 between 2000 and 2015. For example, Bulgaria has a very high correlation, 0.87, with Hungary. Similarly, India's FX rate has a high correlation with that of Turkey (0.83), South Africa (0.82), and Indonesia (0.79). As discussed earlier, India is one of the major trading partner

Table 2 Correlation Matrix between FX rates

	Brazil	Bulgaria	Hungary	India	Indonesia	Mexico	Peru	South Africa	Turkey
Brazil	1								
Bulgaria	0.35	1.00							
Hungary	0.34	0.87	1.00						
India	0.36	0.01	0.43	1.00					
Indonesia	0.36	0.01	0.35	0.79	1.00				
Mexico	0.13	−0.49	−0.07	0.71	0.65	1.00			
Peru	0.55	0.72	0.43	−0.35	−0.17	−0.62	1.00		
South Africa	0.43	0.21	0.50	0.82	0.80	0.51	−0.13	1.00	
Turkey	0.43	−0.28	0.09	0.83	0.75	0.83	−0.44	0.73	1.00

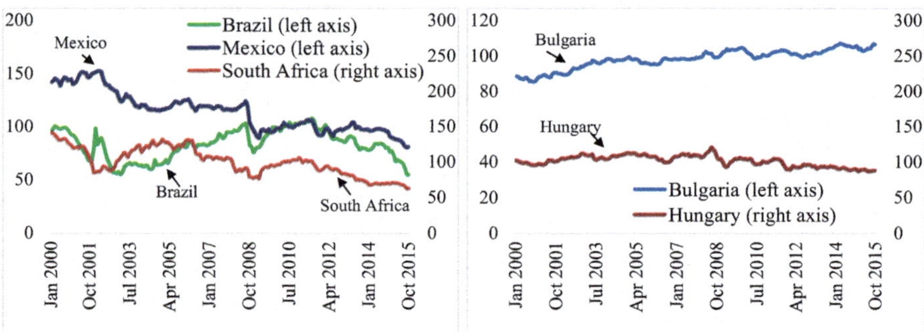

Fig. 6 Nominal effective exchange rate (NEER). Data source: IMF

(#5) of Indonesia. Brazil's currency has a high correlation with the currencies of Peru (0.55), South Africa (0.43), and Turkey (0.43). Such correlations indicate a large bilateral trade between these countries.

In addition to analyzing nominal FX rate against the US Dollar, we can compare FX rates against the currencies of other countries, especially those engaged in trade. Monthly NEER[2] is shown in Fig. 6 between 2000 and 2015 for selected countries. We observe a decreasing trend in NEER, especially over the last decade for Brazil, Hungary, Mexico, and South Africa. This fall in NEER implies a depreciation of the local currency against the currencies (in terms of weighted basket) of the trading partners.

Next, a REER[3] based on the consumer price index (CPI) is illustrated in Fig. 7. For most countries, REER has appreciated until the recent financial crisis of 2007–2008. After a large depreciation during the crisis period, it has started appreciating

[2]According to the IMF, a currency's value under NEER is measured against a weighted average of foreign currencies.

[3]For the calculation of REER, we divide NEER by a price deflator.

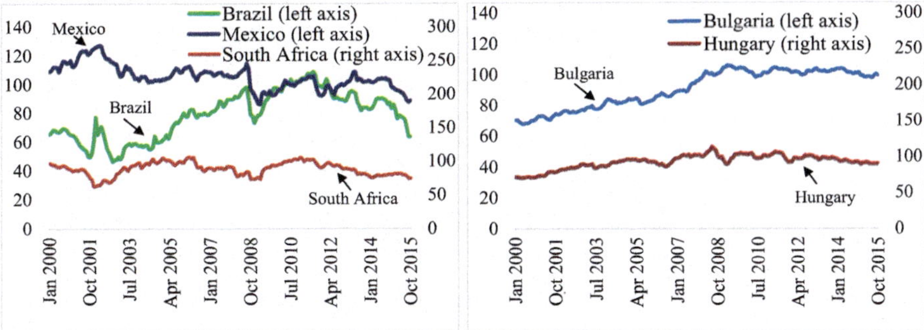

Fig. 7 Real effective exchange rate (REER). Data source: IMF

again. However, in the past few years it started depreciating again, especially for Latin American countries such as Brazil, Mexico, and other developing countries like South Africa. This decrease in REER may imply a boost in international competitiveness as exports become cheaper and import prices increase.

3.4 *External Debt*

External debt as percentage of GDP is shown in Fig. 8 for different countries between 1970 and 2014. For Latin American countries, these ratios reached their peaks in the late 1970s and started declining sharply during debt crises, when these countries defaulted or restructured their debt.[4] Another peak in external debt ratios occurred in the late 1980s and again in the mid or late 1990s, for example, Peru (120 %), Ecuador (80 %), Mexico (80 %), and Brazil (50 %). Similarly, Indonesia's external debt/GDP ratio increased to around 60 % in the late 1980s; however during the Asian Financial Crisis, this ratio hit a record high over 150 %. For the recent period, external debt/GDP ratios continue to be very high in Hungary (over 120 %) and Bulgaria (60 %).

A nation's external debt position can be derived from its International Investment Position (IIP) data. According to the guide for external debt statistics compilation by IMF (2013), this is calculated by deducting all equities (equity capital, equity shares, and others), investment fund shares, financial derivatives, and employee stock options from the total liability in the IIP as shown in the following Table 3.[5]

[4] See the example of Mexico's default in 1982 in Sect. 2.

[5] For details, see IMF's External Debt Statistics: Guide for Compilers and Users by The Task Force on Finance Statistics (IMF 2013).

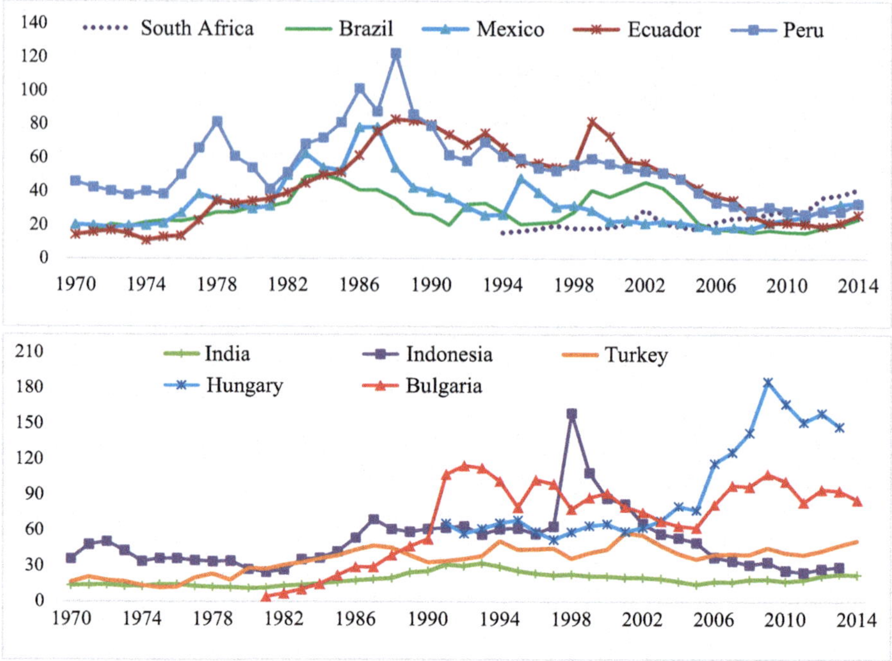

Fig. 8 External debt as percentage of GDP. Data source: The World Bank

Table 3 Deriving gross external debt from IIP

Gross External Debt	=	International Investment Position (Total Liability)							
		Total Liability	–	All Equity (e.g., equity shares)	–	Investment Fund Shares	–	Financial Derivatives	–	Employee Stock Options

3.5 *Credit Default*

Spreads on credit derivatives, especially for fixed income instruments, e.g., bonds and Credit Default Swaps (CDS) can serve as a credit risk indicator. For sovereign CDSs, an issuer of the bond is the State or Government. Sovereign 5-year CDS spreads for different countries are shown in Fig. 9. Levels of CDSs in emerging markets have increased largely during the financial crisis, 2007–2008; interestingly, the CDSs declined as the sentiments of the investors gradually improved. But again, in the past few years (especially 2014–2015), CDSs have risen significantly, mainly triggered by the sell-offs caused by losses in investor confidence. These concerns were affected not only by the global economic and financial changes and collapse of commodity prices, but were

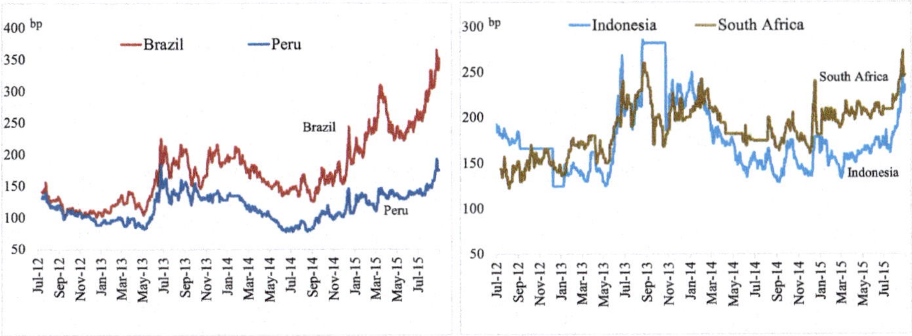

Fig. 9 Credit default swap USD (Price) spread (2012.7–2015.8). Data source: Bloomberg

also downgraded due to credit and other domestic issues, e.g., political and corruption scandals. In particular, sovereign CDS spreads (5-year) for Brazil have shown some increase since 2014 and surged to 500 bp, according to end-of-2015 Bloomberg data (Bloomberg Terminal 2016). This happened when Brazil's sovereign credit rating was lowered to BB+ by Standard & Poor's in September of 2015.

4 Theoretical Model

We present a stochastic growth model in an open economy with the following objective function and state variables. It maximizes the expected Present Value (PV) of utility which is a function of consumption, and denoted as $U(C)$, thus the control variables are consumption (C) and investment (I). The economic growth is constrained with the dynamics of capital (K) and external debt (D). While capital increases with a rise in investment, external debt increases with the interest (r) payments on the external debt but decreases with a rise in the trade balance (or net exports). This is derived from the macroeconomic accounting, where it is defined as Y-C-I; here Y is output or GDP. This standard model is sketched in numerous articles and books, for example in Blanchard (1983).

$$\max_{C,I} \int_0^\infty U(C_t)\, e^{\vartheta t} dt \tag{1}$$

$$dK_t = I_t dt \tag{2}$$

$$dD_t = -(Y_t - C_t - I_t)\, dt + r_t D_t dt \tag{3}$$

A stochastic version of this growth model is solved by Stein (2006) with the production function determined as $Y_t = g_t K_t$, where the return on investment denoted by g is defined as $g = (\Delta Y/Y) / (I/Y)$. The utility function can be a power utility as $U = C_t^{\gamma}/\gamma$, with a positive risk aversion of $(1 - \gamma)$. He emphasizes the importance of the Net Worth (NW), which is defined as $(NW)_t = K_t - D_t$, and solves the debt to the net worth ratio. Using Dynamic Programming and Tobin's Mean and Variance method, optimal external debt to GDP ratio (d^*) is derived as follows[6]:

$$d^* = \left(g \left(1 + \frac{1}{\left(\sigma_{g_t}^2 / \sigma_{g_t - r_t}^2\right) \left(\rho_{g,\, r} \sigma_r / \sigma_g - 1\right) + (g - r) / \left((1 - \gamma)\, \sigma_{g_t - r_t}^2\right)} \right) \right)^{-1}$$

where σ^2 is variance, $\rho_{g,\, r}$ is a correlation between g and r, $1 - \gamma$ denotes risk aversion,

But now we modify the model and introduce the FX rate (E) which affects both the external debt and the interest rate paid on them. External debt denominated in foreign currency can be converted to the local currency using FX rates. Now, the real interest rate will take into account the rate of change of the FX rate: we use $r^{real} + \Delta E/E$. In the next section, we calculate the optimal external debt to GDP ratios and compare it with the actual external debt to GDP ratio and FX rate volatility.

5 Estimation of the Model

Based on the solutions of the theoretical model sketched in Sect. 4, optimal ratios are calculated and empirical findings are presented for selected developing countries using data between 1970 and 2014; this is done for Brazil, Ecuador, India, Indonesia, Peru, and Turkey. The same is done using data between 1981 and 2014 for Bulgaria, data between 1991 and 2014 for Hungary, and data between 1994 and 2014 for South Africa. Different data sources were used: fixed capital formation or fixed investment, GDP, inflation, external debt stock, and FX rates data from the World Bank Databank, interest rate data, and 10-year treasury constant maturity rates from the Board of Governors of the Federal Reserve System were used (U.S. Board of Governors of the Federal Reserve System 2014). Normalized values[7] for the rate of change in FX rates (ΔFX), actual (d), and optimal external debt to GDP (d^*) ratios are illustrated for these countries.

In this section, we analyze Latin American countries starting with Brazil (Fig. 10). In the 1980s, during the debt crisis that covered Latin American countries, the actual external debt to GDP ratio exceeded its optimal ratio; this occurred again in both the early and late 1990s during the Asian Financial Crisis when most emerging economies were affected with increased volatility in FX rates. A similar

[6]For detailed explanation and derivation of the solution method see Stein (2006) and Nyambuu and Bernard (2015).

[7]Normalized values show a deviation of the variable away from its mean.

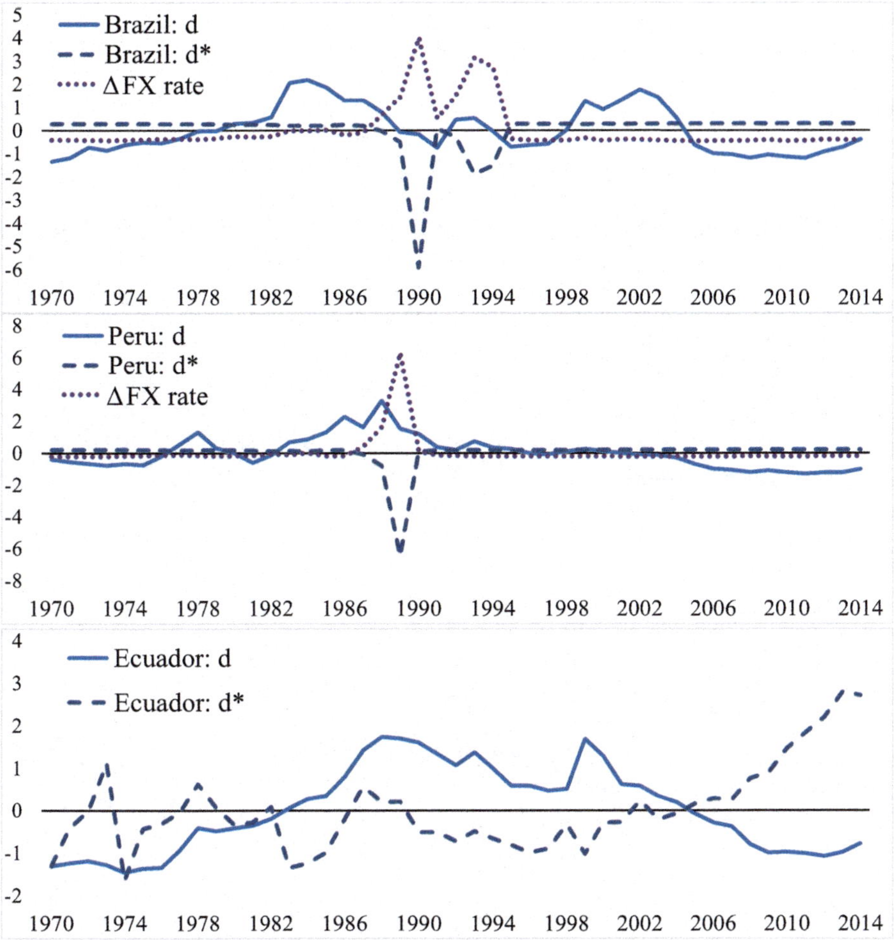

Fig. 10 Optimal debt/GDP ratio and FX rates in Latin America

trend was visible for Peru in the late 1970s, but deviations occurred in the late 1980s due to large debt restructuring issues. These deviations were much larger than expected, and coincided with a sharp swing in the FX rate. As for Ecuador, the country has practiced a full dollarization since 2000 and its currency regime is classified as an arrangement with 'No separate legal tender,' using US Dollar as their own currency. Ecuador heavily depends on oil revenues; thus, changes in oil prices affect its economy through export and fiscal revenues. In addition to volatile resource prices, other external risks include weak external demand coming from trade partners, instability in world finance, and the appreciation of the US Dollar.[8]

[8]For the recent economic situation and external risks see the IMF Country Report on Ecuador (IMF 2015a).

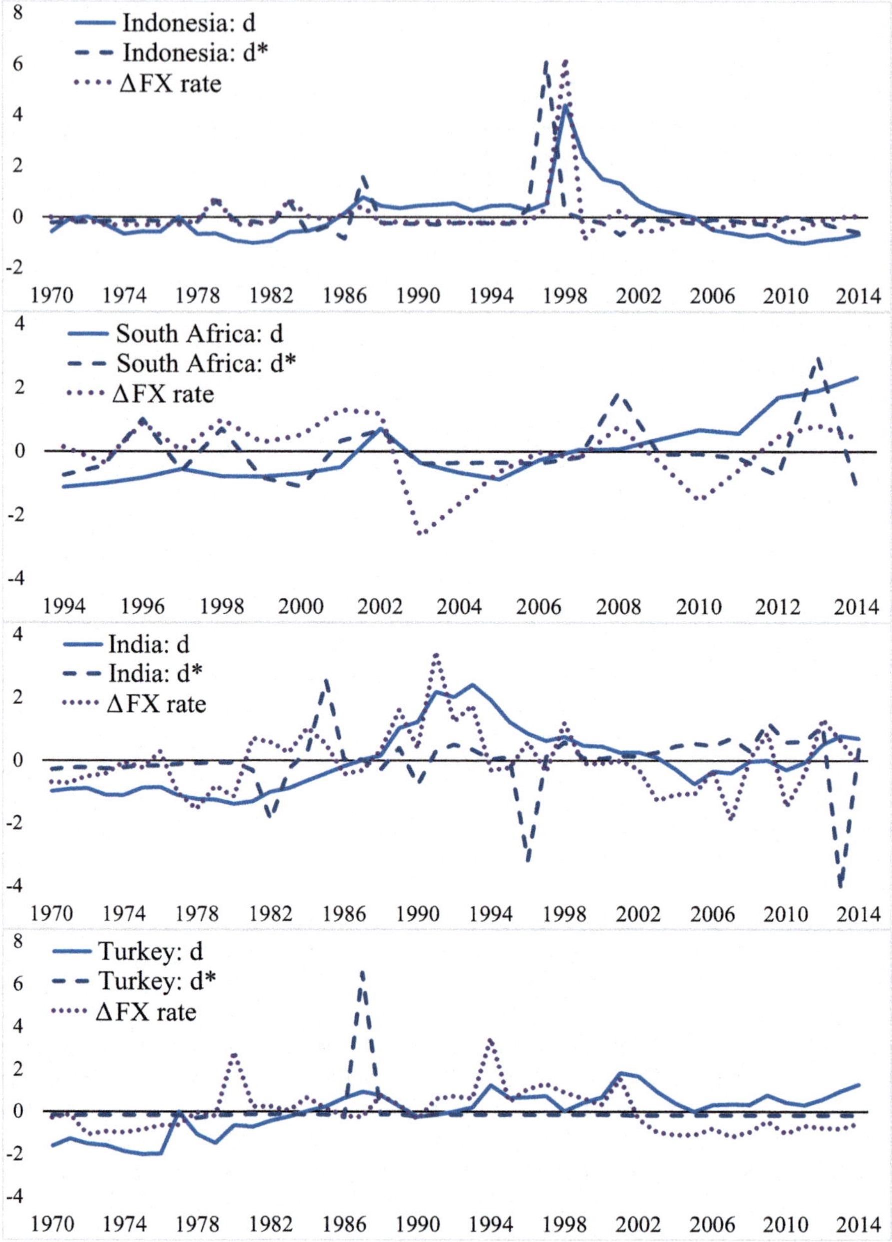

Fig. 11 Optimal debt/GDP ratio and FX rates in Indonesia, South Africa, India, and Turkey

Next, we study Indonesia and South Africa as shown in Fig. 11. Empirical findings highlight the Asian Financial crisis of 1997–1998, when Indonesia had a massive excessive external debt (4 standard deviations away from the historical mean) and a very high swing in the FX rate (6 standard deviations away from its mean rate). A previous rise in excess debt in the 1980s was much smaller. As for South Africa, there was some excess debt in the early 2000s, but the persistent increase in the debt deviations reached a high point (2 standard deviations) in 2014, indicating the risk potential of an unsustainable debt, with a large gap between the actual and optimal debt ratio, as well as higher deviations in the FX rates.

Furthermore, we study India and Turkey, as illustrated in the same Fig. 11. India had large swings in both external debt ratios and FX rates with 3–4 standard deviations for the mean of these ratios in the early through mid-1990s. As for Turkey, debt ratios showed some deviations in the late 1980s, mid-1990s, and early-2000s, with volatile FX rates, and again in 2014, the gap between actual and optimal debt ratios widened.

Finally, we investigate the debt and FX rate trends in Eastern Europe as shown in Fig. 12. Both Bulgaria and Hungary had very volatile FX rates, especially in the 1990s. Bulgaria's external debt had some small swings in the 1990s and again in the late 2000s. But for Hungary, the actual external debt to GDP ratio has been deviating significantly from its optimal ratio since the late 2000s. Hungary, as compared to Bulgaria, seems to have more excess debt problems.

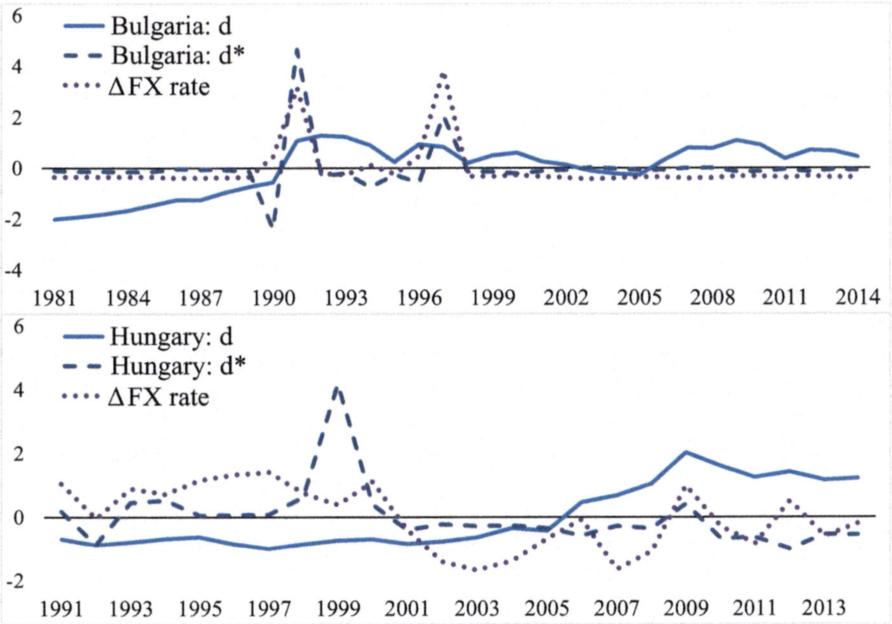

Fig. 12 Optimal debt/GDP ratio and FX rates in Bulgaria and Hungary

6 Conclusion

As globalization has spread deeper in developing countries, exchanges of goods and services, capital, and labor with foreign countries have expanded immensely. All these transactions involve exchanges between domestic and foreign currencies using FX rates, which are observed to be very volatile. In the late 1970s and early 1980s, as countries opened up to the world, both government and private corporations of the developing countries borrowed from foreign lenders, through loans and bond issuing, to finance their consumption and investment projects. Besides sovereign debt, corporate debt (non-financial firms) in emerging economies dominated by countries like China, Turkey, Brazil, India have increased sharply and reached close to $18 trillion or almost 75 % of world GDP in 2014 (IMF 2015b). Changes in FX rates have a direct impact on the interest and principal payments on external debt denominated in foreign currencies. In this paper, we showed how volatile FX rates during a crisis or challenging down-turn, or a sharp and persistent depreciation of the local currency, may contribute to the unsustainability of a country's external debt.

In this paper, a dynamic growth model for a developing country with external debt was considered. The model took FX rates into account since debt issued in a foreign currency needs to be converted into the local currency. In this way, the interest rate payments reflected the changes in the FX rates. Based on the solution of the stochastic version of the standard growth model purposed by Stein (2006) and applied in Nyambuu and Bernard (2015), we calculated the optimal total external debt to GDP ratios, but including the effect of the FX rate movements. Empirical findings for different developing and emerging economies covered in this paper suggest the presence of excess external debt as we have defined it, i.e., an increased gap between the actual and optimal external debt to GDP ratio. This effect was particularly prominent during the debt crises of the 1980s and during the Asian Financial Crisis of 1997–1998. It was also noted when studying individual sovereign default and debt restructuring periods accompanied by increased fluctuations in FX rates. As a result of recent global instability, including slower growth in China with lower demand for raw material, external shocks, e.g., plummeting oil prices, and geo-political events, emerging country currencies have already lost significant value; consequently, borrowers from the emerging world have shown signs of struggle with regard to their debt payments. In addition, a rise in CDS spreads for emerging sovereign bonds as well as corporate issues is attributed to investors' fear of default risk. Thus, over the next few years, we would expect more FX rate volatility and higher borrowing costs; this will, in turn, make it harder to obtain new financing from abroad, especially for countries thought to be at increased risk for credit default, e.g., Brazil.

References

Blanchard, O. J. (1983). Debt and the current account deficit in Brazil. In P. A. Armella, R. Dornbusch, & M. Obstfeld (Eds.), *Financial Policies and the World Capital Market: The problem of Latin American Countries*. Chicago: University of Chicago Press.

Bloomberg Terminal. Statistics. Accessed 1 Feb 2016.

Calvo, G. A., Leiderman, L., & Reinhart, C. M. (1993). Capital inflows and real exchange rate appreciation in Latin America. *IMF Staff Papers, 40*(1), 108–151.

Catão, L., & Kapur, S. (2006). Volatility and the debt-intolerance paradox. *IMF Staff Papers, 53*(2), 195–218.

Das, S.D., Papaioannou, M.G., & Trebesch, C. (2012). Restructuring sovereign debt: Lessons from recent history. In *IMF Seminars, Conferences and Workshops. Financial Crises: Causes, Consequences, and Policy Responses*. https://www.imf.org/external/np/seminars/eng/2012/fincrises/

Davis, J. S. (2015). The macroeconomic effects of debt-and equity-based capital inflows. *Journal of Macroeconomics, 46*, 81–95.

Edwards, S., & Rigobon, R. (2009). Capital controls on inflows, exchange rate volatility and external vulnerability. *Journal of International Economics, 78*, 256–267.

Eichengreen, B. (2000). Taming capital flows. *World Development, 28*(6), 1105–1116.

Eichengreen, B., & Hausmann, R. (1999). Exchange rates and financial fragility. *NBER Working Paper*, 7418.

IMF. (2013). The task force on finance statistics. External debt statistics: Guide for Compilers and Users. Accessed 10 Dec 2015. http://www.tffs.org/edsguide.htm

IMF. (2015a). IMF Country Report, 15/289. https://www.imf.org/external/pubs/ft/scr/2015/cr15289.pdf

IMF. (2015b). Global financial stability report. Vulnerabilities, legacies, and policy challenges risks rotating to emerging markets. October 2015. https://www.imf.org/external/pubs/ft/gfsr/2015/02/

IMF. (2016). E-Library Data. http://www.elibrary.imf.org/. Accessed 10 Jan 2016.

Kolb, R. W. (2011). Sovereign debt: Theory, defaults, and sanctions. In R. W. Kolb (Ed.), *Sovereign debt: From safety to default* (pp. 3–15). Hoboken, NJ: Wiley.

Krugman, P. R. (2014). Currency regimes, capital flows, and crises. *IMF Economic Review, 62*(4), 470–493.

Manasse, P., & Roubini, N. (2009). Rules of thumb for sovereign debt crises. *Journal of International Economics, 78*(2), 192–205.

Mellios, C., & Paget-Blanc, E. (2011). The impact of economic and political factors on sovereign credit ratings. In R. W. Kolb (Ed.), *Sovereign debt: From safety to default* (pp. 3–15). Hoboken, NJ: Wiley.

Nyambuu, U., & Bernard, L. (2015). A quantitative approach to assessing sovereign default risk in resource-rich emerging economies. *International Journal of Finance and Economics, 20*(3), 220–241. doi:10.1002/ijfe.1512.

Reinhart, C. M., & Rogoff, K. (2008). *This time is different: Eight centuries of financial folly*. Princeton: Princeton University Press.

Reinhart, C. M., & Kenneth, R. S. (2014). Financial and sovereign debt crises: Some lessons learned and those forgotten. In S. Claessens, M. A. Kose, L. Laeven & F. Valencia (Eds.), *Financial crises: Causes, consequences, and policy responses*. Washington D.C.: International Monetary Fund.

Stallings, B. (2007). The globalization of capital flows: Who benefits? *The ANNALS of the American Academy of Political and Social Science, 3*(610), 201–216.

Stein, L. J. (2006). *Stochastic optimal control, international finance, and debt crises*. New York, NY: Oxford University Press.

Stein, L. J. (2012). *Stochastic optimal control and the U.S. financial debt crisis*. Berlin: Springer.

Stiglitz, J. (2000). Capital market liberalization, economic growth, and instability. *World Development, 28*(6), 1075–1086.

Stiglitz, J. (2002). *Globalization and its discontents*. New York/London: W.W. Norton.

The World Bank. Databank. http://databank.worldbank.org/data/home.aspx. Accessed 5 Jan 2016.

Standard & Poor's (2006). Sovereign defaults and rating transition data: 2006 Update. Available at http://www2.standardandpoors.com/spf/pdf/fixedincome/KR_sovereign_SovDefaultsTransitionData2006.pdf

Sturzenegger, F., & Zettelmeyer, J. (2006). *Debt defaults and lessons from a decade of crises*. Cambridge, MA: MIT Press.

U.S. Board of Governors of the Federal Reserve System (2014). http://www.federalreserve.gov/. Accessed 5 Nov 2015.

Keynes' Microeconomics of Output and Labor Markets

Duncan K. Foley

JEL codes: D4, D6, D7, E2, E4

1 Introduction

The first readers of John Maynard Keynes' *General Theory* (*GT*) on the whole recognized the empirical importance of Keynes' macroeconomic conclusions, but largely failed to grasp the microeconomic logic supporting those conclusions. This failure led to the gradual replacement of Keynes' original ideas by various "syntheses" such as those proposed by Hicks (1937) and Modigliani (1944) based on the idea that Keynes' macroeconomics could be derived from Walrasian market-clearing microeconomics with the addition of the hypothesis of money wage "stickiness". In the 1970s the intellectual and policy failures of this "bastard Keynesianism" (in Joan Robinson's terms) led to the eclipse of Keynesian theoretical work in macroeconomic theory and the emergence of one or another version of "real business cycle" theory as the hegemonic paradigm of macroeconomic thinking following the path of Robert Lucas and Thomas Sargent.

Parallel to the high drama of macroeconomic paradigm wars important innovations in microeconomic theory made major progress after the 1970s. These innovations centered on "post-Walrasian" theories based on assumptions of contractual incompleteness, strategic complementarity, increasing returns to scale, and quantal responses of economic decision makers. (One synthetic statement that addresses many of these innovations is Samuel Bowles' *Microeconomics*.)

D.K. Foley (✉)
Leo Model Professor of Economics, New School for Social Research
e-mail: duncan.foley@gmail.com; foleyd@newschool.edu

L. Bernard, U. Nyambuu (eds.), *Dynamic Modeling, Empirical Macroeconomics, and Finance*, DOI 10.1007/978-3-319-39887-7_8

A careful reconsideration of Keynes' *GT* reveals that the microeconomic foundations of his thinking are assumptions closely akin to these microeconomic innovations. Keynes insisted on: (a) the monetary character of economic interactions; (b) the pervasiveness of increasing returns to scale due to fixed costs in labor supply and production decisions; and, as a consequence, (c) the need to analyze economic equilibrium through crowding of markets rather than clearing of markets. In this paper I will focus on the second two of these major aspects.

I use the terms "market crowding" and "market clearing" to describe two concepts of market equilibrium. In market clearing equilibria, producers are *price-takers* who perceive no impact of their production decisions on price, and, as a consequence, choose an output level where price is equal to marginal cost: they therefore are indifferent at the margin as to whether they gain or lose demand. In market crowding equilibria producers perceive a negative impact of increased production on price, and, as a consequence, choose an output level where marginal revenue is equal to marginal cost, price is higher than marginal cost: they would prefer higher demand, if they could attract it without lowering their price through advertising, coupons, or "miles". Market clearing equilibria can occur only when producers are operating on convex regions of their production sets and average cost is constant or increasing in output. Market crowding equilibria, in contrast, typically occur when producers are operating on non-convex regions of their production sets (due, for example, to the existence of significant fixed costs), and average cost is declining in output. Market crowding equilibria are enforced in the short run by independent uncoordinated output decisions of existing competitors and in the long period by free entry and exit of competing firms (or households) from segmented industry markets in response to signals of excess or deficient profitability. Market crowding equilibria are Cournot-Nash type equilibria, and in general fail to be Pareto-efficient.

The consistent application of these ideas to Keynes' problems clarifies the roots of the important elements of what he put forward as a general theory of employment, interest and money, particularly the possibility of chronic excess capacity in product markets and involuntary employment in labor markets.

The aim of this paper is to explain in detail how these market-crowding considerations (many of which have been examined piecemeal in the economics literature) together support Keynes' claim to have arrived at a general theory, of which Walrasian equilibrium based on the assumption of market clearing rather than market crowding is indeed a special case.

2 Crowding Equilibrium

In order to provide a well-defined context for the discussion, let us look at a particular model of the product market of a competitive industry.

2.1 Industry Equilibrium Through Market Crowding

Consider an industry in which n identical firms with fixed cost f, constant marginal cost c and a capacity constraint $x_{\max}$ compete. The industry demand curve is

$$p[X] = p_0 - p_1 X$$

where X is the industry output. The demand curve for an individual firm, assuming the other firms are producing on average $\bar{x}$, so that $X = (n-1)\bar{x} + x$ is the total supply, is

$$p\left[x + (n-1)\bar{x}\right] = p_0 - p_1(n-1)\bar{x} - p_1 x$$

The dependence of the firm's price on its output is characteristic of crowding equilibria.

The short-run profit of the firm, before fixed cost, is

$$\pi = P\left[x, \bar{x}, n\right] x - cx = \left(p_0 - p_1(n-1)\bar{x} - p_1 x\right) x - cx = \left(p_0 - c - p_1(n-1)\bar{x}\right) x - p_1 x^2$$

The firm maximizes profit by choosing the best-response output level

$$x^{\dagger}\left[\bar{x}\right] = \begin{cases} x_{\max} & \bar{x} \leq \frac{p_0 - c}{p_1(n-1)} - \frac{2}{n-1} x_{\max} \\ \frac{1}{2}\frac{p_0 - c}{p_1} - \frac{n-1}{2}\bar{x} & \frac{p_0 - c}{p_1(n-1)} - \frac{2}{n-1} x_{\max} \leq \bar{x} \leq \frac{p_0 - c}{p_1(n-1)} \\ 0 & \frac{p_0 - c}{p_1(n-1)} \leq \bar{x} \end{cases}$$

Since all the firms are identical, the Cournot(-Nash) equilibrium level of output, x^C, requires:

$$x^C = x^{\dagger}\left[x^C\right]$$

The short-run equilibrium, which can be thought of as a function of the level of demand as represented by p_0, is:

$$x^C\left[p_0\right] = \begin{cases} \frac{p_0 - c}{p_1(n+1)} & p_0 \leq c + p_1(n+1)x_{\max} \\ x_{\max} & c + p_1(n+1)x_{\max} \leq p_0 \end{cases}$$

The equilibrium price is:

$$p^C\left[p_0\right] = \begin{cases} c + \frac{p_0 - c}{n+1} & p_0 \leq c + p_1(n+1)x_{\max} \\ p_0 - p_1 n x_{\max} & c + p_1(n+1)x_{\max} \leq p_0 \end{cases}$$

The corresponding profit of the typical firm (or profit rate if the size of the investment required is 1) is:

$$\pi^C[p_0] = \begin{cases} \frac{(p_0-c)^2}{p_1(n+1)^2} & p_0 \leq c + p_1(n+1)x_{\max} \\ (p_0 - c - nx_{\max})\, x_{\max} & c + p_1(n+1)x_{\max} \leq p_0 \end{cases}$$

Thus there are two equilibrium regimes, depending on the strength of demand as measured by p_0. If $p_0 \leq c + p_1(n+1)x_{\max}$, there is a crowding equilibrium, in which the equilibrium price is above marginal cost, and the typical firm operates with unused capacity; if $c + p_1(n+1)x_{\max} \leq p_0$ the equilibrium puts the typical firm at its capacity constraint, where marginal cost for an increase in output becomes vertical, so that price is equal to marginal cost.

Of particular interest from the point of view of Keynes' thinking, we can see that in the crowding regime changes in demand are absorbed partly by changes in market price and partly by changes in output through a change in unused capacity, while in the clearing regime a change in demand affects only prices. These regimes are central to Keynes' thinking, and underly his insistence that his analysis is a general theory, that embraces both cases, while the "classical" view is restricted to the market clearing regime, which is of limited value in analyzing real-world economies, particularly in conditions of falling or stagnant aggregate demand.

Note that this model determines a Kaleckian equilibrium markup on marginal cost, which varies with demand as parameterized by p_0.

2.1.1 Queuing

If the typical firm faces a *user cost*, that is a fixed cost that it pays only if it produces a non-zero output, then if the crowding equilibrium price just covers the user cost, the typical firm will be indifferent between producing and not producing, and market equilibrium will lead a certain proportion of firms not to produce, thus forming a queue of completely unused capacity. The effect of user cost is to put a floor under the market equilibrium price.

> The employment of a given volume of labour by an entrepreneur involves him in two kinds of expense: first of all, the amounts which he pays out to the factors of production (exclusive of other entrepreneurs) for their current services, which we shall call the *factor cost* of the employment in question; and secondly, the amounts which he pays out to other entrepreneurs for what he has to purchase from them together with the sacrifice which he incurs by employing the equipment instead of leaving it idle, which we shall call the *user cost* of the employment in question (Keynes 1951, Chap. 3).

The prospective profit of the typical firm for given $\bar{x}$ is:

$$\pi^{\dagger}[\bar{x}] = \begin{cases} (p_0 - c - p_1(n-1)\bar{x})\, x_{\max} & \bar{x} \leq \frac{p_0-c}{p_1(n-1)} - \frac{2}{n-1}x_{\max} \\ \frac{(p_0-c-p_1(n-1)\bar{x})^2}{4p_1} & \frac{p_0-c}{p_1(n-1)} - \frac{2}{n-1}x_{\max} \leq \bar{x} \leq \frac{p_0-c}{p_1(n-1)} \\ 0 & \frac{p_0-c}{p_1(n-1)} \leq \bar{x} \end{cases}$$

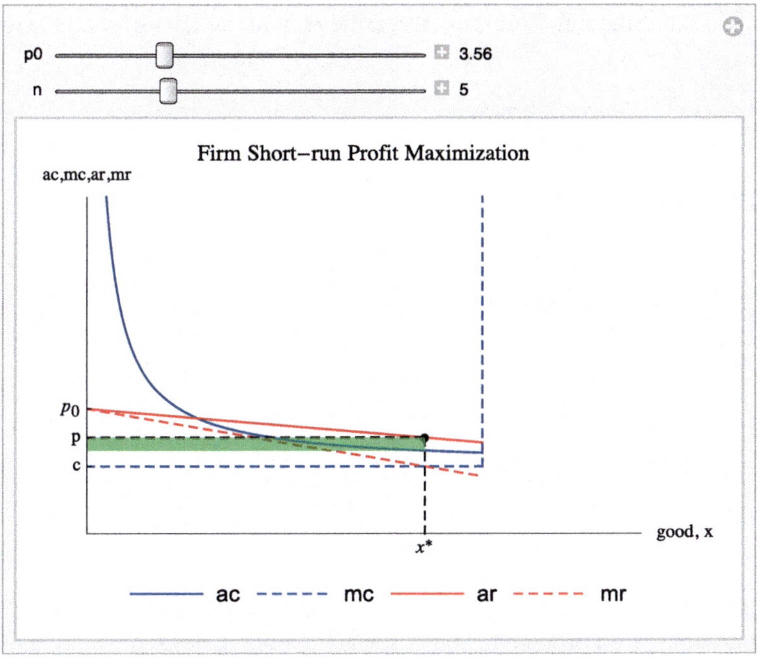

Fig. 1 The short-run profit-maximizing equilibrium of a typical firm with a fixed cost, and constant marginal cost up to a capacity constraint in market-crowding industry equilibrium

The firm will produce only if the prospective short-run profit covers the user cost, denoted $c_0 \geq 0$, or:

$$\frac{(p_0 - c - p_1(n-1)\bar{x})^2}{4p_1} \geq c_0$$

Firms will begin to shut down when average output and price are:

$$\bar{x}^{\text{SD}} = \frac{p_0 - c}{p_1(n+1)} - \frac{2}{n-1}\sqrt{\frac{c_0}{p_1}}$$
$$p^{\text{SD}} = 2\sqrt{c_0 p_1}$$

At this point the typical firm is indifferent between producing $\sqrt{\frac{c_0}{p_1}}$ for profit c_0 or shutting down. Thus the best response of the typical firm when there is a user cost becomes vertical when $\bar{x} = \bar{x}^{\text{SD}}$.

If $\sqrt{\frac{c_0}{p_1}} < x^{SD}$ the equilibrium is the same as without user cost, $x^C = \frac{p_0 - c}{p_1(n+1)}$, $p^N = c + \frac{p_0 - c}{n+1}$.

If $\sqrt{\frac{c_0}{p_1}} \geq x^{SD}$, the equilibrium is

$$x^C = x^{SD} = \frac{p_0 - c}{p_1(n+1)} - \frac{2}{(n-1)}\sqrt{\frac{c_0}{p_1}}$$

$$p^C = c + 2\sqrt{p_1 c_0}$$

$$x^C[p_0, c_0] = \begin{cases} \frac{p_0-c}{p_1(n+1)} - \frac{2}{(n-1)}\sqrt{\frac{c_0}{p_1}} & 0 \leq p_0 \leq c + \frac{(n+1)^2}{n-1}\sqrt{c_0 p_1} \\ \frac{p_0-c}{p_1(n+1)} & c + \frac{(n+1)^2}{n-1}\sqrt{c_0 p_1} \leq p_0 \leq c + p_1(n+1)x_{\max} \\ x_{\max} & c + p_1(n+1)x_{\max} \leq p_0 \end{cases}$$

$$p^C[p_0, c_0] = \begin{cases} c + 2\sqrt{p_1 c_0} & 0 \leq p_0 \leq c + \frac{(n+1)^2}{n-1}\sqrt{c_0 p_1} \\ c + \frac{p_0-c}{n+1} & c + \frac{(n+1)^2}{n-1}\sqrt{c_0 p_1} \leq p_0 \leq c + p_1(n+1)x_{\max} \\ p_0 - p_1 n x_{\max} & c + p_1(n+1)x_{\max} \leq p_0 \end{cases}$$

In this equilibrium there is a queue of non-producing firms since the typical firm makes the same profit producing at the profit-maximizing level or shutting down.

With a user cost the industry market clears through a combination of some firms producing with excess capacity and others completely shutting down.

At the shut-down price, the industry meets the demand by having the necessary proportion of the firms produce at the shut-down output level, and the rest shut down.

With a positive user cost, as p_0 rises, there is a region where industry price remains constant while industry output increases, as shutdown firms come online, then a region where both industry price and industry output increase, and finally a full-capacity region in which price rises while output remains constant.

2.1.2 Labor Markets

This model can also represent a crowded labor(-power) market, in which n households compete to supply a particular type of labor services to a possibly congested market. The marginal cost in this case represents the *marginal disutility* of labor, measured in terms of money. In this case equilibrium at $x_{\max}$ has the natural interpretation of *full employment*, in that each household is working as many hours as it can. The crowding equilibrium corresponds to an *underemployment* equilibrium in which each household would like to have more work if it could get it without lowering its money wage. A user cost would represent some costs of the household actually working at all, such as securing access to transportation, or providing complementary inputs to work such as appropriate clothing, or arranging for childcare. When the user cost of employment is positive, the equilibrium of

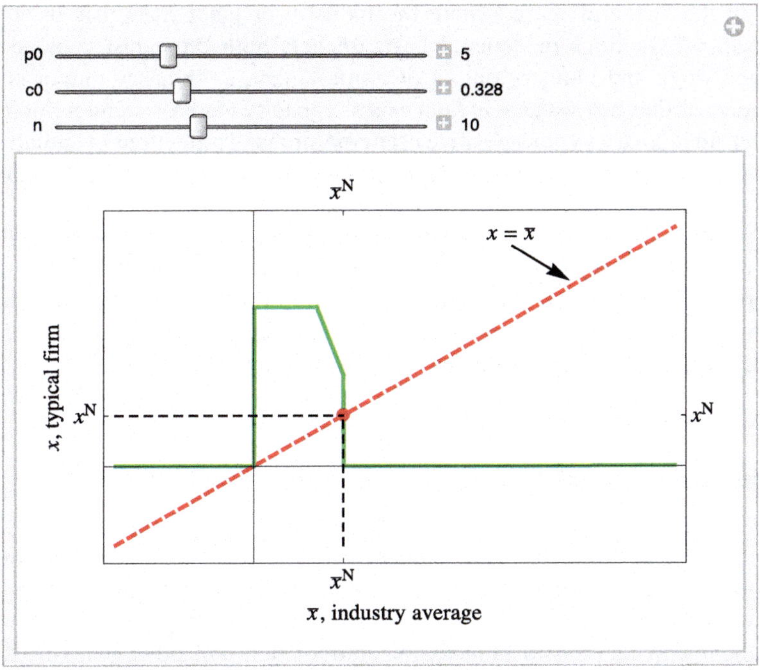

Fig. 2 The equilibrium of an industry of identical firms with a fixed cost, constant marginal cost up to a capacity constraint, and a user cost to producing at all. The equilibrium illustrated here puts the typical firm at the output level where it is indifferent between producing and not producing, so the market equilibrium involves a queue of non-producing firms

the market will exhibit a combination of underemployment and unemployment as a queue of worker household not actually working at all, but willing to work at the equilibrium money wage, forms.

2.2 *Some Macroeconomic Consequences*

In industries or labor markets where crowding and queuing equilibria are possible, it is difficult to lower the money price or wage in the short run by repressing demand. If the industry or labor market is in a market-clearing (or, as Keynes would say, "full-employment") equilibrium, then reductions in demand do have a one-for-one impact on the industry money price or money wage, with no reduction in industry output or level of employment, which remains at the full-capacity or full-employment level. (Or, conversely, increases in demand have a one-for-one impact on industry money price or money wage, with no increase in industry output or employment.)

But in the "general" case, where the industry or labor market is in crowding equilibrium, reductions in demand have impacts both on industry money price or money wage and industry output or employment, as industry output is pulled further and further below capacity and excess capacity increases, underemployment increases. In industries with a positive user cost, demand reductions eventually result in shutting down of some firms or the unemployment of some workers without any impact on industry money price or money wage at all.

It is unlikely that firms in a crowding equilibrium with excess capacity will invest in order to increase capacity, since they already are operating with excess capacity. They do, however, have an incentive to invest to lower their marginal cost, because if a firm anticipates that its competitors will not change their output levels, a lower marginal cost increases the profit-maximizing output and profit of the firm.

3 The Structure of Social Interactions

Crowding equilibrium is an example of a broad class of *social interactions*, in which the actions of other participants (firms or households or individuals) changes the incentives influencing the action of each participant. This type of interaction crops up frequently in relation to fallacies of composition and other ideas in Keynes' economics. (See Bowles 2004, *Microeconomics*, Chap. 4 for a discussion of this class of models.)

It is worth summarizing abstractly the simplest class of these interaction models, in which a large number of identical participants make symmetrical action choices, to clarify the analytical basis of Keynes' ideas, particularly the concept of aggregate demand.

Abstractly then, consider an interacting group of n identical participants, each of whom controls an action variable x. The action of the typical participant is a *response* to the mean level of the actions of the other participants, $\bar{x}$. (Other statistics describing the distribution of the actions of the other participants, such as the median or mode could play the same role, and might be salient in some social situations.) In the simplest case, the action of the typical participant is a linear function of the mean action of the others, subject to a constraint on the extreme values of the action:

$$x[\bar{x}] = \begin{cases} x_{\text{lo}} & \alpha + \beta\bar{x} \leq x_{\text{lo}} \\ \alpha + \beta\bar{x} & x_{\text{lo}} \leq \alpha + \beta\bar{x} \leq x_{\text{hi}} \\ x_{\text{hi}} & x_{\text{hi}} \leq \alpha + \beta\bar{x} \end{cases}$$

In many cases is possible to take $x_{\text{lo}} = 0$, $x_{\text{hi}} = 1$, for example when x is the probability of the typical participant taking some particular action, or, as in the case of an industry, output is measured as a fraction of the typical firm's capacity.

Various nonlinear generalizations of this response function are likely under particular situations, but this bare-bones case is sufficient to understand the qualitative possibilities of this type of social interaction.

It is possible, but not necessary, to derive the response function as a payoff-maximizing *best response*, given a quadratic payoff function of the form:

$$u\left[x,\bar{x}\right] = \alpha x + \beta x\bar{x} + \gamma\bar{x} - \frac{1}{2}x^2 = (\alpha + \beta\bar{x})\,x - \frac{1}{2}x^2 + \gamma\bar{x}$$

In this case $\alpha + \beta\bar{x}$ represents the private monetary return to the typical participant's action, which depends either positively ($\beta > 0$, the case of *strategic complementarity*) or negatively ($\beta < 0$, the case of *strategic substitutability*) on the average action of the other participants. The quadratic term represents the *effort* or cost of the action to the typical individual (and could be replaced by any other concave function of x). The term $\gamma\bar{x}$ does not influence the typical participant's best response, but represents a pure (positive or negative) *external effect* of the average action on the outcome of the typical participant.

The industry equilibrium model is the social interaction model with $\alpha = \frac{p_0-c}{2p_1}$ and $\beta = -\frac{n-1}{2}$.

In this simple version of the social interaction model where all participants are identical (a restriction that can be relaxed in a variety of ways) a Cournot-Nash equilibrium requires that the equilibrium action be a (best) response to itself:

$$x^C = x\left[x^C\right]$$

If $\beta < 1$ there is a unique equilibrium

$$x^C = x^{\dagger} = \begin{cases} x_{\text{lo}} & \frac{\alpha}{1-\beta} \leq x_{\text{lo}} \\ \frac{\alpha}{1-\beta} & x_{\text{lo}} \leq \frac{\alpha}{1-\beta} \leq x_{\text{hi}} \\ x_{\text{hi}} & x_{\text{hi}} \leq \frac{\alpha}{1-\beta} \end{cases}$$

which is stable in the intuitive sense that when the average level of the action rises above the equilibrium level the typical participant's response either falls or rises by a smaller amount, and symmetrically for a fall in the average action level.

When $\beta < 1$ and $\beta \neq 0$, the social interaction model is a *Prisoners' Dilemma* in the sense of a game having a single Cournot-Nash equilibrium that is Pareto-inefficient.

If $\beta > 1$, there is still an equilibrium at $x^{\dagger}$, which is unstable whenever it is interior to the limiting values of the action, $x_{\text{lo}} < x^{\dagger} < x_{\text{hi}}$. In this case there are also stable equilibria at x_{lo} and x_{hi}, and the interior unstable equilibrium $x^{\dagger}$ separates the *basins of attraction* or the two stable extreme equilibria. If $x^{\dagger} = x_{\text{lo}}$ or $x^{\dagger} = x_{\text{hi}}$ it is a unique and stable equilibrium.

When there are two stable equilibria the social interaction model presents what is often called an *Assurance Game*, or a version of the *network externality* based on a strong *strategic complementarity*. When $\beta > 1$ an increase in the proportion of the rest of the participants who choose one of the extreme strategies increases the incentives for the typical participant to choose the same extreme even more. We can be quite sure that the system will converge to one of the two stable equilibria (and equally sure that it cannot maintain itself at the unstable interior equilibrium), but which one will emerge depends on the initial conditions of the system, and therefore is *path-dependent*. Generically, the two stable equilibria are Pareto-comparable.

When $\beta = 1$, the system is in neutral equilibrium for all values of x. At this *bifurcation*, the system intuitively is prone to scale-free fluctuations due to the lack of any stabilizing forces.

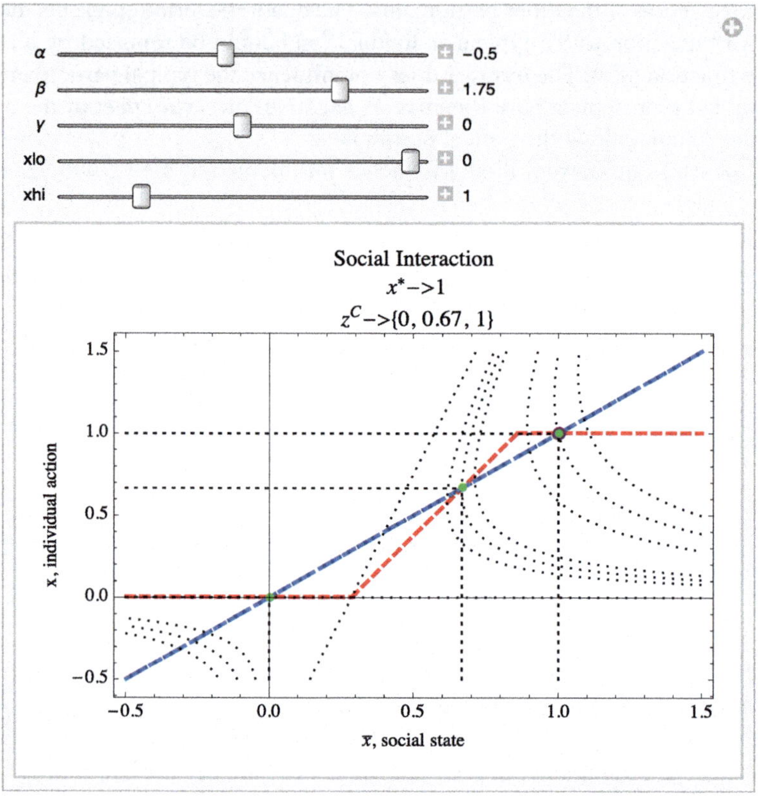

Fig. 3 The social interaction model with a high degree of strategic complementarity ($\beta > 1$). The interior equilibrium is unstable, and separates the basins of attraction of the two stable equilibria at ($x = 0$) and ($x = 1$). In this case the Pareto-efficient equilibrium is ($x = 1$), but depending on path-dependent factors the system might converge on the Pareto-inefficient equilibrium at ($x = 0$). A social interaction system with strong strategic complementarity may undergo a cusp catastrophe in which a stable equilibrium vanishes and forces a catastrophic change to the surviving stable equilibrium

The symmetric Pareto-efficient action level maximizes the typical participant's utility when all the participants are constrained to the same action level:

$$u[x,x] = (\alpha + \gamma)x + \left(\beta - \frac{1}{2}\right)x^2$$

When $\beta < \frac{1}{2}$ this is a concave function with an interior Pareto-efficient action level $\frac{\alpha+\gamma}{1-2\beta}$. When $\beta \geq \frac{1}{2}$ this is a convex function that reaches its maximum at one of the extreme values of the action, depending on whether $(\alpha+\gamma)x_{\mathrm{lo}} + \left(\beta - \frac{1}{2}\right)x_{\mathrm{lo}}{}^2 < (\alpha+\gamma)x_{\mathrm{hi}} + \left(\beta - \frac{1}{2}\right)x_{\mathrm{hi}}{}^2$. In the case where $x_{\mathrm{lo}} = 0$ and $x_{\mathrm{hi}} = 1$ the Pareto-efficient action level x^* is:

$$x^* = \left\{ \begin{array}{ll} x_{\mathrm{lo}} & \beta > \frac{1}{2}\text{and } \alpha + \beta + \gamma < \frac{1}{2} \\ \frac{\alpha+\gamma}{1-2\beta} & \boldsymbol{\beta} < \frac{1}{2} \\ x_{\mathrm{hi}} & \beta > \frac{1}{2}\text{and } \alpha + \beta + \gamma > \frac{1}{2} \end{array} \right.$$

4 Conclusion

This line of thinking (which could easily be extended to other problems, such as the theories of aggregate demand, liquidity preference, and Keynes' views on financial markets) raises two problems, one of mostly scholarly interest, but the other of more pressing practical concern.

The scholar may question whether Keynes himself did think in terms of models equivalent to the class of social interaction models discussed here. I personally think this is likely. Much of the language of *GT* invokes the concept of crowding equilibrium, which was the focus of Robinson's (1932) research on imperfect competition, published during Keynes' development of the *GT* in the *Economic Journal*, of which Keynes was editor, and of Richard Kahn's Cambridge dissertation. Because Keynes was writing in the 1930s before the clarification of the structure of market-clearing general equilibrium models in the work of Kenneth Arrow and Gerard Debreu, and Tjalling Koopmans, the mathematical relation of the two types of equilibrium and the importance of convexity assumptions was not yet fully apparent. Keynes could freely use the term "equilibrium" as it came to him from the Marshall tradition as including market-crowding and market-clearing cases. (One figure who did appreciate the depth of this distinction was Harold Hotelling, whose mathematical insight put him far in advance of the majority of the economics profession in this period.)

Of more practical concern is the question of whether macroeconomics, in its role as a policy science, is wise to sweep the range of issues raised by market-crowding equilibrium and monetary economics under the rug of the sweeping assertion that proper "micro-foundations" for macroeconomics can lie only in the market-clearing, price-taking, Walrasian general equilibrium framework. The empirical evidence for maintained full-employment equilibrium macroeconomics has always been weak, and in the wake of major shocks like the Great Depression or the Great Recession it is difficult to convince ordinary intelligent women and men of its relevance. Unfortunately, as we have witnessed in the period since the 1970s, the internal logical contradictions of "Keynesian" economics (Robinson's "bastard Keynesianism", which appears on the present stage in the guise of "New Keynesian" economics) in the long run leave it vulnerable to erosion and attack from the Walrasian position. In my view this vulnerability lies in the inability of the sticky-wage approach to clarify just what social coordination problem macroeconomic instability reflects, and in particular the way in which money prices themselves, along with systemic liquidity, are a fragile public good. We can do better than recapitulate the sterile controversies of the 1970s and 1980s.

Acknowledgements This paper is a contribution to a festschrift for my long-time friend and colleague, Willi Semmler, with whom I've shared many intellectual adventures and discussions over more than 40 years.

My thinking on Keynes' microeconomics has been shaped in profound ways by the work of Sam Bowles. The usual disclaimer applies. The support of the Institute for New Economic Thinking is gratefully acknowledged.

References

Bowles, S. (2004). *Microeconomics : Behavior, institutions, and evolution*. New York/Princeton, NJ: Russell Sage/Princeton University Press.

Hicks, J. R. (1937). Mr. Keynes and the 'Classics': A suggested interpretation. *Econometrica, 5*(2), 147–159.

Keynes, J. M. (1951). *The general theory of employment, interest and money*. London: Macmillan (8th repr. 1936).

Modigliani, F. (1944). Liquidity preference and the theory of interest and money. *Econometrica, 12*(1), 45. doi:10.2307/1905567.

Robinson, J. (1932). Imperfect competition and falling supply price. *The Economic Journal, 42*(168), 544–554.

“Wavelet-Based” Early Warning Signals of Financial Stress: An Application to IMF’s AE-FSI

Marco Gallegati, Mauro Gallegati, James B. Ramsey, and Willi Semmler

1 Introduction

Following the recent 2008–2009 financial crisis there has been growing interest, from researchers to policy-makers, in developing methods to measure and monitor deteriorating financial conditions. A number of alternative indexes of financial stress have been recently developed with the aim of identifying periods of deteriorating financial conditions and determining whether financial stress is high enough to be a serious concern for financial stability. Such indexes take the form of composite indicators and are constructed by combining a number of indicators representative of different financial markets, i.e. interbank, credit, equity, foreign exchange, etc., in order to capture one or more aspects of systemic financial stress such as information asymmetry, flight to quality, flight to liquidity, increased uncertainty, etc. (see Hakkio and Keeton 2009).[1]

[1]Well known examples are the Financial Stress Indexes of the Federal Reserve Bank of Kansas City (KCFSI), St. Louis (STLFSI), Cleveland (CFSI) and IMF’s Financial Stress Index for Advanced Economies (AE-FSI).

M. Gallegati (✉) • M. Gallegati
DISES and SIEC, Polytechnic University of Marche, Ancona, Italy
e-mail: marco.gallegati@univpm.it; mauro.gallegati@univpm.it

J.B. Ramsey
Department of Economics, New York University, New York, NY, USA
e-mail: james.ramsey@nyu.edu

W. Semmler
Department of Economics, New School for Social Research, New York, NY, USA
e-mail: semmlerw@newschool.edu

L. Bernard, U. Nyambuu (eds.), *Dynamic Modeling, Empirical Macroeconomics, and Finance*, DOI 10.1007/978-3-319-39887-7_9

In addition to favor the creation of such continuous measure of financial stress, the 2008–2009 crisis has also generated renewed interest in the early warning indicators literature since early warning systems can signal financial stress episodes in advance. Several papers have recently addressed this issue with the purpose of investigating to what extent financial stress can be predicted by using a set of early warning indicators (e.g. Misina and Tkacz 2009; Slingenberg and De Haan 2011; Shin 2013; Christensen and Li 2013). Notwithstanding the large set of leading indicators and the wide range of different estimation techniques used in empirical works, there is one point upon which the extensive literature on early warning indicators agrees: the view that only time domain information matters (for a comprehensive review of the literature see Chamon and Crowe 2013). But the analysis in the time domain yields only part of the information embedded in the data, and in order to obtain the remaining part of the information the data need to be analyzed in the frequency domain.

The objective of this paper is to propose a new methodology for the construction of early warning composite indicators that is able to exploit all available information conveyed by individual indicators using the multiresolution decomposition properties of wavelet analysis. Based on the methodology developed in Gallegati (2014a,b), in this paper we construct a "wavelet-based" early warning indicator of financial stress for the IMF financial stress index for advanced economies developed in Cardarelli et al. (2009). Specifically, for each country of the sample we construct a "wavelet-based" early warning indicator of financial stress by aggregating several "scale-based" sub-indices whose components are selected on the basis of their statistical performance at each frequency band. The implicit assumption underlying this alternative approach to composite indicators is that the aspects of financial stress captured by each of these variables can change with the time frame adopted, so that the leading properties of each indicator can be expected to differ across time horizons (Gallegati 2014a,b).

The contribution of the methodology proposed in this paper to the early warning literature is twofold. First, it provides a method that allow to fully exploit all available information conveyed by individual indicators. Second, in contrast to binary variables of the signalling approach (e.g. Kaminsky and Reinhart 1999), it allows the construction of a continuous measures that can provide early warning signals about the evolving conditions of the stress level of the financial system.[2]

[2]Other continuous measures are provided by Brave and Butters' (2012) NFCI's nonfinancial leverage subindex, an early warning index of financial instability that is made up of two nonfinancial leverage measures included in the Chicago Fed's National Financial Conditions Index, and the early-warning composite index proposed by Berti et al. (2012) for early detection of fiscal stress.

The leading properties of each country's "wavelet-based" early warning indicator for its correspondent financial stress index are examined by using univariate statistical criteria and a pseudo out-of-sample forecasting exercise. The empirical results supports the findings reported in Gallegati (2014a) as to the leading indicator properties for financial stress of a "wavelet-based" early warning indicator. Specifically we find that: (1) the "general fit" of the early warning indicators in relation to the financial stress index is rather good, with an average lead of about 2 months; and (2) the "wavelet-based" composite indicator largely outperforms any individual financial variable taken in isolation in predicting financial stress at every horizon, with the gain increasing as the time horizon increases.

The paper is organized as follows. Section 2 explains the motivation for using wavelets in economic applications and the methodology used for constructing a "wavelet-based" composite index. Section 3 describes the main features of the dataset used, i.e. the IMF financial stress index for advanced economies. The methodology of constructing the "wavelet-based" early warning indicator for each country of the sample and the "scale-by-scale" selection of various indicators in construction of the aggregate index. In Sect. 4 we examine the ability of each country's early-warning composite indicator of financial stress to provide early signals of financial stress by performing an out-of-sample exercise. Section 5 concludes.

2 "Wavelet-Based" Composite Indexes: Motivation and Methodology

In the construction of composite indexes the attention of the researcher is focused on finding the best-performing indicators with respect to the phenomenon to be explained (turning points, crisis or stressful periods, etc.), with the timing classification and selection of such "reliable" individual indicators being based on the time domain information content provided by time series aggregate data. However, if the variable realizations depend on different frequency components rather than just one as implicitly stated in the standard time domain approach, any information regarding the different frequency components contained into the original data series will be lost (Masset 2015).[3]

[3]The economic intuition supporting the application of time-frequency domain techniques in macroeconomics and finance is that many economic and financial processes are the results of decisions of agents with different, sometimes very different, time horizons information. For example, in financial markets the presence of heterogeneous agents with different trading horizons may generate very complex patterns in the time-series of economic and financial variables (e.g., Muller et al. 1995; Lynch and Zumbach 2003).

In such a context, joint time-frequency analysis methods combining information from both time-domain and frequency-domain simultaneously, like wavelets, can be very appealing.[4] Wavelets are mathematical functions that allows the decomposition of a signal into its different frequency components and to study the dynamics of each of these components separately.[5] The most important gain from using the multiresolution decomposition properties of wavelet transform for the indicators approach is that it allows to efficiently process all available information by exploiting the different informative content of individual indicators at different time scales. In particular, wavelet analysis can reveal structure to a problem and/or relationships between variables that standard time series analysis is not able to capture. Indeed, the results obtained using wavelet analysis in economic applications clearly show that the separation by time scale decomposition analysis can be of great benefit for a proper understanding of economic relationships that are actually recognized as puzzling (Ramsey 2002). For example, the results obtained in several recent papers provide evidence of different informative content of different interest rate spreads for future output (Gallegati et al. 2014) and of stock and bond market prices for investments (Gallegati and Ramsey 2013, 2014) at different time scales.

2.1 *The Methodology*

The "wavelet-based" methodology differs from the traditional "indicators approach" as to the construction and variable selection procedures in that the overall composite index is obtained by aggregating several sub-indices, each corresponding to a different frequency band. In particular, each sub-index is obtained by combining those indicators displaying the best statistical cross-correlations performance on a scale-by-scale basis. The result is a composite indicator that, by combining the best-performing individual components at each scale, is able to process all available information efficiently and parsimoniously. Moreover, since the wavelet transform allows the decomposition of the observed time series into a sum of structural and random error components it can also deal with the problem of false signals by reducing the incidence of irregular movements.

[4]The analysis in the time and frequency domain is able to produce more accurate information on the presence of (highly localized) patterns and dominant scales of variation in the data such as characteristic scales, and on the relationships between variables as well (e.g. Ramsey and Lampart 1998a,b; Kim and In 2003; Aguiar-Conraria et al. 2008; Aguiar-Conraria and Soares 2011; Gallegati et al. 2011).

[5]A brief introduction to wavelets in general, and the maximal overlap discrete transform in particular, is provided in the Appendix.

The methodology to construct a "wavelet-based" composite indicator requires going through several steps:

- first, each individual variable is decomposed into its time scale components by applying the Discrete Wavelet Transform (multiresolution decomposition analysis);
- second, a set of variable is chosen at each scale level on the basis of previously selected statistical criteria;
- third, at each scale level the selected variables are aggregated into a composite "scale-based" sub-index using a weighting methodology;
- fourth, the various composite sub-indexes are aggregated by simple sum into a "wavelet-based" composite index.

Let I_1, I_2,..., I_N be the set of N "reliable" indicators available for the construction of the composite index. By applying a J-level multi resolution decomposition analysis we can provide a complete decomposition of each individual indicator I_i into a smoothed version of the original signal and a set of detail information at different scales:

$$I_i \approx S_J[I_i] + D_J[I_i] + \ldots.. + D_j[I_i] + \ldots + D_2[I_i] + D_1[I_i] \tag{1}$$

where $S_J[I_i]$ represents the "smooth component" of the signal, and $D_j[I_i]$, with $j = 1, 2, ..J$, the detail signal components at ever-increasing level of detail.[6] The scale-by-scale analysis of the relationships between each variable and the reference series allows identification of the "best-performing" variables on a " scale-by-scale" basis. These "best-performing" individual variables at each scale level can then be used to construct a "scale-based" composite sub-index CI_{D_j} for each decomposition level $j, j = 1, 2, \ldots, J$, by means of weighted aggregation. Thus, with $k \leq$n statistically significant "reliable" indicators at the scale level j, the composite sub-index CI_{D_j} will be

$$CI_{D_j} = \omega_{1,j}D_j[I_1] + \omega_{2,j}D_j[I_2] + \omega_{3,j}D_j[I_3] + \ldots + \omega_{k,j}D_j[I_k] \tag{2}$$

where ω_{kj} is the weight of each indicator k at scale j. Finally, by aggregating the j "scale-based" composite sub-indexes CI_{D_j} we can obtain the "wavelet-based" composite index CI^W, that is

$$CI^W = CI_{S_J} + CI_{D_J} + \ldots\ldots.. + CI_{D_j} + \ldots\ldots. + CI_{D_1}. \tag{3}$$

[6]The wavelet transform uses a flexible time-scale window that narrows when focusing on small-scale features and widens on large-scale features. By using short windows at high frequencies and long windows at low frequencies the wavelet transform has good frequency and time localization properties, that is it displays good time resolution and poor frequency resolution at high frequencies, and good frequency resolution but poor time resolution at low frequencies (see Kumar and Foufoula-Georgiou 1997).

Two different applications of this methodology have been recently provided in Gallegati (2014a,b). In the first case a "wavelet-based" early warning composite indicator of financial stress has been constructed and its ability to provide early warning signals of financial distress evaluated for the US. The results indicate that the "wavelet-based" early warning indicator largely outperforms any individual financial variable taken in isolation in predicting financial stressful periods in the US at every horizon and that the gain tends to increase as the time horizon increases (see Gallegati 2014a). In the latter a "wavelet-based" leading indicator for the US business cycle is created. The comparison with the OECD-CLI and its derived measures indicate that the "wavelet-based" composite index tends to provide early signals of business cycle turning points well in advance of the OEDC-CLI (Gallegati 2014b).

In what follows we apply the "wavelet-based" methodology to the IMF financial stress index for advanced economies (AE-FSI) developed in Cardarelli et al. (2009). In particular, for each country of the sample we construct a "wavelet-based" early warning indicator of financial stress by selecting, among the individual indicators used in the construction of the IMF financial stress index, those indicators displaying the best leading performance on a "scale-by-scale" basis. Then, we examine the leading properties of each country's early warning composite indicator for the detection of financial stress periods by using univariate statistical criteria and a pseudo out-of-sample forecasting exercise.

3 "Wavelet-Based" Early Warning Composite Indicators for the IMF's AE-FSI

Although the literature on early warning systems for currency and banking crises has been extensive,[7] developed economies have generally received limited attention mainly because of the relatively rare occurrence of financial crises in these countries. Still less attention has been devoted to develop early warning composite indicators providing early signs of financial instability.[8]

There are several reasons that explain why IMF financial stress index for advanced economies (AE-FSI) represents an outstanding candidate for evaluating the robustness of such a "wavelet-based" early warning indicator of financial stress. First of all, although the FSI is constructed for each of the 17 countries in the sample by using a uniform minimum set of indicators, it is quite robust in capturing the

[7]Early warning signal models can be based on composition of leading indicators as in Kaminsky (1998), and Kaminsky and Reinhart (1999), or on probit/logit models, as in Berg and Pattillo (1999).

[8]A rare example is the Kaminsky's (1998) leading indicator approach to early warning system where the warning signals have to be transformed into a binary variable.

main financial stress episodes documented in the literature.[9] Second, by including such a large set of countries for quite a long period of time it is possible to provide evidence about a very large number of past episodes of financial stress that are thus likely to capture the whole spectrum of events potentially able to originate periods of stress in financial markets. Indeed, for each advanced economy the AE-FSI is a weighted average of seven indicators[10] capturing events in banking (banking-sector stock price volatility, the spread between interbank rates and the yield on treasury bills, and the slope of the yield curve), securities (corporate bond spreads, stock market returns, and stock return volatility), and exchange markets (exchange rate volatility).

3.1 IMF Financial Stress Index for Advanced Economies (AE-FSI)

The international dataset developed by Cardarelli et al. (2009) includes 17 advanced economies covering about 80 % of advanced economy GDP. Specifically, the dataset provides a market based Financial Stress Index (AE-FSI) for Australia, Austria, Belgium, Canada, Denmark, Finland, France, Germany, Italy, Japan, Netherlands, Norway, Spain, Sweden, Switzerland, United Kingdom, and United States.

Each country's IMF's AE-FSI is composed of the following variables[11]:

- the 'banking-sector beta', i.e. the standard capital asset pricing model (CAPM) beta, defined as:

$$\beta_{ij} = cov(r_{ij}^{M}, r_{ij}^{B})/\sigma_{i,M}^{2},$$

  where r represents the year-over-year banking or market returns, computed over a 12-month rolling window (Beta);
- TED or interbank spread, defined as the 3-month LIBOR or commercial paper rate minus the government short-term rate (Ted);

[9]Although it would be possible to get more reliable early warning indicators in the considered countries by investigating alternative indicators as to the one included in the IMF sample, in order to emphasize the ability of wavelet multi resolution analysis to provide an efficient data reduction technique we use for the construction of the "wavelet-based" composite index the same individual financial indicators that at aggregate level are able to provide information on the current level of financial stress.

[10]All components are available in monthly frequency but for different time periods. The aggregate index is only computed when data for all components were available, thus countries' samples can be slightly different in length.

[11]For data sources and definitions see Balakrishnan et al. (2009). We use the dataset version updated to march 2012.

- Corporate bond yield spread, defined as corporate bond yield minus long-term government bond yield (Corps);
- Inverted term spread, defined as the government short-term rate minus government long-term rate (Invts);
- Monthly stock market returns, computed as the month-over-month change in the stock index multiplied by minus one, so that a decline in equity prices corresponds to increased securities-market-related stress (Stockd);
- Stock market volatility, measured as the 6-month (backward looking) moving average of the squared month-on-month returns (RSQ ret);
- the foreign exchange market volatility, measured as the 6-month (backward looking) moving average of the squared month-on-month growth rate of the exchange rate (RSQ reer).[12]

Similarly to Kaminsky and Reinhart (1999) the aggregation of these subindices is based on a variance-equal weighting, so that each component is computed as a deviation from its mean and weighted by the inverse of its variance. The FSI is robust to other weighting schemes, including those based on principal components.[13]

After standardizing each variable, that is demeaning (using the arithmetic mean) and dividing by its standard deviation, the aggregate FSI for each country is obtained by summing up the seven components:

AE-FSI = 'banking-sector beta' + TED spreads + inverted term spreads + corporate debt spreads + stock market returns + stock market volatility + exchange market volatility

A value of zero of the AE-FSI implies neutral financial market conditions on average across the subindices, while positive values imply financial stress (i.e. prices are on average above means or trends). A value of 1 indicates a one-standard deviation from average conditions across subindices, where values higher than 1 have in the past been associated with a crisis.

3.2 Constructing the "Wavelet-Based" Composite Indicator

Since the components of the composite indicator are selected on a "scale-by-scale" basis, the procedure for calculating a "wavelet-based" composite index starts with decomposing all variables in the IMF's AE-FSI into their time-scale components

[12]Both volatility measures replaces the previously used GARCH(1,1) specification.

[13]This approach is the most common weighting method in the literature. Previous research have shown that variance-equal weighting performs as well in signaling stress episodes as weighting based on economic fundamentals (Illing and Liu 2006). Moreover, robustness tests indicate that equal-variance weights are very similar to weights identified by a principal components analysis of the stress subindices (Kappler and Schleer 2013).

through the maximum overlap discrete wavelet transform (MODWT).[14] In this way each variable is partitioned into a set of different time scale components, each corresponding to a particular frequency range. However, before performing wavelet analysis a number of decisions must be made: what type of wavelet transform to apply, which family of wavelet filters to use, and how boundary conditions at the end of the series are to be handled. For our purposes we use the Daubechies' (1992) least asymmetric $LA(8)$ wavelet filter of length $L = 8$, the most widely used filter in economic applications, with reflecting boundary conditions.[15]

The application of the MODWT with a number of levels $J = 6$ produces seven individual crystals[16]: one vector of smooth coefficients s_6, representing the underlying smooth behavior of the data at the coarse scale, and six vectors of details coefficients $d_6, d_5, d_4, d_3, d_2, d_1$, representing progressively finer scale deviations from the smooth behavior.[17] Using monthly data scale 1 wavelet coefficients are associated to 2–4 month periods, while scales 2–6 are associated to 4–8, 8–16, 16–32, 32–64 and 64–128 month periods, respectively. Hence, detail levels from d_1 to d_3, represent the very short-run dynamics of a signal (and contains most of the noise of the signal), levels d_4, d_5 and d_6 roughly correspond to the standard business cycle time period (Stock and Watson 1999).

Just as the usual unconditional cross-correlation coefficients provide a measure of the degree of association between variables in the time domain, wavelet cross-correlation coefficients can provide information on the relationship between two processes at different scales. In particular, the results from wavelet cross-correlation analysis can provide very useful insights into the timing behavior of different variables at different frequency bands, especially as to their leading behavior. This kind of information can be usefully exploited to construct a parsimonious and

---

[14]Because of the practical limitations of DWT in empirical applications the *maximal overlap discrete wavelet transform* (MODWT). The MODWT is a non-orthogonal variant of the classic discrete wavelet transform (DWT) that, unlike the DWT, is translation invariant, as shifts in the signal do not change the pattern of coefficients, can be applied to data sets of length not divisible by 2^J and returns at each scale a number of coefficients equal to the length of the original series is applied.

[15]In order to calculate wavelet coefficient values near the end of the series boundary conditions are to be assumed. The series may be extended in a periodic fashion (periodic boundary condition) or in a symmetric fashion (reflecting boundary condition). With reflecting boundary condition the original signal is reflected at its end point to produce a series of length $2N$ which has the same mean and variance as the original signal. The wavelet and scaling coefficients are then computed by using a periodic boundary condition on the reflected series, resulting in twice as many wavelet and scaling coefficients at each level.

[16]A limit to the level of decomposition is given by the number of observations, $J = log_2 N$.Each set of wavelet transform coefficients is called a crystal in wavelet terminology.

[17]As the MODWT wavelet filter at each scale j approximates an ideal high-pass filter with passband $f \in [1/2^{j+1}, 1/2^j]$, the level j wavelet coefficients are associated to periods $[2^j, 2^{j+1}]$. On the other hand scaling coefficients are associated with frequencies $f \in [0, 1/2^{j+1}]$.

"informationally efficient" composite leading index.[18] Indeed, given that our aim is to construct a composite indicator providing early warning signals of financial stress we are interested in finding those variables that at each frequency band are leading the AE-FSI series for each country. Hence, after decomposing the regression variables into their different time scale components using the MOWDT, we examine the cross-correlation of each component series with the corresponding FSI at different time scales using wavelet cross-correlation coefficients.

In Fig. 1 we show, as an example, wavelet cross correlation results between stock returns declines and AE-FSI for the UK and the US.[19] The MODWT-based wavelet cross-correlation coefficients (solid lines), along with the corresponding approximate confidence intervals (dotted lines), against time leads and lags up to 12 months for all scales are shown. Figure 1 provides clear evidence of leading behavior for stock returns decline over AE-FSI at scale levels 4 and 5, that is at frequency bands corresponding to business cycles periodicities. The statistical criterium used for selecting the component series of the wavelet based composite indicator is that of cyclical conformity, where correlation analysis is used to verify whether the cyclical profiles of the individual component and the reference series, the AE-FSI in this case, are highly related.[20] Therefore, on the basis of the above mentioned criterium the wavelet detail vectors D_6, D_5 and D_4 of stock market returns can be included into their corresponding early warning sub-indexes.[21]

In Table 1 the wavelet details vectors for which there is evidence of a leading behavior at peak-correlation with the corresponding financial stress index at different scales are reported. They represent the time scale components selected for the construction of each country's "wavelet-based" early warning composite index on the basis of the results of wavelet cross-correlation analysis. There are striking differences as to the leading behavior of the different time scale components among variables and across scales. The "scale-by-scale" cross-correlation results indicate that TED or interbank spread (Ted), Inverted term spread (Invts) and stock returns declines (Stockd) display the most notable leading properties for the AE-FSI. Two main features emerge in the composition of a "wavelet-based" composite leading index from the different results across scales. First, each "scale based" composite sub-index displays just a subset, sometimes a very small subset, of all indicators, and in this sense it can be considered a parsimonious index. Second, since the leading behavior is mostly concentrated at higher scales, the low scales sub-indexes,

---

[18]Parsimony and efficiency are related to the ability of reducing redundant information in the construction of the composite index.

[19]Full results are not presented here for brevity, but are available on request by the author.

[20]In the indicators approach to business cycles cross-correlation analysis is complemented by turning point analysis since forecasting turning points is one of the main objectives of the leading indicators approach.

[21]The wavelet details vectors D_6, D_5, ..., D_1 are obtained from the corresponding wavelet coefficients d_6, d_5, ..., d_1 through the synthesis or reconstruction process. The synthesis or reconstruction process reconstructs the original signal from the wavelet coefficients using the inverse DWT.

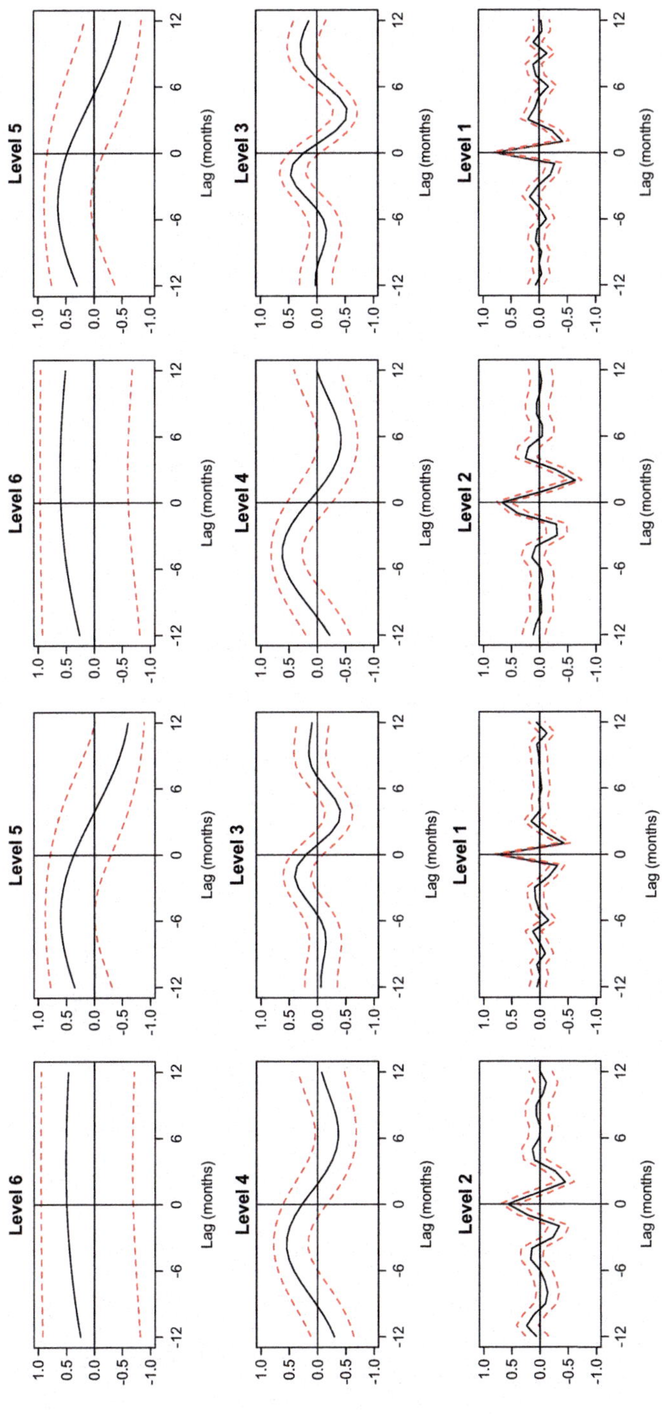

Fig. 1 Wavelet cross correlation results between stock returns declines and AE-FSI at different time scales for the UK (*left panel*) and the US (*right panel*)

Table 1 Wavelet detail vectors included in each country's "wavelet-based" early warning index of financial stress EWI_c^W

	Beta	Ted	Inv. TS	Corp. spr.	Stock Dec.	RSQ ret.	RSQ reer
EWI_{AUT}^W	D_5	D_6	D_3, D_4, D_5	D_3, D_4	D_4	–	D_6
EWI_{AUS}^W	D_5	D_6	D_5, D_6	–	D_3, D_4, D_5	D_4	–
EWI_{BEL}^W	–	D_3, D_4, D_5	D_4, D_5, D_6	D_6	D_4, D_5	–	–
EWI_{CAN}^W	D_5	D_3, D_4, D_5, D_6	D_4, D_5, D_6	–	D_3, D_4, D_5, D_6	–	–
EWI_{DEN}^W	D_5	D_4, D_5	D_3, D_5, D_6	D_3	D_3, D_4, D_5	–	D_5
EWI_{FIN}^W	D_6	D_3, D_5, D_6	D_6	D_6	D_4, D_5	–	–
EWI_{FRA}^W	–	–	D_3, D_4	D_3, D_5	D_3, D_4, D_5	D_4	–
EWI_{GER}^W	–	D_3	D_5, D_6	–	D_3, D_4, D_5	–	D_5
EWI_{ITA}^W	–	D_3, D_4, D_5	–	–	–	D_4, D_6	–
EWI_{JAP}^W	–	D_6	–	D_5	D_3, D_4, D_5, D_6	–	–
EWI_{NED}^W	–	–	D_4, D_5, D_6	D_6	D_3, D_4, D_5	–	D_3
EWI_{NOR}^W	D_5	–	D_4, D_6	D_6	D_3, D_4, D_5	D_6	D_5
EWI_{SPA}^W	–	D_3, D_5	D_3, D_5, D_6	–	D_4, D_5	–	D_5
EWI_{SWE}^W	D_3, D_5	–	D_4, D_5, D_6	D_4, D_5, D_6	D_3, D_4, D_5, D_6	–	–
EWI_{SWI}^W	D_4, D_5	D_3	D_5, D_6	–	D_3, D_4, D_5	–	D_5
EWI_{UK}^W	–	D_3, D_4, D_5, D_6	D_4, D_5, D_6	–	D_3, D_4, D_5	–	–
EWI_{US}^W	D_5	D_3, D_4, D_5, D_6	D_5, D_6	–	D_3, D_4, D_5	–	–

Table 2 Leading behavior by variable and time scale: a summary

Scale	1 (2–4)	2 (4–8)	3 (8–16)	4 (16–32)	5 (32–64)	6 (64–128)
Beta	–	–	21 %	12 %	47 %	12 %
Ted	–	–	47 %	41 %	47 %	41 %
Invts	–	–	29 %	52 %	**82 %**	**82 %**
Corps	–	–	21 %	12 %	21 %	29 %
Stockd	–	–	**70 %**	**94 %**	**94 %**	18 %
RSQ ret	–	–	6 %	18 %	–	18 %
RSQ reer	–	–	6 %	–	21 %	6 %

Note: The numbers represent the percentage of countries for which each specific time scale component is included in the corresponding sub-index
The bold values refer to components selected in more than 2/3 of the times (countries)

corresponding to higher frequency components, can be excluded from the overall composite index without losing informative content. Therefore, since the noise component is generally restricted to scales corresponding to higher frequencies (see Ramsey et al. 2010; Gallegati and Ramsey 2012), the "wavelet-based" composite index is likely to be smoother than the corresponding traditional composite index.

The results presented in Table 1 are summarized in Table 2. The numbers in Table 2 represent the percentage of countries for which each specific time scale component is included in the corresponding sub-index. For example, stock returns declines at scale levels 4 and 5 have leading properties for the financial stress index in 16 countries out of 17, that is in about 94 % of the cases.

The results in Table 2 indicate that the leading behavior of the inverted term spread at scales 5 and 6, corresponding to frequencies between 3 and 10 years, and of stock returns declines at scales 3, 4 and 5, corresponding to frequencies between slightly less than 1 and 5 years, is common to almost all advanced countries in the sample. The inverted term spread at scale 4, the beta of bank stock index at scale 5 and TED spread at scales from 3 to 6 display a leading behavior for about half of the countries in the sample (from 7 to 9 countries). For the remaining variables, as well as for the variables previously mentioned at other scales, the evidence of a leading behavior is sparse and generally limited to a few number of countries. Finally, at shorter scales, that is at scale levels 1 and 2, there is no evidence of any leading behavior (the peak correlation is generally at zero lag).

Figure 2 shows for the G6 countries the "wavelet-based" early warning indicators EWI_c^W (smooth line) constructed using the individual indicators indicated in Table 1 along with their corresponding IMF's AE-FSI index (irregular line).

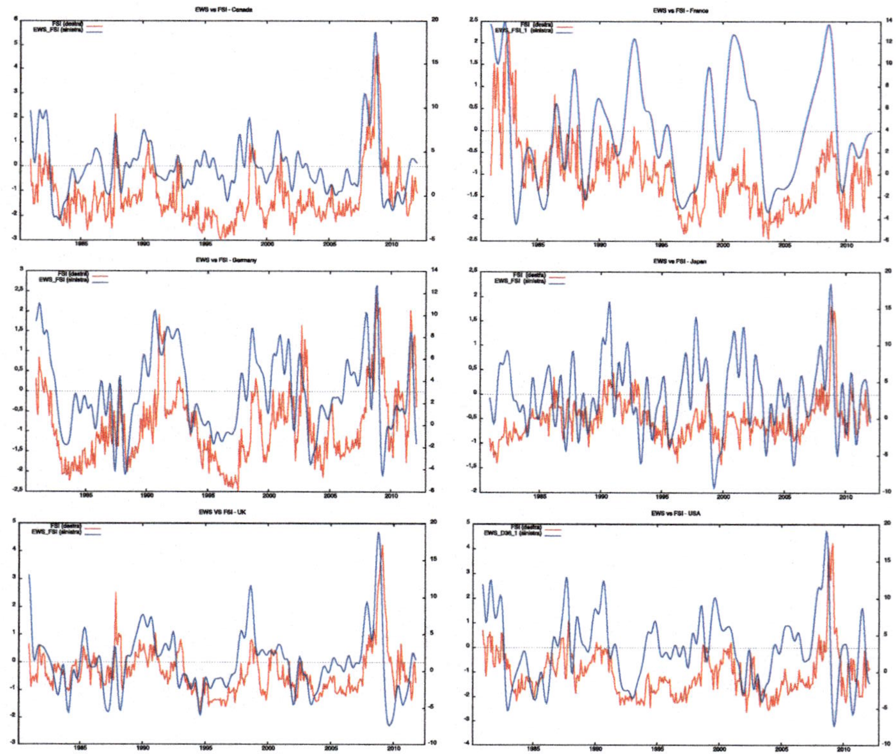

Fig. 2 Wavelet-based early warning indicators EWI_c^W (*smooth line*) and AE-FSI (*irregular*) for the G6 countries

4 Evaluation of the "Wavelet-Based" Early Warning Composite Indicator of Financial Stress

4.1 Cross-Correlation Results

In order to test the ability of the "wavelet-based" early warning composite indicator to give information about future financial stress episodes we evaluate its performance using both statistical and econometric methods. In particular, for each country we use the cross correlation function between the early warning composite indicator, EWI_c^W, and the financial stress index, AE-FSI. In Table 3 the months lead at which the maximum correlation occurs between the early warning indicator and its corresponding financial stress index (column two) and the correlation coefficient at the peak value (column three) are presented. The number of months lead gives the average lead of the "wavelet-based" composite indicator over the AE-FSI and indicate the phase shift between the two series. The mean lead of the early warning

Table 3 Leading time and peak correlation value between EWI_c^W and AE-FSI

Country	Lead time	Peak value
Austria	2	0.6498
Australia	2	0.5289
Belgium	2	0.6825
Canada	2	0.6824
Denmark	1	0.6045
Finland	5	0.5332
France	2	0.5229
Germany	3	0.6123
Italy	4	0.5941
Japan	2	0.4471
Netherland	3	0.6677
Norway	1	0.6625
Spain	2	0.4950
Sweden	2	0.5881
Switzerland	1	0.6556
UK	3	0.6197
US	3	0.593
Mean	2.3	0.5964
Median	2	0.6045

indicator is close to 2 months. Indeed, although the leading time seem to be quite variable, ranging from 1 (Denmark, Norway and Switzerland) to 5 months (Finland), for most of the countries the peak correlation is observed at lead-time two (eight countries) and three (four countries). The correlation coefficient at the peak provides a linear measure of the extent to which the pattern of the early warning indicator and the financial stress index resemble each other. The "general fit" of the early warning indicators in relation to the financial stress index is rather good, since most of the countries have correlation coefficients very close or above 0.60. Exceptions are Australia, Finland, France, Japan and Spain whose cross-correlation coefficient values, between 0.44 and 0.53, are lower than average. In sum, the evaluation results presented in Table 3 indicate that the "wavelet-based" early warning composite indicator provides reliable information about the future evolution of the financial stress index, a finding also confirmed by the slight difference between the mean and the median values of the statistical criteria considered.

4.2 Forecasting Financial Stress

Financial stress indexes, as measures of current financial stress conditions, can also enter early warning models as predicted variables and be used to evaluate the ability of alternative indicators to provide early warning signals of financial instability. In order to evaluate the ability of the "wavelet-based" composite indicator EWI_c^W to provide early warning signals of financial stress we follow the forecasting framework of Misina and Tkacz (2009) and Hatzius et al. (2010) by performing a pseudo out-of-sample forecasting exercise.[22] We compare a simple autoregression model where the current FSI is a function of its k-quarter lagged value

$$FSI_t = \alpha + \beta FSI_{t-k} + \epsilon_{1,t} \tag{4}$$

against an augmented model in which several additional explanatory variables are individually added to the baseline autoregressive specification

$$FSI_t = \alpha + \beta FSI_{t-k} + \gamma X_{t-k} + \epsilon_{2,t} \tag{5}$$

[22]For the forecasting experiment we use quarterly data obtained by converting monthly data through averaging.

where X include the set of indicators used in the AE-FSI and the "wavelet-based" composite indicator EWI_c^W.

The forecasting exercise is conducted by splitting the whole sample into two subsamples as in Christensen and Li (2013): the first, from 1981:Q1 to 2007:Q3 is used to estimate the parameters of the benchmark and augmented models, the latter, from 2007:Q4 to 2011:Q4, is used to generate forecasts of the AE-FSI. These forecasted values of the financial stress index are obtained on a recursive basis by adding one observation to the sample and then repeating the estimation and the forecast at the different horizons k of 1, 2 and 4 quarter(s). This forecasting framework attempts to replicate the production of forecasts in real time since forecasts are provided since using all information that is available in each time period.[23]

For an evaluation criterion of the forecasting performances of the two models we use the differences between the forecasted financial stress index originating from models (4) and (5) and the actual values of the financial stress index, Specifically, we use the relative root mean squared error (RMSE) expressed as the ratio of the RMSE of the augmented model relative to that of the benchmark model. This ratio provides a test of model fit, so that a value below 1 indicates an improvement in forecast accuracy relative to the benchmark model, while when the ratio is above 1.0 this indicates a worsening of the forecasting performance relative to the baseline autoregressive model.

Table 4 presents the relative RMSE's ratios for this forecasting exercise where the "wavelet-based" composite indicator EWI_c^W and the different financial variables used in the construction of the IMF financial stress index for advanced economies are individually added as explanatory variables to the augmented model. The main findings is that the "wavelet-based" composite indicator EWI_c^W largely outperforms any individual financial variable taken in isolation in predicting financial stress at every horizon. Moreover, the gain tends to increase as the horizon increases.[24]

---

[23]Moreover, the issue of the impact of data revisions does not apply in this case since the observations of our financial variables are not revised.

[24]The only exception is given by the TED spread for Canada at 1 and 2 quarters horizons, and for Finland at 2 and 4 quarters horizons.

Table 4 Out-of-sample forecasting results—k=1

	Australia			Austria			Belgium			Canada		
var	1	2	4	1	2	4	1	2	4	1	2	4
Beta	0.9747	0.9654	0.9530	1.0248	1.0582	1.0835	1.0287	1.0543	1.0737	1.0508	1.1088	1.1515
TED	0.9822	0.9963	1.0123	1.0742	1.0683	1.0076	1.0096	0.9742	0.9424	0.7582	0.6716	0.6392
Invts	0.9620	0.9242	0.8899	0.8898	0.8668	0.9646	0.9970	0.9930	0.9850	1.0607	1.0589	1.0198
Corps	0.9881	0.9553	0.9107	1.0238	0.9841	0.9252	1.0393	0.9909	0.9123	1.0332	1.0704	1.1229
Stockd	1.0382	1.0672	1.1078	1.0717	1.1445	1.2044	1.0610	1.1136	1.1925	0.9921	0.9797	0.9585
RSQret	0.9739	1.0014	1.0276	1.0430	1.0810	1.0999	1.0007	1.0144	1.0119	0.9155	0.9374	1.0146
RSQreer	0.9326	0.9430	0.9796	0.9725	0.9784	1.0062	1.0004	1.0708	1.2345	0.9976	1.0042	1.0336
EWI^W	0.9097	0.8547	0.8209	0.7660	0.6880	0.6244	0.8323	0.7181	0.6332	0.8547	0.7398	0.6691
	Denmark			Finland			France			Germany		
var	1	2	4	1	2	4	1	2	4	1	2	4
Beta	1.0318	1.1358	1.2703	1.0019	1.0196	1.0467	1.0022	1.0853	1.1757	0.9715	1.0141	1.0845
TED	0.9626	0.9759	0.9622	0.9087	0.8288	0.7339	0.9989	0.9993	0.9985	1.0039	1.0122	1.0534
Invts	0.9688	0.9904	0.9951	0.9958	0.9939	0.9588	0.9810	0.9785	0.9590	1.0212	1.0169	0.9988
Corps	0.9677	0.9971	0.9973	0.9524	0.9307	0.8595	0.9854	1.0036	0.9786	0.9878	0.9489	0.9007
Stockd	0.9698	0.9999	0.9957	0.9982	1.0705	1.0597	0.9994	1.0467	1.0451	0.9911	0.9763	0.9828
RSQ ret	0.9695	0.9861	0.9911	0.9956	0.9938	1.0039	1.0048	1.0095	1.0125	0.9899	0.9922	0.9892
RSQ reer	0.9882	1.0334	1.0567	0.9888	0.9809	1.0308	0.9918	0.9537	0.9950	0.9846	0.9892	1.0053
EWI^W	0.9233	0.9130	0.8938	0.9011	0.8353	0.8653	0.9590	0.9186	0.8801	0.9392	0.8705	0.8550

Notes: The table displays the relative ratio of the root mean squared error (RMSE) of model (5) relative to that of model (4). Rows headings denote the variable added to the baseline model. The columns under each country's panel denote the forecast horizon (k) measured in quarters beyond the end of the sample

Table 5 Notes: see Table 4

	Italy			Japan			Netherland			Spain		
var	1	2	4	1	2	4	1	2	4	1	2	4
Beta	0.9191	1.0007	1.1862	1.0125	1.0277	1.0450	1.0181	1.0505	1.1023	0.9243	0.9486	0.9843
TED	0.9985	0.9757	0.9434	1.0100	1.0122	1.0327	0.9457	0.9650	0.9916	0.9759	0.9569	0.9437
Invts	0.9915	1.0076	1.0812	1.0109	1.0240	1.0498	0.9544	0.8889	0.8269	0.8959	0.8686	0.8174
Corps	1.0106	1.0161	1.0113	1.0204	1.0539	1.0925	0.9873	0.9125	0.8398	0.9940	0.9911	0.9895
Stockd	1.1282	1.1983	1.2835	0.9950	0.9766	0.9544	1.0002	1.0019	1.0156	1.1447	1.1892	1.2544
RSQ ret	1.0442	1.0549	1.0493	1.0100	1.0058	0.9901	1.0122	1.0622	1.1314	0.9597	0.9680	0.9696
RSQ reer	0.9995	1.0122	1.0389	1.1253	1.2867	1.1857	1.0017	1.0245	1.0629	1.0009	0.9930	0.9870
EWI^W	0.8071	0.6686	0.5082	0.9459	0.8967	0.8554	0.8587	0.7418	0.6911	0.8542	0.8051	0.7590
	Sweden			Switzerland			UK			US		
var	1	2	4	1	2	4	1	2	4	1	2	4
Beta	1.0161	1.0138	1.0142	0.9962	0.9827	0.9925	0.9946	1.0061	1.0413	1.0033	0.9971	0.9861
TED	1.0056	1.0127	1.0128	1.0128	0.9966	0.9924	0.9806	0.9700	0.9458	0.9690	0.9368	0.9001
Invts	0.9884	0.9675	0.9949	1.0493	1.0259	1.0012	1.0153	0.9871	0.9435	1.0138	1.0121	0.9972
Corps	1.0096	0.8447	0.8016	1.0134	1.0091	0.9925	1.0595	1.1082	1.1466	1.0106	1.0244	1.0348
Stockd	1.0177	1.0415	1.0850	1.0124	1.0695	1.1172	1.0051	1.0104	1.0243	1.0044	1.0151	1.0304
RSQ ret	0.9604	0.9439	0.9155	0.9288	0.9442	0.9067	1.0013	1.0216	1.0522	0.9722	1.0236	1.1876
RSQ reer	0.9026	0.9094	0.9885	0.9761	0.9900	1.0186	0.9368	0.9555	0.9871	0.9742	0.9882	1.0279
EWI^W	0.9018	0.8214	0.7736	0.9213	0.8451	0.8045	0.7971	0.7292	0.6613	0.9389	0.8106	0.7202

Notes: The table displays the relative ratio of the root mean squared error (RMSE) of model (5) relative to that of model (4). Rows headings denote the variable added to the baseline model. The columns under each country's panel denote the forecast horizon (k) measured in quarters beyond the end of the sample

5 Conclusions

In this paper we apply a "wavelet-based" methodology for the early detection of periods of financial distress for 17 developed countries for which the IMF's has constructed a financial stress index (Cardarelli et al. 2009). The proposed methodology takes advantage of the ability of wavelet analysis to fully exploit the whole information present in the data and differs from the traditional one as to the construction and data selection procedure. The composite leading indicator is obtained by aggregating several "scale-based" sub-indexes whose components are selected on the basis of their cross-correlations properties at each frequency band.

The "wavelet-based" early warning composite indicator of financial stress is evaluated using univariate statistical criteria and a pseudo out-of-sample forecasting exercise. The overall performance of the early warning indicator is quite good. The average leading time is about 2 months and its "general fit" is satisfactory. Moreover, the findings of the out-of-sample forecasting exercise indicate that the 'wavelet-based' composite indicator largely outperforms at every horizon any individual financial variable taken in isolation in predicting financial stress for each country of the sample and that the gain tends to increase as the horizon increases.

These findings, given the number of developed economies included in the IMF's financial stress index, provide evidence on the robustness of the proposed methodology and on the usefulness of exploiting all information available on the time and frequency domain for the construction of composite indicators with leading properties.

Acknowledgements The research leading to these results has received funding from the European Union, Seventh Framework Programme FP7/2007-2013 Socio-economic Sciences and Humanities under Grant Agreement No. 320278 - RASTANEWS.

Appendix: Basic Notions of Wavelets

Given a stochastic process $\{X\}$, if we denote with $H = (h_0, \ldots, h_{L-1})$ and $G = (g_0, \ldots, g_{L-1})$ the impulse response sequence[25] of the wavelet and scaling filters h_l, and g_l, respectively, of a Daubechies compactly supported wavelet (with L the width of the filters), when $N = L^2$ we may apply the orthonormal discrete wavelet transform (DWT) and obtain the wavelet and scaling coefficients at the *jth* level

[25]The impulse response sequence is the set of all filter coefficients. The filter coefficients must satisfy three properties: zero mean ($\sum_{l=0}^{L-1} h_l = 0$), unit energy ($\sum_{l=0}^{L-1} h_l^2 = 1$) and orthogonal to its even shifts ($\sum_{l=0}^{L-1} h_l h_{l+2k} = 0$).

defined as[26]

$$w_{j,t} = \sum_{l=0}^{L-1} h_{j,l} X_{t-l}$$

$$v_{J,t} = \sum_{l=0}^{L-1} g_{j,l} X_{t-l},$$

where $h_{j,l}$ and $g_{j,l}$ are the level j wavelet and scaling filters and, due to downsampling by 2^J, we have $\frac{N}{2^J}$ scaling and wavelet coefficients.[27]

The DWT is implemented via a filter cascade where the wavelet filter h_l is used with the associated scaling filter g_l[28] in a pyramid algorithm (Mallat 1989) consisting in an iterative scheme in which, at each iteration, the wavelet and scaling coefficients are computed from the scaling coefficients of the previous iteration.[29]

However the orthonormal discrete wavelet transform (DWT), even if widely applied to time series analysis in many disciplines, has two main drawbacks: the dyadic length requirement (i.e. a sample size divisible by 2^J),[30] and the fact that the wavelet and scaling coefficients are not shift invariant due to their sensitivity to circular shifts because of the decimation operation. An alternative to DWT is represented by a non-orthogonal variant of DWT: the maximal overlap DWT (MODWT).[31]

In the orthonormal Discrete Wavelet Transform (DWT) the wavelet coefficients are related to nonoverlapping differences of weighted averages from the original observations that are concentrated in space. More information on the variability of the signal could be obtained considering all possible differences at each scale, that is considering overlapping differences, and this is precisely what the maximal overlap algorithm does.[32] Thus, the maximal overlap DWT coefficients may be considered the result of a simple modification in the pyramid algorithm used in computing DWT coefficients through not downsampling the output at each scale and inserting zeros

[26]The expressions used for DWT (and MODWT) wavelet and scaling coefficients refer to functions defined over the entire real axis, that is $t \in \Re$ as in this case $X_t = X_{t \bmod N}$ when $t < 0$.

[27]At the *jth* level the inputs to the wavelet and scaling filters are the scaling coefficients from the previous level $(j-1)$ and the output are the *jth* level wavelet and scaling coefficients.

[28]The wavelet and scaling filter coefficients are related to each other through a quadrature mirror filter relationship, that is $h_l = (-1)^l g_{L-1-l}$ for $l = 0, \ldots, L-1$.

[29]The only exception is at the unit level ($j = 1$) in which wavelet and scaling filters are applied to original data.

[30]This condition is not strictly required if a partial DWT is performed.

[31]The MODWT goes under several names in the wavelet literature, such as the "non-decimated DWT", "stationary DWT" (Nason and SIlverman 1995), "translation-invariant DWT" (Coifman and Donoho 1995) and "time-invariant DWT".

[32]Indeed, the term maximal overlap refers to the fact that all possible shifted time intervals are computed. As a consequence, the orthogonality of the transform is lost, but the number of wavelet and scaling coefficients at every scale is the same as the number of observations.

between coefficients in the wavelet and scaling filters.[33] In particular, the MODWT wavelet and scaling coefficients $\tilde{w}_{j,t}$ and $\tilde{w}_{j,t}$ are given by

$$\tilde{w}_{j,t} = \frac{1}{2^{j/2}} \sum_{l=0}^{L-1} \tilde{h}_{j,l} X_{t-l}$$

$$\tilde{v}_{J,t} = \frac{1}{2^{j/2}} \sum_{l=0}^{L-1} \tilde{g}_{J,l} X_{t-l},$$

where the MODWT wavelet and scaling filters $\tilde{h}_{j,l}$ and $\tilde{g}_{j,l}$ are obtained by rescaling the DWT filters as follows[34]:

$$\tilde{h}_{j,l} = \frac{h_{j,l}}{2^{j/2}}$$

$$\tilde{g}_{j,l} = \frac{g_{j,l}}{2^{j/2}}.$$

The MODWT wavelet coefficients $\tilde{w}_{j,t}$ are associated with generalized changes of the data on a scale $\lambda_j = 2^{j-1}$. With regard to the spectral interpretation of MODWT wavelet coefficients, as the MODWT wavelet filter $h_{j,l}$ at each scale j approximates an ideal high-pass with passband $f \in [1/2^{j+1}, 1/2^j]$,[35] the λ_j scale wavelet coefficients are associated to periods $[2^j, 2^{j+1}]$.

MODWT provides the usual functions of the DWT, such as multiresolution decomposition analysis and variance analysis based on wavelet transform coefficients, but unlike the classical DWT it

- can handle any sample size;
- is translation invariant, as a shift in the signal does not change the pattern of wavelet transform coefficients;
- provides increased resolution at coarser scales.[36]

[33] The DWT coefficients may be considered a subset of the MODWT coefficients. Indeed, for a sample size power of two the MODWT may be rescaled and subsampled to obtain an orthonormal DWT.

[34] Whereas DWT filters have unit energy, MODWT filters have half energy, that is $\sum_{l=0}^{L-1} \tilde{h}_{j,l}^2 = \sum_{l=0}^{L-1} \tilde{g}_{j,l}^2 = \frac{1}{2^j}$.

[35] On the other hand at scale λ_J the scaling filter $g_{J,l}$ approximates an ideal low-pass filter with passband $f \in [0, 1/2^{j+1}]$.

[36] Unlike the classical DWT which has fewer coefficients at coarser scales, it has a number of coefficients equal to the sample size at each scale, and thus is over-sampled at coarse scales.

In addition, MODWT provides a larger sample size in the wavelet variance and correlation analyses and produces a more asymptotically efficient wavelet variance estimator than the DWT.

In addition to the features stated above wavelet transform is able to analyze the variance of a stochastic process and decompose it into components that are associated to different time scales. In particular, given a stationary stochastic process $\{X\}$ with variance σ_X^2 and defined the level j wavelet variance $\sigma_X^2(\lambda_j)$, the following relationship holds

$$\sum_{j=1}^{\infty} \sigma_X^2(\lambda_j) = \sigma_X^2$$

where $\sigma_X^2(\lambda_j)$ represent the contribution to the total variability of the process due to changes at scale λ_j. This relationship says that wavelet variance decomposes the variance of a series into variances associated to different time scales.[37] By definition, the (time independent) wavelet variance for scale λ_j, $\sigma_X^2(\lambda_j)$, is defined to be the variance of the j-level wavelet coefficients

$$\sigma_X^2(\lambda_j) = var\{\tilde{w}_{j,t}^2\}.$$

As shown in Percival (1995), provided that $N - L_j \geq 0$, an unbiased estimator of the wavelet variance based on the MODWT may be obtained, after removing all coefficients affected by the periodic boundary conditions,[38] using

$$\tilde{\sigma}_X^2(\lambda_j) = \frac{1}{\tilde{N}_j} \sum_{t=L_J}^{N} \tilde{w}_{j,t}^2$$

[37] The wavelet variance decomposes the variance of certain stochastic processes with respect to the scale $\lambda_j = 2^{j-1}$ just as the spectral density decompose the variance of the original series with respect to frequency f, that is

$$\sum_{j=1}^{\infty} \sigma_X^2(\lambda_j) = varX = \int_{-1/2}^{1/2} S_X(f)df$$

where $\sigma_X^2(\lambda_j)$ is wavelet variance at scale λ_j and $S_{(.)}$ is the spectral density function.

[38] As MODWT employs circular convolution, the coefficients generated by both beginning and ending data could be spurious. Thus, if the length of the filter is L, there are $(2^j - 1)(L - 1)$ coefficients affected for 2^{j-1}-scale wavelet and scaling coefficients, while $(2^j - 1)(L - 1) - 1$ beginning and $(2^j - 1)(L - 1)$ ending components in 2^{j-1}-scale details and smooths would be affected (Percival and Walden 2000).

where $\tilde{N}_j = N - L_j + 1$ is the number of maximal overlap coefficients at scale j and $L_j = (2_j - 1)(L - 1) + 1$ is the length of the wavelet filter for level j.[39] Thus, the jth scale level j wavelet variance is simply the variance of the nonboundary or interior wavelet coefficients at that level (Percival 1995; Serroukh et al. 2000). Both DWT and MODWT can decompose the sample variance of a time series on a scale-by-scale basis via its squared wavelet coefficients, but the MODWT-based estimator has been shown to be superior to the DWT-based estimator (Percival 1995).

Starting from the spectrum $Sw_{X,j}$ of the scale j wavelet coefficients it is possible to determine the asymptotic variance V_j of the MODWT-based estimator of the wavelet variance (covariance) and construct a random interval which forms a $100(1-2p)\%$ confidence interval.[40]

The formulas for an approximate $100(1 - 2p)\%$ confidence intervals MODWT estimator robust to non-Gaussianity for $\tilde{\sigma}^2_{X,j}$ are provided in Gençay et al. (2002).[41]

Similar to their classical counterparts, we can define the wavelet covariance between two processes X and Y at wavelet scale j as the covariance between scale j wavelet coefficients of X and Y, that is $\gamma_{XY}(\lambda_j) = Cov(w^X_{j,t} w^Y_{j,t})$, and the wavelet correlation between two time series $\rho_{XY}(\lambda_j)$ as the ratio of the wavelet covariance, $\gamma_{XY}(\lambda_j)$, and the square root of their wavelet variances $\sigma_X(\lambda_j)$and $\sigma_Y(\lambda_j)$ (see Whitcher et al. 1999, 2000). The wavelet correlation coefficient $\rho_{XY}(\lambda_j)$ provides a standardized measure of the relationship between the two processes X and Y on a scale-by-scale basis and, as with the usual correlation coefficient between two random variables, $|\tilde{\rho}_{XY}(\lambda_j)| \leq 1$. Specifically, given the unbiased estimators of the wavelet variances, $\tilde{\sigma}_X(\lambda_j)$ and $\tilde{\sigma}_Y(\lambda_j)$, and covariance, $\tilde{\gamma}_{XY}(\lambda_j)$, the unbiased estimator of the wavelet correlation for scale j, $\tilde{\rho}_{XY}(\lambda_j)$, may be obtained by

$$\tilde{\rho}_{XY}(\lambda_j) = \frac{\tilde{\gamma}_{XY}(\lambda_j)}{\tilde{\sigma}_X(\lambda_j)\tilde{\sigma}_Y(\lambda_j)}. \tag{6}$$

---

[39]The quantity estimated in equation (2) is time-independent even in case of nonstationary processes but with stationary dth-order differences, provided that the length L of the wavelet filter is large enough to make the wavelet coefficients $\tilde{w}_{j,t}$ a sample of stationary wavelet coefficients (Serroukh et al. 2000). This is because Daubechies wavelet filters may be interpreted as generalized differences of adjacent averages and are related with difference operator (Whitcher et al. 2000).

[40]For a detailed explanation of how to construct the confidence intervals of wavelet variance, see Gençay et al. (2002), pp. 254–256.

[41]The empirical evidence from the wavelet variance suggest that $N_j = 128$ is a large enough number of wavelet coefficients for the large sample theory to be a good approximation (Whitcher et al. 2000).

References

Aguiar-Conraria, L., Azevedo, N., & Soares, M. J. (2008). Using wavelets to decompose the time-frequency effects of monetary policy. *Physica A: Statistical Mechanics Applications, 387*(12), 2863–2878.

Aguiar-Conraria, L., & Soares, M. J. (2011). Oil and the macroeconomy: Using wavelets to analyze old issues. *Empirical Economics, 40*(3), 645–655.

Balakrishnan, R., Danninger, S., Elekdag, S., & Irina, T. (2009). The transmission of financial stress from advanced to emerging economies. IMF Working Papers 09/133. Washington: International Monetary Fund.

Berg, A., & Catherine, P. (1999). Predicting currency crises: The indicators approach and alternative. *Journal of International Money and Finance, 18*, 561–586.

Berti, K., Salto, M., & Mathieu, L. (2012). An early-detection index of fiscal stress for Eu countries. Economic Papers, 475. Brussels: European Commission.

Brave, S., & Butters, R. A. (2012). Detecting early signs of financial instability. Chicago Fed Letter No. 305. Chicago: The Federal Reserve Bank of Chicago.

Cardarelli, R., Elekdag, S., & Lall, S. (2009). Financial stress, downturns, and recoveries. IMF Working Papers 09/100, Washington: International Monetary Fund.

Chamon, M., & Crowe, C. W. (2013). Predictive indicators of financial crises. In G. Caprio (Ed.), *The evidence and impact of financial globalization*. San Diego: Academic Press.

Christensen, I., & Fuchun, L. (2013). A semiparametric early warning model of financial stress events. Working Paper 2013–13. Ottawa: Bank of Canada.

Coifman, R. R., & Donoho, D. L. (1995). Translation-invariant De-noising. In A. Antoniadis & O. Georges (Eds.), *Wavelets and statistics*. New York: Springer.

Daubechies, I. (1992). *Ten lectures on wavelets*. CBSM-NSF Regional Conference Series in Applied Mathematics, Vol. 61. Philadelphia: SIAM.

Gallegati, M. (2014a). Early warning signals of financial stress: A "wavelet-based" composite indicators approach. In S. Frauke (Ed.), *Advances in non-linear economic modeling: theory and applications*. Berlin: Springer.

Gallegati, M. (2014b). Making leading indicators more leading: A wavelet-based method for the construction of composite leading indexes. *OECD Journal of Business Cycle Measurement and Analysis, 1*, 67–84.

Gallegati, M., Gallegati, M., Ramsey, J. B., & Semmler, W. (2011). The US wage Phillips curve across frequencies and over time. *Oxford Bulletin of Economics and Statistics, 73*(4), 489–508.

Gallegati, M., & Ramsey, J. B. (2012). Errors-in-variables and the wavelet multiresolution approximation approach: A Monte Carlo study. In B. H. Baltagi, R. C. Hill, W. K. Newey & H.L. White (Eds.), *Essays in honor of Jerry Hausman*. Advances in Econometrics, Vol. 29. Bingley, UK: Emerald Group Publishing Limited.

Gallegati, M., & Ramsey, J. B. (2013). Bond vs stock market's Q: Testing for stability across frequencies and over time. *Journal of Empirical Finance, 24*, 138–150.

Gallegati, M., & Ramsey, J. B. (2014). How *forward looking* are equity and bond markets for aggregate investments? *Journal of Economics and Business, 75*, 1–24.

Gallegati, M., Ramsey, J. B., & Semmler, W. (2014). Time scale analysis of interest rate spreads and output using wavelets. *Computational Statistics & Data Analysis, 76*, 283–290.

Gençay, R., Selçuk, F., & Whitcher, B. (2002). *An introduction to wavelets and other filtering methods in finance and economics*. San Diego: Academic Press.

Hakkio, C. S., & Keeton, W. R. (2009). Financial stress: What is it, how can it be measured, and why does it matter? *Federal Reserve Bank of Kansas City Economic Review*, Second Quarter, pp. 5–50.

Hatzius, J., Hooper, P., Mishkin, F., Schoenholtz, K. L., & Watson, M. W. (2010). Financial conditions indexes: A fresh look after the financial crisis. NBER Working Paper No. 16150. Cambridge, MA: National Bureau of Economic Research.

Illing, M., & Liu, Y. (2006). Measuring financial stress in a developed country: An application to Canada. *Journal of Financial Stability, 2*(3), 243–265.

Kaminsky, G. L. (1998). Currency and banking crisis: A composite leading indicator. Working Paper, Board of Governors of the Federal Reserve System.

Kaminsky, G. L., & Reinhart, C. M. (1999). The twin crises: The causes of banking and balance-of-payments problems. *American Economic Review, 89*(3), 473–500.

Kappler, M., & Schleer, F. (2013). How many factors and shocks cause financial stress? Discussion Paper No. 13–100. Mannheim: ZEW.

Kim, S., & In, F. H. (2003). The relationship between financial variables and real economic activity: Evidence From spectral and wavelet analyses. *Studies in Nonlinear Dynamics & Econometrics, 7*(4), 1–16.

Kumar, P., & Foufoula-Georgiou, E. (1997). Wavelet applications in geophysics. *Reviews of Geophysics, 35*(4), 385–412.

Lynch, P. E., & Zumbach, G. O. (2003). Market heterogeneities and the causal structure of volatility. *Quantitative Finance, 3*(4), 320–331.

Mallat, S. (1989). A theory for multiresolution signal decomposition: The wavelet representation. *IEEE Transactions on Pattern Analysis and Machine Intelligence, 11*, 674–693.

Masset (2015). Analysis of financial time series using wavelet methods. In C. F. Lee & J. C. Lee (Eds.), *Handbook of Financial Econometrics and Statistics*. New York: Springer.

Misina, M., & Tkacz, G. (2009). Credit, asset prices, and financial stress. *International Journal of Central Banking, 5*(4), 95–122.

Muller, U. A., Dacorogna, M. M., Dave, R. D., Pictet, O. V., Olsen, R. B., & Ward, J. (1995). Fractals and intrinsic time - a challenge to econometricians. Working Papers 16 Aug 1993. Olsen and Associates. URL http://ideas.repec.org/p/wop/olaswp/_009.html.

Nason, G. P., & SIlverman, B. W. (1995). The stationary wavelet transform and some statistical applications. In A. Anestis & G. Oppenheim (Eds.), *Wavelets and statistics*. New York: Springer.

Percival, D. B. (1995). On estimation of the wavelet variance. *Biometrika, 82*(3), 619–631.

Percival, D. B., & Walden, A. T. (2000). *Wavelet methods for time series analysis*. Cambridge: Cambridge University Press.

Ramsey, J. B. (2002). Wavelets in economics and finance: Past and future. *Studies in Nonlinear Dynamics & Econometrics, 6*(3), 1–29.

Ramsey, J. B., Gallegati, M., Gallegati, M., & Semmler, W. (2010). Instrumental variables and wavelet decomposition. *Economic Modeling, 27*(6), 1498–1513.

Ramsey, J. B., & Lampart, C. (1998a). The decomposition of economic relationships by time scale using wavelets: Money and income. *Macroeconomic Dynamics, 2*(1), 49–71.

Ramsey, J. B., & Lampart, C. (1998b). The decomposition of economic relationship by time scale using wavelets: Expenditure and income. *Studies in Nonlinear Dynamics & Econometrics, 3*(1), 23–42.

Serroukh, A., Walden, A. T., & Percival, D. B. (2000). Statistical properties of the wavelet variance estimator for non-Gaussian/non-linear time series. *Journal of the American Statistical Association, 95*(449), 184–196.

Shin, H. S. (2013). Procyclicality and the search for early warning indicators. IMF Working Paper no. 258. Washington: International Monetary Fund.

Slingenberg, J. W., & de Haan, J. (2011). Forecasting financial stress. DNB Working Paper No. 292. Amsterdam: De Nederlandsche Bank.

Stock, J. H., & Watson, M. P. (1999). Business cycle fluctuations in US macroeconomic time series. In J. B. Taylor & M. Woodford (Eds.), *Handbook of macroeconomics*. Amsterdam: Elsevier North-Holland.

Whitcher, B., Guttorp, P., & Percival, D. (1999). Mathematical back- ground for wavelet estimators of cross-covariance and cross-correlation. Technical Report 38, National Research Center for Statistics and the Environment, Boulder, Seattle, USA.

Whitcher, B., Guttorp, P., & Percival, D. B. (2000). Wavelet analysis of covariance with application to atmospheric series. *Journal of Geophysics Research Atmosphere, 105*(D11), 14941–14962.

Corporate Liquidity under Financial Constraints and Macroeconomic Uncertainty

Daniel Samaan and Immo Schott

1 Introduction

Brealey and Myers (2002) and Kim et al. (1998) have referred to the value of liquidity as one of the ten unsolved problems in finance. Indeed, no comprehensive explanation exists as to why and when firms clinge to large amounts of cash, an asset with zero nominal and mostly negative real return. Advances in information technologies as well as computerized management accounting systems have facilitated financial transactions and cash management. People can pay with smart phones or credit cards. Firms can track worldwide sales, inventory and their cash flows practically in real-time. Together with more sophisticated financial instruments and interlinked global banking systems these factors should have significantly reduced the amount of cash held by firms. However, the opposite is the case: The slogan "Cash is dead! Long live cash" (see Williams 2012) is not only valid for the U.S. economy. Our analysis shows a secular increase of average cash holdings by non financial corporations for most regions and countries across the World (see Fig. 1). While differences exist in terms of their magnitude and their specific progression over time, the main trend of cash-to-asset ratios appears to be similar for all regions except for India.

D. Samaan (✉)
Research Department, International Labor Organization, Geneva, Switzerland
e-mail: samaan@ilo.org

I. Schott
Department of Economics, University of Montreal, Montreal, QC, Canada
e-mail: immo.schott@umontreal.ca

L. Bernard, U. Nyambuu (eds.), *Dynamic Modeling, Empirical Macroeconomics, and Finance*, DOI 10.1007/978-3-319-39887-7_10

Fig. 1 Median cash holdings over time

In the U.S., for example, the cash ratio increases from about 0.04 in 1990 to about 0.12 in the late 2000s.[1] Very similar ratios and evolutions of ratios can be observed for other advanced countries, like Germany or Canada. In the UK, a significant increase in cash holdings can be observed but changes in cash ratios of British firms tend to be less strong, starting at about 0.055 in 1990 and peaking at around 0.095 in 2005. The cash ratios of Chinese firms tend to be higher and increase from about 0.1 in 1995 to about 0.2 in 2010.[2] It also seems to be the case for most regions that cash ratios have come down from their overall peaks during the economic crisis. The evolution of cash ratios for the world as a whole shows a similar pattern to the individual regions. Main differences can only be found in the Indian sample where cash ratios remain basically flat in the long term and fluctuate around a ratio of about 0.03.

As can be seen from Fig. 1, cash holdings have risen to unprecedented levels with the outbreak of the financial and economic crisis in 2007 and 2008. Using the median as a notion of average has the advantage of being more robust towards

[1]Note: These ratios are comparable with findings in Bates et al. (2009), see Table I, p. 1991.

[2]We ignore the first observations for China due to the small sample size. For years 1990–1992 only between six and ten firms report cash holdings and we consider this number too small to be representative. After 1995 the sample size in China reaches 80–100 firms and increases to more than 1000 firms after 1999.

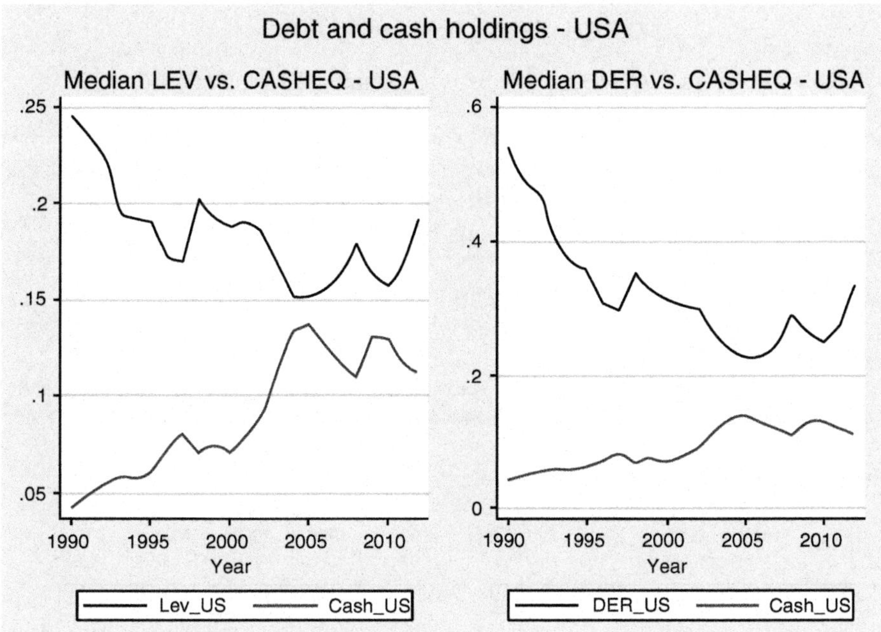

Fig. 2 Firm leverage and cash holdings, 1990–2012

outliers, but the evolvement of cash holdings in terms of means gives a similar picture (see for example Figs. 2 and 3 for a comparison of means and medians in the United States). Means are generally higher than medians but the shape of the curves look very similar. This rise of cash holdings during and in the aftermath of the crisis also sparked new interest in the possible motives of firms' liquidity demand (see for example Bates et al. 2009), and in particular in the relation of cash holdings and firms' debt (see Acharya et al. 2007; Denis and Sibilkov 2010; Acharya et al. 2012). In this article, we argue that firms seek to hold cash because of certain "market frictions" (Stiglitz and Weiss 1984): they are exposed to an unknown future about their sales and about (re-) financing. Cash holdings then reflect anticipated uncertainty in the macroeconomy, as well as uncertainty on the firm and industry level concerning business prospects and financing possibilities. At the same time, such individual firm behavior can have undesirable results on the aggregate level. Hoarding large asset shares in the form of cash can lead to lower aggregate demand when financial assets are not eventually transformed into real investment (see Fazzari et al. 1988).

One of the first theoretical treatments of economic motives for holding cash dates back to Keynes (1936) who distinguished the transaction motive, the precautionary motive and the speculative motive as differing factors for economic agents' desire to hold cash and to demand liquidity. Cash held for transaction purposes serves to

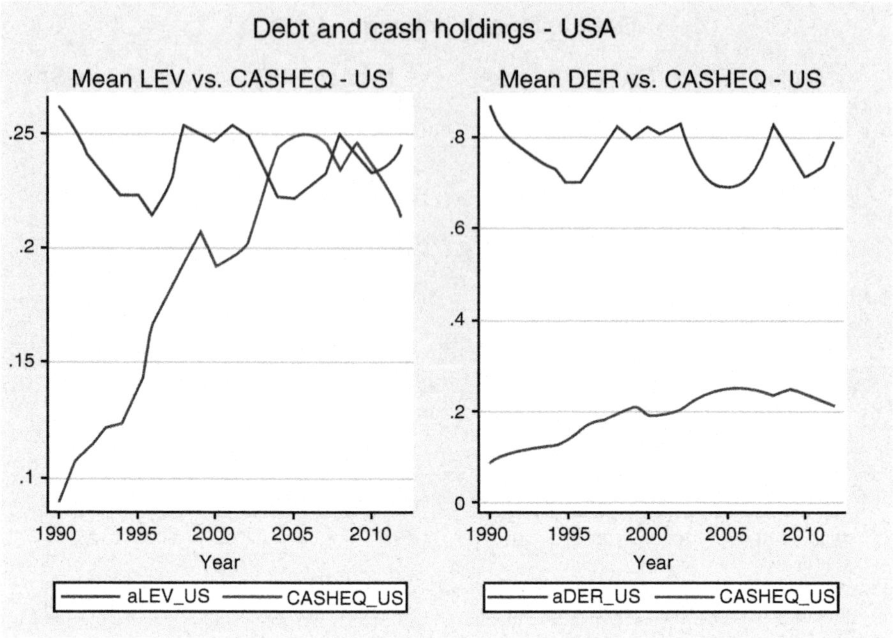

Fig. 3 Firm leverage and cash holdings, 1990–2012

bridge the interval between the receipt of income and its disbursement. In the context of business operations, the transaction motive arises from the time interval between incurring normal operating costs and sales receipts. Baumol (1952) showed in a formal setting that the optimal (cost minimizing) demand for cash for transaction purposes depends on the anticipated value of transactions per time period, and transaction costs like banking fees and the interest rate. The interest rate can either be an actual cost incurred or be an opportunity cost like a foregone alternative investment.

The other two motives arise from the fact that agents have to take decisions in an environment of uncertainty. Keynes sees the purpose of the precautionary motive in providing for contingencies requiring sudden expenditures and for unforeseen opportunities of advantageous purchases. Holding cash has also the advantage that its nominal asset value in terms of money can never decline in the future, which is not the case for any other asset. The precautionary motive according to Keynes captures both, the desire to wait for better investment opportunities in the future and the fear of unforeseen negative events. Miller and Orr (1966) expand the Baumol model by introducing stochastic operational cash flows. In this case, the (known) mean and variance of operational cash flows influence the optimal amount of cash holdings. The Miller-Orr Model covers some aspects of the precautionary motive but remains closely linked to the transaction motive since the stochastic shocks on the cash balance are i.i.d. shocks with known mean and variance. The speculative motive in the sense of Keynes has a very specific meaning and refers to motives of bond market investors' anticipation of future bond prices and aggregate money supply. It appears to us that this motive does largely not apply to financial management considerations of non-financial corporations and will therefore be disregarded.

Clearly, in the absence of any market frictions, perfect foresight of economic agents, and a positive interest rate, none of these motives is convincing. Companies would reduce their liquidity to zero. It follows in reverse, that the demand for cash arises from transaction costs, uncertainty or risk about the future, or from principal agency relations and information asymmetries that lead to agency costs, or from a combination of these factors. Relevant principal agency problems in this context concern the relation between managers and shareholders, whereby the former may have an incentive to keep more liquidity under their control than would be desirable for shareholders. Jensen (1986) called this agency costs of free cash flows and suggested high leverage and high dividend payout ratios could reduce agency costs from excessive cash holdings. Opler et al. (1999)[3] speak of agency costs of managerial discretion.

Another principal agency relationship that influences cash holding behavior concerns borrowers and creditors. If financing is provided by creditors they face a credit risk which arises from the principal-agency relationship and the ignorance of the contracting party's future actions. Creditors may therefore decide to restrain or even freeze credit which may bring firms into situations in which debt financing is restricted or not available. Note that such refinancing risk for companies exists in *addition* to the company's operational risk and solely arises from the information asymmetries between creditors and debtors. Only the operational risk is a true social cost. (An entrepreneur without external creditors does not face that type of risk.) But both types of risk can be a driver for cash holdings of firms. The exact reasons for imposed credit constraints are manifold and can range from problems in the financial system like an overall loss of confidence to firm specific factors. Jensen and Meckling (1976) and Myers (1977) discuss agency cost of debt and their implications for investment and the capital structure of firms.

While we separate credit risk from operational risk conceptionally, it is of course clear that the two are linked and, for example, higher business risk can lead to higher credit risk: Lower sales of a company's products may change its actual or perceived willingness to meet its debt obligations. These considerations of credit risk apply to external equity financing analogously and could be more broadly subsumed under the term "external financing risk".

Based on the previous theoretical considerations, we can broadly distinguish three theories of cash holdings and financial structure. In the first view, there exist a marginal benefit and marginal cost of cash holdings, basically built on the argument of transaction and debt financing costs. Internal financing is cheaper but liquidity does not bear a return so that there is a tradeoff between holding much and holding

[3]OPSW.

little cash. Opler et al. (1999) refer to this theory as the "transaction cost model". A second view, not necessarily conflicting with the transaction cost model, emphasizes the agency costs of managerial discretion. In both cases, firms would have a target cash ratio, maximizing shareholder value in the first case, and higher than optimal cash ratios in the second case (maximizing the utility of management).

An alternative theory is the pecking order or financial hierarchy model. Information asymmetries between managers and investors lead to a preference of internal funds over debt and of debt over equity as means of financing (Myers and Majluf 1984). According to this view, the issuance of stock can be interpreted as a signal to old and new investors and lead to adverse selection. In the pecking order model, firms adjust their leverage (net debt[4] exogenously given capital expenditures. There is no optimal amount of cash. Firms finance investment with internally created cash and accumulate any surplus to repay debt when due. If the internally created cash is insufficient to finance investment the firm issues debt, assuming it can always do so at low cost.

Since none of the two motives for liquidity preference (transaction motive and precautionary motive) nor agency costs are directly observable, we use firm level accounting data and ratio analysis to infer on the magnitude of each of the above discussed factors.

2 Data, Sample Selection and Construction of Uncertainty Indicators

2.1 Overview of Main Variables

We construct our sample of firms from the *FactSet*-Database[5] using observations for the period 1990–2012. The database consists of financial accounting and complementary firm level data on publicly listed companies from a large group of countries. Our sample includes surviving and non-surviving firms. From the available data, we construct a panel with the following variables that are subsequently employed in the econometric analysis. We can thus index each variable by an i for the individual firm ($i = 1, \dots, N$) and by a t for the time period ($t = 1990, \dots, 2012$). To simplify the notation we usually drop the index unless required for purposes of clarity.

We measure our main variable of interest, firm liquidity, as cash and cash equivalents (CASHEQ). This is a broader concept of liquidity than cash only (CASH) and is typically employed in the literature.[6] For numbers to be comparable across firms we calculate all variables as shares of total assets unless otherwise

[4]Net debt equals debt minus cash.

[5]http://www.factset.com/.

[6]On the macroeconomic level CASHEQ would roughly correspond with M1 while CASH would be the equivalent of M0. The two measures are also highly correlated (0.77).

stated. Firms' investment behaviour is captured by several stock and several flow variables stemmin from firms' balance sheets, as well as income and cash flow statements. Annual investment behaviour and growth opportunities can be measured through capital expenditures (CAPEX), R&D expenses (RD), and the market-to-book ratio (MTB). R&D expenditures can be interpreted as investments into future opportunities (see for example Opler et al. 1999). Opler and Titman (1994) point out that R&D expenditures as a form of investment typically bring about high information asymmetries between management and capital providers. Accordingly, and because R&D activities result mostly in intangible assets (if at all) which can hardly be used as collateral, investments into R&D are more costly to finance via external capital than ordinary capital expenditures (see also Bates et al. 2009). The market-to-book value is a proxy for Tobin's q and indicates growth opportunities for the firm, i.e. the likelihood that a firm will have positive net present value projects in the future (see for example Jung et al. 1996 or Smith and Watts 1992). Everything else equal, the opportunity costs of a lack of liquidity for such firms would by higher than for low-growth firms. Most authors suggest that a company may consider non-cash liquid assets as substitutes for cash since they can be converted into cash quickly at relatively low cost (see for example Ozkan and Ozkan 2004 or Bates et al. 2009). Measures of such fungible assets are the value of current assets (CA), consisting of cash, financial investments, accounts receivable and inventory, or working capital (WK), which is equal to current assets minus current liabilities, or liquidity (LIQ), equaling working capital minus cash only or inventories (INV).

Several variables are available to measure the degree of indebtedness of firms, in particular leverage (LEV), debt-equity-ratio (DER) and equity (EQ). The three variables are highly correlated (> 0.65). Based on the available number of observations and our preference to express all balance sheet variables on the same scale, we choose LEV as our measure of indebtedness. Retained earnings (Ret) are a subaccount of equity in which undistributed profits are accumulated. We ignore this latter variable in our analysis. A non balance sheet variable of indebtedness is a company's rating (RAT). It includes an evaluation of firms' credit risk. Since RAT is only available for a limited number of firms, and mainly only for large firms, we exclude this variable from further analysis. An indicator of a firm's cost of debt financing its the effective interest rate (EIR), calculated as current interest expense on debt divided by the sum of short-term and long-term debt.

Cash distributions to shareholders are captured by the four variables dividends over total assets (DIV), dividend payout ratio (DPR), dividend yield (DY) and share-buy-backs (SBB). In our analysis, we regard such occurrences as a sign of available financial means and thus an indicator of a financially unconstrained firm. DIV and DPR are positively correlated and we prefer DIV since it is measured in terms of total assets. DY includes the stock market valuation of a firm's shares, which is of less interest to us. SBB, as an alternative form of cash distributions, are only used by a small proportion of companies and the variable is therefore discarded.

We calculated several variables to grasp the profitability of a firm: The profit rate (PR), return on assets (ROA) and the operating profit margin (OPM). The main relation to cash holdings that we see is that profitability can signal to outside financiers the firm's potential to pay back loans or equity. We use ROA as one component to construct a variable that indicates financial constraints. Another possible use for these variables is that high profitability combined with high cash holdings may be an indication of agency costs (The Jensen argument).

Two measures for the cash flow are available, cash flow from operations (CFO) is provided from the FactSet Database and is equal to funds from operations minus extraordinary items minus funds from changes in working capital ("net cash from operating activities"). Funds from operations are derived from company's cash flow statements and are typically calculated as: net Income plus depreciation plus deferred taxes plus other funds. We also measure cash flow (CF) in a broader sense by adding depreciation to net income.

Firm size can be measured through the number of employees (EMP) or the logarithm of total assets (LTA) measured in real 2000 dollars. Since LTA is much more broadly available we focus on this measure of company size. The age of a firm can be proxied by the number of years passed since its initial public offering (IPO). Strictly speaking IPO does not measure the overall age of the firm but the time since its first listing on capital markets. Both variables are likely to be correlated though. IPOd is a dummy variable that indicates whether the firm's IPO has occurred within the last 5 years. Unsurprisingly, Bates et al. (2009) show that cash holdings of firms with a recent IPO are higher.

Another rational for cash holdings that has been brought up in the literature is anticipated tax payments (Foley et al. 2007). We collected information on overall taxes paid (TAX) and foreign taxes paid (FTAX). Related to these two variables is foreign income (FINC) that can be used to distinguish domestic income from income abroad. We basically ignore these variables since we are not convinced by the argument and do not find empirical evidence for it.

2.2 Uncertainty and Risk: Idiosyncratic, Industry-Wide, and Aggregate Risk

The economics literature tends to distinguish between risk and uncertainty, whereby "risk" is usually referred to a situation of uncertainty faced by an economic agent in which subjective or objective probabilities can be assigned to future states of the world (see e.g. Savage 1954). A situation in which no such probabilities can be assigned to possible future states of the word is generally referred to as a decision under "uncertainty". We will not engage in the discussion whether or not most unknown future economic events should be regarded as "risk" or as happening in an environment of fundamental uncertainty as Keynes (1936) suggested. We consider "uncertainty" and "risk" as synonyms in this article and broadly understand risk as firms' exposure to an known future concerning key economic variables of interest.

We propose a range of variables to measure the risk to which firms are exposed: One broad, and commonly used indicator of a firm's risk is the volatility of its cash flow (stdCF). The firms' main purpose can be reduced to the creation of cash flows that can be distributed to stakeholders. Higher volatility of these cash flows are a measure of the firms exposure to business risks. We use a firm's annual revenue (REV) to construct several additional measures of risk. We first compute the firm-specific standard deviation of cash flows. This is done over the whole sample and as a rolling 3-year period. The variables are called stdCF and stdCF3. Following the literature on investment and dynamic factor adjustment (see Cooper and Haltiwanger 2006), we can estimate firm-level production functions to recuperate an idiosyncratic shock component from our data. We assume that the production function is of the form $Y_{it} = A_{it}K_{it}^{\alpha}L_{it}^{\gamma}$, where $A_{it} = A_t \cdot \varepsilon_{it}$, consists of aggregate components and a firm-specific one. We use the value of total assets as a firm's capital stock input and the total number of employees for our labor input. Our output measure is revenue. All variables are in logs.We carry out a GMM estimation with complete time dummies where we instrument for capital with its first lag we compute the residuals, which correspond to A_{it} (in logs). We then decompose the residual into the idiosyncratic and aggregate component. This step amounts to measuring the aggregate shock as the mean of A_{it} for each year. The idiosyncratic shock component is identified as the deviation from the year-specific mean. As an alternative specification we allow for industry-specific shocks and measure the industry-wide aggregate shock as the mean of A_{it} for each year and each industry.[7] The idiosyncratic shock is then measured as the deviation from the year-industry mean.

For each firm we now compute the standard deviation of the idiosyncratic shock component over time. This is the variable called riskidio. The variable riskidio3 computes a 3-year rolling standard deviation. We can use the aggregate component of the shock to back out industry-specific risk. We devise several measures. The first one simply computes the annual standard deviation of the idiosyncratic risk of firms within the industry (riskindustry). The second measure uses the annual mean of the idiosyncratic risk (riskindustry2). The third is based on the average of the rolling standard deviation of cash flows.

To obtain an aggregate risk measure we compute the standard deviation of GDP growth for each country. GDP figures were obtained as PPP adjusted from the IMF.

Since we are particularly interested in the relationship of cash holdings and risk, we also construct an alternative risk measure following Campbell et al. 2001.

2.3 *Financial Constraints Indicator*

We use an indicator function to classify firms into financially constrained and unconstrained firms. We follow many authors (see for example Almeida et al. 2009 or Fazzari et al. 1988) by assuming that non-payment of dividends is an indication of "financial constraints". The rationale is that firms in need of liquidity would

[7] An industry is classified by the first two digits of the industry code.

not distribute cash from funds within the firm. However, a non-dividend paying firm is not necessarily constrained as high-growth firms also tend to pay no or low dividends as long as they have sufficient investment opportunities available. Many fast growing firms in the technology sector, for example Apple, have paid no or little dividends in the past—without being constrained in the sense that they may have problems obtaining external financing. In order to exclude such firms, we create a variable FINCSTR which takes the value of one in a firm-year if and only if the pay-out is zero and the return on assets (ROA) is non-positive.

Financial constraints, measured in this sense, do not necessarily mean that a firm is unable to obtain new credit or new equity financing. Our variable FINCSTR is—more broadly—an indicator for firms in which management is concerned about "not having enough liquidity" at its disposal to carry out current and future activities. Splitting the population into constrained and non constrained firm-years and taking averages reveals the following descriptive results (see Fig. 5).

2.4 The Sample of 'Normal Firms'

In order to construct a feasible panel of 'normal firms', we clean the data in the following manner: In a first step we exclude all financial firms. This is done because we suspect that financial firms hold cash for different reasons than non-financial firms, e.g. reserve requirements.This constitutes a standard procedure in the literature. We then apply a routine to detect duplicates in the FactSet Database, that is firms that are listed on more than one stock market or that enter the database twice with variations in the name. We exclude observations for cash ratios (to total assets) higher than 1 as these values are clearly implausible. Importantly, we require that firms have at least four consecutive time series observations of several key variables during the sample period.We find that during this step also a large amount of extreme outliers are removed from database. Companies that are listed for only very few consecutive periods often do not represent 'normal' firms.

In a next step, we follow Baum et al. (2006) and remove extreme outliers from the database where these can with a high probability be attributed to mergers and acquisitions. Specifically, firms that have experienced a very high change in their balance sheet totals at any time during the sample period are excluded. For each country we compute the distribution of year-to-year growth in total assets and remove firms below the 1 % and above the 99 % quartiles. Baum et al. (2006) remove a much larger fraction of their data via this procedure than we do (they cut off at the 10 % and 90 % values). On the other hand, they do not require that a firm be in the sample several consecutive times. Our decision to set the cutoff values at the 1st, respectively 99th percentile is motivated by an analysis of those firms that would not remain in the dataset had we done the same type of correction as Baum et al. (2006). The firms that would be removed show a very different composition than the overall sample. In particular, many of the to-be-removed firms belong to the technology sector, and are smaller than the remaining firms. They show much higher cash holdings (about 24 % vs. 16 %). A closer look at individual firms from this subset furthermore reveals that asset growth often exceeds 50 % p.a. but that the absolute values of total assets are small, rendering this much more plausible. In short, if we were applying the exact same criterion as Baum et al. (2006) we feel that we would be introducing a significant sample-bias.

A last step in the correction concerns deviations in a firm's cash holdings over time within each country. If these vary by more than three standard deviations from the mean cash holdings in that country, the firm is removed from our sample. The summary statistics of the full sample after applying all cleaning procedures ("Modified Baum Correction") are listed in Tables 8–14). It is possible that country-specific business practices, differing legal and accounting frameworks and varying overall macroeconomic conditions lead to differences in firms' cash holding strategies. We therefore present the summary statistics of the following (sub-) panels:

1. World (full sample)
2. United States
3. United Kingdom
4. European Union (27)
5. G7
6. China
7. India

3 Determinants of Firms' Cash Holdings

3.1 Pooled OLS Regressions with Cluster-Robust Standard Errors

Before turning to a detailed evaluation of the role of financial constraints and debt, we evaluate empirically which factors determine firms' cash holdings. We employ a variety of panel regressions to determine which financial variables and firm characteristics are important. Using $CASHEQ_t$ as the dependent variable, we largely follow the literature on the "transaction cost model" (Opler et al. 1999), but we add our set of risk variables and the financial constraints variable as dependent variables. We keep most of the variables that are usually identified as the determinants of cash holdings in the literature on the transaction cost model (see also Bates et al. 2009) as control variables.

Table 1 shows the results of various ordinary least squared (OLS) models whereby we regress our risk variables on CASHEQ, subsequently adding control variables and using the whole data set. The results of the six models show that

Table 1 OLS regression comparisons

	(1)	(2)	(3)	(4)	(5)	(6)
	OLS1	OLS2	OLS2a	OLS3	OLS4	OLS4_US
CAPEX	−0.3370*** (0.0171)	−0.2626*** (0.0159)	−0.2432*** (0.0127)	−0.2411*** (0.0127)	−0.3107*** (0.0130)	−0.4365*** (0.0293)
DIV	0.2126*** (0.0534)			0.3037*** (0.0416)	0.2705*** (0.0408)	0.1546 (0.1169)
LEV	−0.3381*** (0.0070)	−0.3371*** (0.0073)	−0.3261*** (0.0059)	−0.3194*** (0.0059)	−0.3043*** (0.0057)	−0.3945*** (0.0123)
ROA	0.0598*** (0.0086)	0.0286** (0.0094)	0.0383*** (0.0071)	0.0459*** (0.0079)	0.0391*** (0.0078)	0.0581*** (0.0105)
CF	−0.1220*** (0.0076)	−0.0890*** (0.0070)	−0.0804*** (0.0060)	−0.0817*** (0.0060)	−0.0607*** (0.0059)	−0.0948*** (0.0101)
std_CF	0.2807*** (0.0267)	0.1895*** (0.0264)	0.2064*** (0.0225)	0.2076*** (0.0224)	0.1875*** (0.0223)	0.1443*** (0.0356)
MTB	0.0053*** (0.0004)	0.0063*** (0.0004)	0.0051*** (0.0003)	0.0049*** (0.0003)	0.0042*** (0.0003)	0.0048*** (0.0005)
risk_idio	0.0736*** (0.0075)	0.0851*** (0.0070)	0.0657*** (0.0061)	0.0665*** (0.0060)	0.0605*** (0.0059)	0.0876*** (0.0110)
risk_industry	0.0661*** (0.0197)	0.0287 (0.0187)	0.0645*** (0.0164)	0.0612*** (0.0164)	−0.0027 (0.0159)	0.0272 (0.0294)
RD	0.6315*** (0.0284)					
IPOd	0.0364*** (0.0032)	0.0275*** (0.0029)	0.0319*** (0.0026)	0.0324*** (0.0026)	0.0295*** (0.0025)	0.0313*** (0.0044)
EMP	−0.0002*** (0.0000)	−0.0001*** (0.0000)	−0.0001*** (0.0000)	−0.0002*** (0.0000)	−0.0002*** (0.0000)	−0.0004*** (0.0001)
DIVd		−0.0189*** (0.0025)	−0.0101*** (0.0020)			
RAT		−0.0022 (0.0024)				
RD2		0.0265*** (0.0033)	0.0178*** (0.0028)	0.0175*** (0.0027)	0.0217*** (0.0027)	0.0325*** (0.0060)
RD3		0.1258*** (0.0068)	0.1163*** (0.0060)	0.1168*** (0.0060)	0.1153*** (0.0057)	0.1233*** (0.0082)
RD4		0.2419*** (0.0119)	0.2202*** (0.0106)	0.2186*** (0.0106)	0.2133*** (0.0104)	0.2077*** (0.0126)
fin_cstr				0.0206*** (0.0028)	0.0190*** (0.0027)	0.0210*** (0.0044)
fin_cstr_crisis				−0.0124*** (0.0037)	−0.0104** (0.0036)	0.0016 (0.0058)

(continued)

Table 1 (continued)

	(1)	(2)	(3)	(4)	(5)	(6)
	OLS1	OLS2	OLS2a	OLS3	OLS4	OLS4_US
INV					−0.2405***	−0.3037***
					(0.0077)	(0.0155)
_cons	0.4372***	0.3749***	−0.0751***	0.1249***	0.1902***	0.2071***
	(0.0135)	(0.0144)	(0.0132)	(0.0087)	(0.0089)	(0.0159)
N	67,427	62,753	100,035	100,035	99,578	31,285
r2	0.4262	0.4714	0.4147	0.4162	0.4408	0.5062

Standard errors in parentheses
* $p < 0.05$, ** $p < 0.01$, *** $p < 0.001$

most variables consistently have the expected signs. In particular, investment and R&D activities lead to lower cash holdings; firms that pay dividends hold more cash (non-constraint firms), and higher leverage, i.e. more debt financing seems to lead to lower cash holdings. This latter result supports the pecking order theory that we will discuss further below. A main focus of the public debate on high corporate cash holdings has been the alleged link between private investment and cash holdings. The discussion centers on the idea that capital expenditures could be stimulated by increasing a firm's cost of holding cash. This would increase the opportunity costs of cash holdings and might channel cash reserves into then-profitable investments. The validity of such an argument relies crucially on the assumption that cash holdings are high because of a lack of investment opportunities for firms. If, on the other hand, firms hold cash because they want to avoid a situation in which they find themselves to be financially constrained a very different policy recommendation follows.

Higher operational cash flows lead on average to lower cash holdings. In line with this finding, the firm specific risk variables, cash flow volatility and the idiosyncratic risk are significant and confirm our hypothesis that cash is being held out of risk considerations. Financially constraint firms also hold more cash, confirming our second hypothesis. However, our industry risk variable is not significant in all specifications.

The presence of financial constraints emerges as one of the key explanatory factor for the changes in corporate cash holdings. As our analysis of various univariate test suggests, being a financially constrained firm in a developed country is a good predictor of a significantly higher cash-to-asset ratio. Focussing on the subsample of U.S. and Chinese firms, for example, we find that in the U.S. financially constrained firms hold more cash than unconstrained firms, while in China the opposite is true. The dynamics of firm cash holdings give us a good clue as to what is driving those differences. For the constrained firms we see a steady rise of cash reserves prior to the beginning of the financial crisis: an increase of around 45 % between 2001 and 2007. Starting in 2008, however, there is a rapid fall of cash reserves.

An interpretation that comes to mind is that these financially constrained firms burned significant amounts of cash in a period where external financing was more difficult to obtain. A very different picture emerges for the non-constrained firms.

Cash holdings actually decreased between 2004 and 2008, and began to increase only *after* the beginning of the financial crisis. This may have occurred as the result of precautionary thinking. Capital expenditures may have been deferred to a future date, the cash for which is in the meantime held in liquid form. For China there is no specific trend with the exception of an increase in cash holdings by non-constrained firms starting in 2010. The correlation between cash holdings by constrained and unconstrained groups is 0.37 for the U.S., −0.36 for U.K., −0.26 for G7, 0.17 for EU27, and 0.49 for China, 0.76 for India, 0.64 for Indonesia. So, it is generally much higher for developing countries. The difference in cash holdings between constrained and unconstrained firms for the G7 countries is positively correlated with GDP growth, 0.36. For non-G7 countries it is −0.25. For India it is −0.41 but for China it is 0.60. The U.S. has 0.57 and UK 0.70 correlation, EU27 0.24, and non-EU27 has −0.59.

CFO is now significant and has a negative sign for constrained firms in the US. For the unconstrained firms the sign is positive and it is significant for the UK. For China both numbers are positive but only significant for the unconstrained firms. A simple interpretation could be that more sales lead to more cash for China and all unconstrained firms. Constrained firms in US/UK are using the additional money for something else, e.g. paying down debt, investment or R&D activities.

INV is almost always significant and there is an important difference between countries. In the U.S. and the U.K. the sign is about 2–4 times larger for constrained firms.

The idiosyncratic risk variable is never significant for China or India. In the U.S. and the U.K. it is much larger for the constrained group. In the U.K. it is even insignificant for the unconstrained group. This is interesting because it supports our main arguments that cash holdings are primarily determined by financial constraints and firms' exposure to risk. In the U.S. and the U.K., stdCF is only significant for the unconstrained firms. This is in general similar to the variable for idiosyncratic risk variable.

3.2 OLS Regression for a Sub-sample of Advanced Countries

In light of the apparent differences between advanced and developing countries and the larger amount of available variables, we set up a similar regression as in the previous section for advanced countries only. The results are shown in Table 2 and show that, indeed, risk on the industry level is not significant, while firm specific risk and macroeconomic risk are.

Table 2 OLS regression comparisons

	(1)
	ADVolsfull
CAPEX	−0.3963***
	(0.0153)
INV	−0.3217***
	(0.0090)
DIV	0.1771**
	(0.0582)
ROA	0.0403***
	(0.0048)
fin_cstr	0.0271***
	(0.0028)
fin_cstr_crisis	−0.0148***
	(0.0038)
LEV	−0.3264***
	(0.0060)
CF	−0.0768***
	(0.0058)
std_CF	0.1778***
	(0.0193)
MTB	0.0039***
	(0.0003)
risk_idio	0.0411***
	(0.0050)
risk_industry	0.0114
	(0.0102)
risk_agg	0.0257**
	(0.0079)
RD	0.5552***
	(0.0203)
IPOd	0.0418***
	(0.0027)
EMP	−0.0003***
	(0.0000)
GDPindex	−0.0003**
	(0.0001)
_cons	0.1054
	(0.0608)
N	83,255
r2	0.4339

Standard errors in parentheses
* $p < 0.05$, ** $p < 0.01$, *** $p < 0.001$

3.3 Panel Models

We also tested the validity of our main regression results in Sect. 3.1 by exploiting the special characteristics of the panel data set. We tested the determinants of cash holdings with a variety of panel models (see Table 3) but did not find any significantly different results as from the simple OLS regressions.

3.3.1 Fixed Effects Estimator

The fixed effect estimator removes a potential firm fixed-effect by subtracting the corresponding model for individual means. This model implicitly assumes that x_{it} is correlated only with the time-invariant component of the error, α_i, and not with the time-varying component ε_{it}. This desirable property of consistent parameter estimation in the FE model is tempered, however, by the inability to estimate the coefficients or a time-invariant regressor. Also the within estimator will be relatively imprecise for time-varying regressors that very little over time. The results of the estimation can be found in Table 4. The first column labeled FE performs the regression on levels variables only. The second column labeled FE SQ also includes square terms.

3.3.2 First-Difference Estimator

Instead of mean-differencing we can also first-difference in order to obtain the within estimator. This alternative has the advantage of relying on weaker exogeneity assumptions that become important in dynamic models. Since we do not find a significant time trend in the data, we leave out the intercept of this model specification. The results can be found in columns 3 and 4 of Table 4.

3.3.3 Random Effects Estimator

This estimator is the FGLS estimator (feasible generalized least squares) for the model $y_{it} = x_{it}'\beta + (\alpha_i + \varepsilon_{it})$ under the assumption that the random effect α_i is iid (and also the idiosyncratic error, of course). It is inconsistent if the FE model is appropriate. The RE estimator can be obtained by OLS in the transformed model $(y_{it} - \hat{\theta}_i\bar{y}_i) = (1 - \hat{\theta}_i)\alpha + (x_{it} - \hat{\theta}_i\bar{x}_i)'\beta + \{(1 - \hat{\theta}_i)\alpha + (\varepsilon_{it} - \hat{\theta}_i\bar{\varepsilon}_i)\}$, where $\hat{\theta}_i$ is a consistent estimate of $\theta_i = 1 - \sqrt{\sigma_\varepsilon^2/(T_i\sigma_\alpha^2 + \sigma_\varepsilon^2)}$. The RE estimator uses both between and within variation and has special cases of pooled OLS ($\hat{\theta}_i = 0$) and within estimation ($\hat{\theta}_i = 1$). The RE estimator approaches the within-estimator as T gets large and as σ_α^2 gets large relative to σ_ε^2, because in those cases $\hat{\theta}_i \rightarrow 1$.

Table 3 Model comparison

	(1) FE	(2) FE SQ	(3) FD	(4) FD SQ	(5) RE	(6) RE SQ
LEV	−0.1276***	−0.2730***	−0.1091***	−0.2159***	−0.1425***	−0.2983***
	(0.0108)	(0.0247)	(0.0145)	(0.0287)	(0.0090)	(0.0220)
LIQ	−0.2041***	−0.2081***	−0.2552***	−0.2235***	−0.1848***	−0.1669***
	(0.0133)	(0.0133)	(0.0141)	(0.0148)	(0.0110)	(0.0112)
CF	0.1206***	0.1145***	0.0937***	0.0944***	0.1201***	0.1114***
	(0.0167)	(0.0167)	(0.0140)	(0.0139)	(0.0164)	(0.0162)
IK	−0.1078***	−0.2807***	−0.1241***	−0.2259***	−0.1229***	−0.3239***
	(0.0178)	(0.0469)	(0.0151)	(0.0349)	(0.0168)	(0.0433)
DIV	0.0008**	0.0008**	0.0006*	0.0006*	0.0005	0.0005
	(0.0003)	(0.0003)	(0.0002)	(0.0002)	(0.0003)	(0.0003)
lta	−0.0090***	−0.0088***	−0.0179***	−0.0545**	−0.0073***	−0.0214***
	(0.0026)	(0.0026)	(0.0053)	(0.0201)	(0.0011)	(0.0060)
SAL	−0.1007***	−0.0898**	−0.1134***	−0.1165***	−0.0183	−0.0114
	(0.0285)	(0.0284)	(0.0336)	(0.0329)	(0.0146)	(0.0144)
mtb	0.0019***	0.0041***	0.0007	0.0016	0.0021***	0.0050***
	(0.0006)	(0.0010)	(0.0006)	(0.0009)	(0.0005)	(0.0009)
LEV2		0.2709***		0.2137***		0.2988***
		(0.0366)		(0.0438)		(0.0333)
IK2		0.8939***		0.5277***		1.0569***
		(0.1931)		(0.1470)		(0.1799)
mtb2		−0.0002**		−0.0001		−0.0003***
		(0.0001)		(0.0001)		(0.0001)
vola			0.0004*	0.0004*	0.0003**	0.0003*
			(0.0002)	(0.0002)	(0.0001)	(0.0001)
LIQ2				−0.1208**		−0.0788**
				(0.0396)		(0.0290)
lta2				0.0030*		0.0011**
				(0.0013)		(0.0004)
_cons	0.1821***	0.1944***			0.1506***	0.2066***
	(0.0195)	(0.0197)			(0.0115)	(0.0235)
N	18189	18189	13383	13383	18189	18189
r2	0.1421	0.1537	0.1747	0.1860		
r2_o	0.0896	0.1096			0.1345	0.1493
r2_b	0.0905	0.1126			0.1467	0.1714
r2_w	0.1421	0.1537			0.1363	0.1523
sigma_a	0.0776	0.0765			0.0668	0.0658
sigma_e	0.0500	0.0497			0.0500	0.0495
rho	0.7065	0.7035			0.6407	0.6385

Standard errors in parentheses
* $p < 0.05$, ** $p < 0.01$, *** $p < 0.001$

Table 4 Model comparison

	(1)	(2)
	AB 2SLS	AB GMM
L.CASH	0.3266***	0.3079***
	(0.0340)	(0.0369)
L2.CASH	0.0387**	0.0389**
	(0.0145)	(0.0139)
LEV	−0.0995***	−0.0966***
	(0.0120)	(0.0121)
LIQ	−0.3268***	−0.3181***
	(0.0131)	(0.0127)
CF	0.1332***	0.1313***
	(0.0151)	(0.0148)
IK	−0.1820***	−0.1665***
	(0.0148)	(0.0144)
DIV	0.0006*	0.0005*
	(0.0002)	(0.0002)
lta	−0.0188***	−0.0164***
	(0.0031)	(0.0029)
SAL	−0.0149	−0.0089
	(0.0243)	(0.0217)
mtb	−0.0001	0.0001
	(0.0006)	(0.0006)
vola	0.0006***	0.0006***
	(0.0002)	(0.0001)
_cons	0.1948***	0.1807***
	(0.0208)	(0.0193)
N	22,503	22,503

Standard errors in parentheses
* $p < 0.05$, ** $p < 0.01$, *** $p < 0.001$

The results for the RE estimator are also reported in columns 5 and 6 of Table 4. As with the two previous specifications the last column includes square terms that highlight that non-linear effects exist for some variables. Both the FE and the RE models show a slightly larger estimated standard deviation for α_i than for ε_{it}. This underlines the importance of the individual-specific component of the error (the random effect). The output labeled *rho* equals the intraclass correlation of the error $\rho_u = \text{Corr}(u_{it}, u_{is}) = \sigma_\alpha^2/(\sigma_\alpha^2 + \sigma_\varepsilon^2)$, for all $s \neq t$.

3.3.4 Hausman Test

This tests the null hypothesis that individual effects are random. The result is a strong rejection of this hypothesis. The difference in the size coefficient is not significant but most others are and the test strongly rejects equality of the coefficients. A shortcoming of the standard Hausman test is that it requires the RE estimator to be efficient. This in turn requires that the α_i and ε_{it} are i.i.d., an invalid assumption if cluster-robust standard errors for the RE estimator differ substantially from default standard errors, which they do in our case.

A robust Hausman test could test the $H_0 : \gamma = 0$ in the auxiliary OLS regression $(y_{it} - \hat{\theta}_i \bar{y}_i) = (1 - \hat{\theta}_i)\alpha + (x_{it} - \hat{\theta}_i \bar{x}_i)'\beta_1 + (x_{it} - \hat{\theta}_i \bar{x}_i)'\gamma + v_{it}$ where x_1 denotes only the time-varying regressors. A Wald test of $\gamma = 0$ can be shown to be asymptotically equivalent the standard test when the RE estimator is fully efficient under H_0. In the more likely case that the RE estimator is not fully efficient, Woolridge (2010) proposes performing the Wald test using cluster-robust standard errors. Using the xtoverid command in Stata leads to the same conclusion, though, that we should use FE.

3.3.5 Arellano-Bond Estimator

We can allow for CASH in period t to depend on CASH in previous periods. However, OLS with a lagged dependent variable and serially correlated error leads to inconsistent parameter estimates. We still potentially have a fixed effect, which we eliminate by first differencing. Consistent estimators can be obtained by IV estimation of the parameters in the first-difference model, using appropriate lags of regressors as the instruments. The model we are using now is an AR(p):

$$y_{it} = \gamma_1 y_{i,t-1} + \cdots + \gamma_p y_{i,t-p} + x_{it}'\beta + \alpha_i + \varepsilon_{it}$$

with $t = p + 1, \ldots, T$. There are several different reasons for correlation in y over time: (1) directly through y in preceding periods, called true state dependence; (2) directly through observables x, called observed heterogeneity; and (3) indirectly through the time-invariant effect α_i, called unobserved heterogeneity.

Now the model is first-differenced and we assume that ε_{it} are serially uncorrelated (even for a given individual). This can be tested and relaxed using xtdpd. The idea here is to observe that $\Delta y_{i,t-1}$ is correlated with the error $\Delta\varepsilon_{it}$, even if the errors are serially uncorrelated (because $\Delta\varepsilon_{it} = \varepsilon_{it} - \varepsilon_{it-1}$). At the same time $\Delta\varepsilon_{it}$ is *uncorrelated* with $\Delta y_{i,t-k}$ for $k \geq 2$. This means that we can use IV estimation using lagged variables as instruments. We now have to distinguish between exogenous regressors (no special treatment), weakly exogenous (or predetermined regressors) and contemporaneously endogenous regressors. For the second category we assume that they are correlated with past errors but are uncorrelated with future errors. We can then instrument them in the same way that $y_{i,t-1}$ is instrumented using subsequent lags of $y_{i,t-1}$. For the contemporaneously endogenous regressors $x_{i,t-1}$ is no longer a valid instrument and we have to use $x_{i,t-2}$.

We can obtain two different IV estimators. The 2SLS estimator and the GMM. The model is overidentified so GMM is more efficient. For the specification of the model we include two lags for CASH. A third lag turns out to be insignificant. The results are presented in Table 4. The first two rows show results for the

Arellano-Bond 2SLS and GMM estimator under the assumption that only CASH is endogenous and needs to be instrumented. The estimators are almost identical so there is little efficiency gain here from GMM estimation.

For the given specifications we can make some tests. The first one is about the serial correlation of the residuals. The Arellano-Bond test for zero autocorrelation in first-differenced errors tests the H_0 that there is no autocorrelation. If the ε_{it} are serially uncorrelated we expect to reject at order 1 but not at higher orders. This is the case for both specifications. The second test is a test of overidentifying restrictions. For the cases discussed we use 33 instruments to estimate 12 parameters, so we have 21 overidentifying restrictions. The H_0 that the overidentifying restrictions are valid is rejected in both cases ($p = 0.0000$ and $p = 0.0001$). This rejection indicates that at least one of the instruments is not valid.

4 Cash Holdings and the Leverage of the Firm

4.1 Pecking Order Theory

We showed in the previous part of the paper that cash holdings of corporations are too high to be just explained by the trade-off model. Absolute validity of the trade-off model would mean that transaction costs had been continuously rising and also went up during the crisis. While "classical" transaction costs such as brokerage fees, the availability of other fungible inventory or the opportunity costs of available profitable investment do play a role, they cannot explain the secular increase on cash holdings across the board. Instead, we saw that various measures of risk on the firm and economy wide level are critical for explaining firms' preference for cash. The other important factor that we have emphasized and demonstrated is firms' access to finance or the existence of financial constraints. We do not consider any of these factors as "classical" transaction costs but they represent macroeconomic costs and the existence of information asymmetries that have risen over the last 20 years.

An alternative theory of cash holdings, however, claims that focussing on cash holdings is completely irrelevant. If the pecking-order theory (financial hierarchy) applies, no optimal cash-level exists. According to the pecking-order theory a firm's leverage reacts passively to changes in the firms' internal funds (see for example Opler et al. 1999). Accumulating cash leads to a reduction of the leverage (net debt) because cash is considered negative debt. For a given investment opportunity set, we would expect cash balances to be simply a buffer between cash-inflows from operations and cash-outflows for investment (and other purposes). Thus, managers simply observe the evolvement of internal liquidity generated by the firm and compare this liquidity with planned investment (capital expenditures). If the latter are higher then managers resort to debt financing (and eventually equity financing). This means that in this case, i.e. low internally generated liquidity, leverage (LEV) has to go up (see Shyam-Sunder 1999). Eventually, if no more debt can be raised, the firm would try to raise equity, which would then lower the leverage again. Other measures of the debt ratio like the debt-to-equity ratio (DER) would behave in the same manner, that is DER would increase, once a firm issues debt in response to low levels of internally generated liquidity.

These empirical implications do, however, crucially depend on what one is willing to assume to be decision variables of the firm's management. If the investment opportunity is given exogenously (capital expenditures plus change in working capital), payouts to shareholders are also exogenously determined, one can attempt to test the applicability of the pecking order theory by measuring the relation of the internal deficit and the change in debt as proposed by Shyam and Sunder (1999). But are capital expenditures that we observe in period t a good measure of a firm's a priori investment opportunity set or do managers realize just those investments that they deem viable in light of available internal and externable funds? (see also Fazzari et al. 1988) Similar doubts may be raised about the exogeneity of dividend payments.[8] In fact, we take the view that payment or not-payment of dividends in t is a (financing) decision at discretion of management and indicates (together with other factors) whether or not a managers considers a firm to be short of liquidity or not.

We observe a tendency of declining CFO worldwide since the 1990s, measured in terms of median and mean. In the U.S., the median CFO drops from about 0.1 in 1990 to −0.2 during the economic crisis (2008) and then recovers a bit to about −0.12 in 2012. The fall in mean is less severe where CFO drops from about 0.1 in 1990 to 0.05 in 2012. Declining averages can also be found for the other regions examined, except for China (increasing CFO) and India (inconclusive fluctuations).

Using two notions of indebtedness, LEV and DER,[9] we find that leverage and debt-to-equity ratios in the U.S. have significantly decreased between 1990 and 2012 if measured by the median. Figure 4 depicts the decrease in debt and the mirroring increase in cash holdings for the United States. We can also see from the chart that the increase in cash holdings has slowed down since 2005 and indebtedness started to reverse its downward movement at the same time. Hence, the relationship between cash holdings and debt appears to be inverse with higher cash ratios being held as debt decreases and vice versa.

---

[8]Shyam and Sunder (1999) measure the internal deficit process via the variable DEF_t in their Eqs. (1) and (2). The regression proposed in Eq. (2) is not if, for example, the amount of capital expenditures in t or the amount of dividends in t also depends on the overall amount of the deficit. In other words, if managers cut dividends or reduce investment in order to reduce the deficit and thus the amount of necessary financing, we would face an endogeneity problem.

[9]While both variables, LEV and DER, measure indebtedness, their accounting definitions in FactSet reveal some minor differences: Total debt (td) is the sum of short term and long term debt (credits). the debt-to-equity ratio (DER) is then defined as total debt over common equity. The leverage (LEV) is defined as total debt over common equity plus total liabilities (= total assets). Total liabilities include total debt but also several other positions such as accounts payable or provisions.

Unfortunately, these results are not as clear and unambiguous as one might wish: Fig. 4 shows the same time series chart if averages are measured by the mean. While mean cash holdings show a similar pattern, indebtedness seems to be quite stable over time exhibiting only minor fluctuations around some time average. Our results thus confirm the findings of Bates et al. (2009) for the United States who also find decreasing medians for LEV but relatively stable means. Although, if we consider all regions in our sample, we find that means and medians in most cases convey a uniform message. Indebtedness clearly decreases over time for the World, G7, Canada, the United Kingdom and India. Indebtedness remains stable—exhibiting some fluctuations—in Germany and China. For the EU27, we find similar results as for the U.S. with decreasing medians for LEV and DER and rather stable means for the same variables. We do not find any indication for an increasing trend of indebtedness in any of the regions examined.

In order to better understand the interdependencies between capital structure and cash holdings we also examine the evolution of net debt (ND). Given the decrease in leverage for most regions (except Germany and China) and the simultaneous increase in cash holdings, we would expect to observe falling net debt ratios. This is indeed the case for all regions, even including Germany. Thus, net debt and leverage tend to move in the same direction. For the U.S., we can see that median LEV and median ND move almost parallel to each other, mirroring the evolution of cash holdings (see Fig. 4). We find again differences between medians and means in the

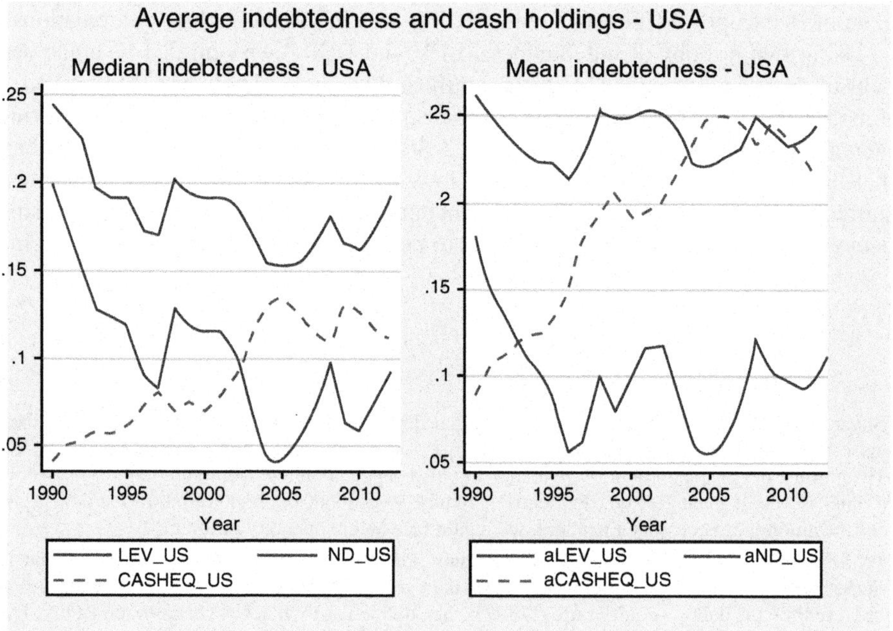

Fig. 4 Firm leverage—medians vs. means, 1990–2012

U.S. sample. Means are shown on the right hand side of Fig. 4.[10] The charts of mean LEV and mean ND also show a very similar progression but none of them declines clearly over time. This result differs with the findings of Bates et al. (2009) who determine a decline of ND in terms of mean and median. Our time series is shorter and does not cover the observations 1980–1990. However, Bates et al. (2009) report hardly any decline for of average ND for this period and the subsequent decline is interrupted by a sharp increase in 1998 that we also observe in our data. In our case, this increase lasts until 2001 and another increase occurs after 2006. We therefore believe that one cannot generally speak of a decline of ND. In summary, we do find a decline in ND and LEV for the whole U.S. sample in terms of the median but find ND and LEV rather stable in terms of the mean.

Observing just one ratio such as LEV over time gives some insight into the financial policy of the firm but not a unique one. Suppose a firm in a pecking-order world would want to raise liquidity through debt financing and subsequently realize investment projects with the obtained cash over time. Such a policy would result in an increase of LEV. The same would occur to ND. But if the firm raises liquidity for other purposes than investment, for example, to hold more cash in anticipation of possible future liquidity problems, LEV would increase, ND would fall and NDER, the net debt-to-equity ratio would fall. It is of main importance to understand whether the observed increased liquidity in firms' balance sheet is merely a reflection of high internal cash flows that are just temporarily "parked" in the firm to be either used for future investments or debt reductions, or if such increase is the outcome of a deliberate increase of asset liquidity in response to certain firm characteristics and other external factors.

4.1.1 Are Cash Holdings Really a Substitute for Debt?

In this section, we employ several tests to analyze the relation between internal cash flow and change in debt. If financial constraints play a role in cash holding behavior, as we claim, we would expect that firms attempt to adjust their cash ratios and debt ratios to optimal levels in order to avoid opportunity costs arising from a lack of financing (tradeoff model). Cash holdings then serve as a buffer between internal cash flow, exogenous payouts (flow of funds deficit) and the portion of the deficit that is financed through external debt.

In absence of financial constraints (and ignoring possible agency considerations), firms have no reason to regard cash or debt ratios as important decision parameters. In this case, debt ratios simply vary over time, reflecting differing states of the flow of funds deficit. Actual cash balances are irrelevant or can be interpreted as negative debt so that not debt ratios but net debt ratios (debt minus cash holdings) vary over time, passively reacting to the flow of funds deficit. This second view is a simple version of the pecking-order theory (Myers and Majluf 1984).

[10]For all other regions, means and medians show the same evolution.

In the previous regressions, we specified both, CFO and LEV, as explanatory variables for cash holdings (see Table 1). However, each of the two theories would predict a different impact of these variables on cash holdings. The pecking-order theory would suggest a positive relation between CFO and cash holdings and a negative relation between cash holdings and debt. The trade-off model would suggest that lower CFO leads to higher cash holdings. The impact of LEV on cash holdings should be negative, at least if a higher level of debt is a sign of non-existing constraints, as is suggested by our variable FINCSTR.

We set up a series of regressions similar to Shyam-Sunder (1999) in which we test if the pecking-order theory has empirical relevance for financially constrained and unconstrained firms. We define three versions of the flow of funds deficit (fof) in period t:

1. $\text{fof1}_t = \text{-cfo}_t$
2. $\text{fof2}_t = \text{-cfo}_t + \text{capex}_t$
3. $\text{fof3}_t = \text{-cfo}_t + \text{capex}_t + \text{pts}_t + \Delta\ \text{wk}_t$

In addition, we also scale each of the three flow of funds definitions by total assets in period t, so that for example $\text{FOF}_{1t} = \text{fof}_{1t}/\text{ta}_t$ for each enterprise i. All three definitions (including the third one) are only proxies for the internal deficit as they ignore for example all cash relevant events that are not related to the firm's operations and changes in working capital that are not financed through immediate use of liquidity. A general model for testing the pecking-order theory can then be specified as follows:

$$\Delta td_{it} = \alpha_i + \beta_i\, fof_{it} + \epsilon_{it} \tag{1}$$

or

$$\Delta TD_{it} = \alpha_i + \beta_i\ FOF_{it} + \epsilon_{it} \tag{2}$$

whereby Δ denotes first differences and ϵ_{it} denotes spherical disturbances: Additional debt assumed (retired) during period t should be approximately equal to the flow of funds deficit (surplus). The simplest version of the pecking-order model would predict $\alpha_i = \alpha_j = \alpha$ and $\beta_i = \beta_j = \beta$ for all firms, so that we have no firm-specific effects and can estimate α and β with pooled OLS. In a perfect pecking-order world, we would find $\alpha = 0$ and $\beta = 1$.

$$\Delta TD_{it} = \alpha + \beta\ FOF_{it} + \epsilon_{it} \tag{3}$$

Alternatively, we could allow for firm-specific financing behaviour, for example, a traditionally higher degree of equity-financing due to industry, management preferences, or other firm-specific characteristics. Such characteristics would be captured in differing α_i.

$$\Delta TD_{it} = \alpha_i + \beta\ FOF_{it} + \epsilon_{it} \tag{4}$$

Table 5 Pooled OLS regressions of Eq. (2) for USA

	firms uc	firms c	firms uc	firms c	firms uc	firms c
FOF1	0.0504**	−0.213				
	(2.82)	(−1.14)				
FOF2			0.106***	−0.160		
			(6.29)	(−1.10)		
FOF3					0.0285	−0.523***
					(1.49)	(−3.80)
Constant	0.0137***	−0.0698	0.0112***	−0.0778	0.00876***	0.0971***
	(7.86)	(−0.92)	(14.19)	(−0.97)	(11.09)	(4.13)
Observations	48,461	27,134	48,337	26,780	45,684	22,117

t-statistics in parentheses
* $p < 0.05$, ** $p < 0.01$, *** $p < 0.001$

We provide estimates for all three versions of the pecking-order model. Table 5 shows the results of pooled OLS regressions using all available observations for FOF1, FOF2 and FOF3 in the United States. The results show that the pecking-order model has some explanatory power for non-constrained firms even though the beta coefficients are much smaller than 1. The constants are relatively close to zero, especially if the flow of funds deficit is approximated by FOF3.

Table 5 shows the same regression with FOF3 as the independent variable for differing numbers of consecutive observations per firm. Thus, the first regression uses only firm-years if at least five consecutive variables of ΔTD and FOF3 are available. The second (third) regression requires at least 10 (15) consecutive observations. The sign of the beta coefficient remains unchanged in each of these estimations but the level of significance varies. Overall the beta coefficient of non-constrained firms lies between + 0.0297 and +0.0757. A different picture emerges for financially constrained firms. The sign of the pecking-order coefficient is negative and its absolute value is considerable larger. The different estimations suggest beta values between −0.119 and −0.523. This result indicates that financially constrained firms do not (cannot?) finance their internal cash deficit through issuing new debt.

These results are confirmed by our individual effects regressions shown in Tables 6 and 7.

4.2 Cash Holdings, Internal Cash and Financial Constraints

This brings us back to the question of internal cash flows. As argued before, we prefer to look at cash flow from operation (CFO) as an indicator of internal cash flows, i.e. without deducting current dividend payments, investment and other

Table 6 Pooled OLS regressions of Eq. (3) for USA

	firms uc (5)	firms c (5)	firms uc (10)	firms c (10)	firms uc (15)	firms c (15)
FOF3	0.0297	−0.351*	0.0679***	−0.119	0.0757**	−0.264
	(1.43)	(−2.53)	(3.74)	(−1.50)	(2.80)	(−1.34)
Constant	0.00894***	0.0552**	0.00774***	0.0134	0.00778***	0.0220
	(12.21)	(2.67)	(12.79)	(1.39)	(11.24)	(1.19)
Observations	43,690	18,391	36,183	11,059	25,715	5240

t-statistics in parentheses
Note: Numbers in brackets refer to the number of consecutive observations used in the regression
$^* p < 0.05$, $^{**} p < 0.01$, $^{***} p < 0.001$

Table 7 Fixed and random effects estimations of Eq. (4) for USA

	RE firms uc	RE firms c	FE firms uc	FE c	FD firms uc	FD firms c
FOF3	0.0382	−0.541***	0.0512*	−0.479***		
	(1.93)	(−4.02)	(2.38)	(−3.41)		
D.FOF3					−0.615***	−0.000882
					(−3.58)	(−0.03)
Constant	0.00656***	0.0983***	0.00801***	0.0887***	−0.0336***	−0.00206**
	(4.48)	(4.53)	(11.21)	(3.31)	(−4.38)	(−3.22)
Observations	45,684	22,117	45,684	22,117	16,788	38,184

t-statistics in parentheses
$^* p < 0.05$, $^{**} p < 0.01$, $^{***} p < 0.001$

payments. We consider it likely that the magnitude of these latter payments may be adjusted by managers in response to high or low CFO in certain circumstances. CFO itself, however, is hardly influenceable by current management decisions and its decline reduces managers' room to maneuver. This "room to maneuver" could even be in fully accordance with shareholders' interest and does not necessarily have to result from agency conflicts à la (Jensen 1986). But since most managerial decisions require monetary transactions, a reduction of CFO may require actions to obtain liquidity from elsewhere.

Firstly, cash flow from operation (CFO) for unconstrained firm is significantly higher than the CFO for constrained firms. This results holds for all regions examined and for all years. It shows that our variable FINCSTR discriminates well between firms that may be in need of liquidity and those which are not. Furthermore, the ratio of CFO for unconstrained firms remains remarkably stable over time with little fluctuations while the CFO of constrained firms declines and shows more fluctuations. Regressing CFO of unconstrained firms on a constant and time yields a zero coefficient. This is also the case for all regions examined even though CFO for unconstrained firms in China and India exhibit some stronger fluctuations compared to the other regions.

In order to further substantiate this claim, we compare the average CFO of constrained firm-years with non constrained ones and find large differences in terms of medians and means in all regions examined. We perform a two-sided Wilcoxon-Mann-Whitney (WMW) rank sum test to analyse the central tendency in the two populations (constrained vs. unconstrained firm-years). The test leads us to reject the null hypothesis of two equal probability distributions at 0.01 % significance for all regions. We conclude therefore that the medians are significantly different. We also perform Mood's median test which rejects the null hypothesis of equal medians at the 0.01 % significance level for all regions.

Secondly, constrained firms tend to hold more cash than unconstrained firms. This second result is not as pervasive as the first one. It clearly holds for most firms in our sample, particularly for firms in advanced economies. Means and medians for constrained versus unconstrained firms show significant differences and we reject again the null hypothesis of the WMW test on CASHEQ except for Germany, where the null cannot be rejected. In China and India, however, where also significant difference exist, constrained firms seem to hold *less* cash, possibly indicating a different behaviour. For German firms, the results are not conclusive. Mean cash holdings of constrained firms are slightly higher than for unconstrained firms but the opposite is true for the medians. The time series charts of average cash holdings for Germany also do not indicate a clear difference of liquidity holdings for constrained versus non-constrained firms. We do also note in this context that the sample of German firms is rather small compared to other advanced economies.

Thirdly, the overall increase in cash holdings is largely driven by constrained firms. Inspection of the time series charts show that most of the increase in cash holdings comes from higher cash holdings of financially constrained firms. Again, this third result is less pervasive than the first two. In China and India, the opposite is true, i.e. increases in cash holdings mainly come from unconstrained firms and in several EU27 countries like Germany no conclusive result can be seen from the charts. A regression of CASHEQ on a constant and time yields higher coefficients for constrained firms than for unconstrained firms for basically all advanced economies examined (for example 3.9 % vs. 3.1 % in the U.S. or 3.7 % vs. 0.1 % for G7 countries).

Fourthly, given the previous three results we conclude that the increase in cash holdings in advanced economies is inversely related to CFO and largely driven by financially constrained firms. In the emerging economies, this is not the case. While financially constrained firms do also have significant lower CFO in emerging economies, those firms do not build up more cash but unconstrained firms with higher CFO do. The overall increase of cash holdings in emerging economies is smaller than in the advanced economies.

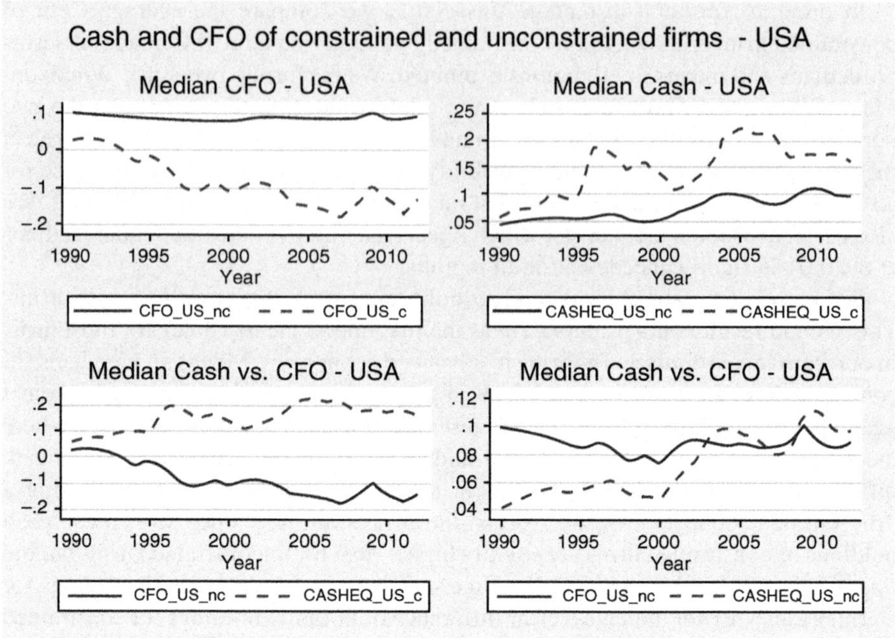

Fig. 5 Cash holdings and cash flow from operation over time for financially constrained and unconstrained firms in the U.S.

Finally, we examine what can be said in terms of firms indebtedness and capital structure. The median leverage of non-constraint firms tends to be higher, or put differently, financially constrained firms hold less debt. We find, however, again differences between firms in advanced economies and India and China.[11] We reject the null hypothesis of the WMW test for the variable LEV at 0.01 % significance for all regions. We also reject the null hypothesis of equal medians based on Mood's test at 0.01 % significance for all regions except for the EU27 where the p value is equal to 0.005. Further examination of median leverages over time show that the median leverage for non-constrained firms hardly changes over time. For example on a world wide scale the leverage of non-constrained firms remains virtually constant at about 0.2 between 1990 and 2012. Similar values are found for the U.S. (median also constant at about 0.2) and other advanced economies. A very different picture evolves for the median of constrained firms as LEV declines over time and exhibits

[11]Findings for Hong Kong, Taiwan and South Africa match those of China and India.

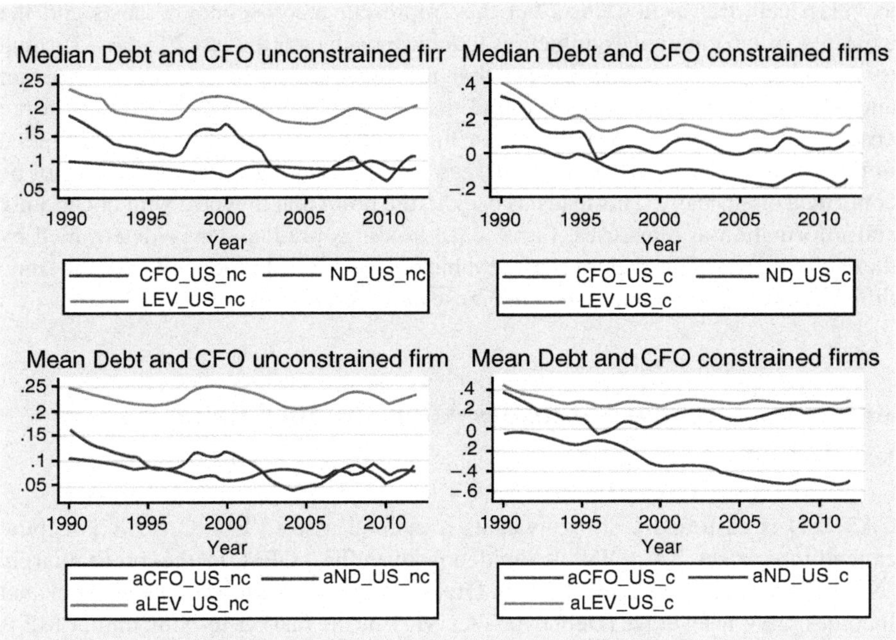

Fig. 6 Debt, net debt and operational cash flow of U.S. firms

strong fluctuations. In China and India, we find that the constrained firms have a higher leverage than unconstrained firms but also observe more fluctuation in the leverage of constrained firms.

5 Conclusion

We demonstrated that companies around the World, especially in developed countries have been increasing their cash ratios. This increase cannot be explained by transaction motives alone. While "classical" transaction costs such as brokerage fees, the availability of other fungible inventory or the opportunity costs of available profitable investment do play a role, they cannot explain the secular increase on cash holdings across the board. Instead, we saw that various measures of risk on the firm and economy wide level are critical for explaining firms' preference for cash. We constructed several indicators to capture risk.The other important factor that we have emphasized and demonstrated is firms' access to finance or the existence of financial constraints. We do not consider any of these factors as "classical" transaction costs but they represent macroeconomic costs and the existence of information asymmetries that have risen over the last 20 years. Turning to the pecking order which claims that the focus on cash is entirely irrelevant and cash is just negative debt, we find that it has some explanatory power in the absence of financial constraints. When financially constrained firms are included in the sample, a different picture emerges and the pecking-order theory cannot be confirmed empirically. This leads us back to the point that in world with uncertainty and information asymmetries, firms' cash holdings predominantly determined by these two factors. This pattern is observable for most OECD economies, while some differences emerge with respect to Chinese and Indian firms.

Appendix: Summary Statistics After Modified Baum Correction

CASHEQ is cash and cash equivalents over total assets (TA). CAPEX is capital expenditures over TA. DIV is dividends over TA. OPM is the profit margin (Net Income over Revenue). DPR is Dividend Payout Ratio (Dividends over Net Income). LEV is leverage (Debt over TA). MTB is the market-to-book ratio. EMP is total employment in 1000. VOLA is the yearly stock price volatility. RET is retained earnings over TA. DER is the debt-equity ratio. SBB is share buybacks over TA. CFO is cash from operations over TA. EQ is equity over TA. REV is revenue over equity. ROA is return on assets (Net Income over TA). EIR is the effective interest rate (interest expense over debt). LTA is the log of TA. CA is current assets over TA. LIQ is Liquidity: (Working Capital minus Cash) over TA. CF is Cash Flow: (Gross income + depreciation) over TA. TAX is total income tax over TA. RAT is the FactSet Rating. DY is Dividend Yield. PR is the Profit Rate (Net income over Equity). RD is R&D expenditures over TA. FINC is foreign income as % of total income. FTAX is income tax paid abroad as % of total income tax. IPO is years since IPO.

See Tables 8, 9, 10, 11, 12, 13 and 14.

Table 8 Summary statistics after Modified Baum Correction

	World						
	Count	Mean	Min	p25	p50	p75	Max
CA	346, 058	0.505	0.000	0.313	0.508	0.692	1.000
CAPEX	348, 324	0.059	0.000	0.012	0.034	0.075	0.488
CASH	259, 194	0.122	0.000	0.017	0.058	0.150	1.000
CASHEQ	358, 936	0.176	0.000	0.036	0.103	0.232	1.000
CF	327, 237	−0.121	−12.320	−0.019	0.051	0.098	0.379
CFO	345, 356	−0.032	−5.127	−0.025	0.044	0.102	0.397
DER	327, 693	0.719	0.000	0.030	0.325	0.881	11.397
DIV	334, 456	0.011	0.000	0.000	0.000	0.012	0.170
DPR	332, 257	0.186	−1.146	0.000	0.000	0.256	3.570
DY	286, 719	1.619	0.000	0.000	0.000	2.500	15.932
EIR	286, 174	10.127	0.000	3.082	5.787	9.215	302.558
EMP	255, 623	5.108	0.000	0.149	0.623	2.477	2640.000
EQ	349, 188	0.435	−4.440	0.303	0.485	0.689	0.991
FINC	114, 223	9.665	0.000	0.000	0.000	1.462	100.000
fin_cstr	358, 112	0.308	0.000	0.000	0.000	1.000	1.000
FTAX	135, 213	14.793	−32.473	0.000	0.000	3.225	134.231
INV	352, 414	0.113	0.000	0.005	0.075	0.174	1.000
IPO	74, 983	3.482	0.000	1.000	3.000	5.000	11.000
LEV	351, 929	0.225	0.000	0.022	0.178	0.350	1.637
LIQ	242, 387	0.027	−3.469	−0.065	0.047	0.189	0.720
LTA	356, 923	4.399	−4.828	3.020	4.482	5.908	13.459
MTB	283, 855	2.574	0.145	0.801	1.423	2.692	50.695
OPM	323, 137	−0.220	−15.500	−0.021	0.029	0.083	1.000
PR	352, 624	0.014	−6.118	−0.033	0.066	0.160	5.132
RAT	171, 378	1.633	1.000	1.333	1.609	1.972	3.000
RD	190, 442	0.038	0.000	0.000	0.002	0.028	0.975
REV	351, 160	0.860	0.000	0.350	0.761	1.216	3.455
ROA	352, 197	−0.159	−13.028	−0.048	0.021	0.062	0.347
SBB	278, 804	0.004	0.000	0.000	0.000	0.000	0.146
TAX	244, 152	0.234	−1.379	0.000	0.097	0.410	3.147
VOLA	239, 847	41.306	0.000	28.589	39.319	51.648	99.931
WK	336, 459	0.140	−3.320	0.006	0.152	0.330	0.905
risk_idio	376, 152	0.388	0.000	0.145	0.244	0.442	6.863
risk_industry	504, 852	0.416	0.250	0.319	0.382	0.476	0.750
std_CF	481, 560	0.283	0.000	0.030	0.061	0.165	8.736
N	504, 852						

Table 9 Summary statistics after Modified Baum Correction

	United States						
	Count	Mean	Min	p25	p50	p75	Max
CA	71,386	0.522	0.000	0.274	0.522	0.770	1.000
CAPEX	71,760	0.051	0.000	0.010	0.029	0.063	0.488
CASH	42,945	0.194	0.000	0.021	0.087	0.247	1.000
CASHEQ	73,338	0.247	0.000	0.028	0.121	0.377	1.000
CF	67,879	−0.455	−12.320	−0.272	0.028	0.090	0.379
CFO	69,420	−0.182	−5.123	−0.141	0.038	0.106	0.397
DER	56,941	0.727	0.000	0.001	0.246	0.849	11.395
DIV	71,567	0.005	0.000	0.000	0.000	0.000	0.170
DPR	72,572	0.085	−1.143	0.000	0.000	0.000	3.565
DY	60,303	0.521	0.000	0.000	0.000	0.000	15.906
EIR	54,117	14.448	0.000	5.135	7.556	11.608	302.489
EMP	64,682	6.340	0.000	0.041	0.371	2.791	2200.000
EQ	68,023	0.285	−4.440	0.201	0.455	0.695	0.991
FINC	44,912	7.367	0.000	0.000	0.000	0.000	100.000
fin_cstr	73,138	0.512	0.000	0.000	1.000	1.000	1.000
FTAX	38,996	20.979	−32.440	0.000	0.000	30.754	134.231
INV	71,222	0.096	0.000	0.000	0.024	0.145	1.000
IPO	13,413	3.678	0.000	1.000	3.000	6.000	11.000
LEV	68,486	0.247	0.000	0.002	0.164	0.375	1.637
LIQ	37,868	−0.034	−3.469	−0.083	0.057	0.251	0.720
LTA	71,429	4.150	−4.828	2.126	4.498	6.476	13.459
MTB	47,179	3.908	0.146	1.239	2.122	3.930	50.659
OPM	62,706	−0.602	−15.500	−0.206	0.011	0.066	1.000
PR	70,668	0.023	−6.117	−0.132	0.076	0.207	5.132
RAT	40,291	1.606	1.000	1.400	1.600	1.833	3.000
RD	51,404	0.078	0.000	0.000	0.010	0.093	0.975
REV	70,402	0.880	0.000	0.262	0.728	1.308	3.455
ROA	69,710	−0.522	−13.028	−0.318	−0.005	0.052	0.347
SBB	69,099	0.007	0.000	0.000	0.000	0.000	0.146
TAX	66,549	0.182	−1.379	0.000	0.000	0.380	3.147
VOLA	52,756	50.323	0.000	33.503	48.515	66.338	99.931
WK	65,380	0.119	−3.320	−0.000	0.181	0.415	0.905
risk_idio	89,688	0.523	0.002	0.167	0.300	0.642	5.289
risk_industry	106,152	0.444	0.250	0.365	0.394	0.476	0.750
std_CF	102,324	0.666	0.000	0.050	0.151	0.663	8.736
N	106,152						

Table 10 Summary statistics after Modified Baum Correction

	United Kingdom						
	Count	Mean	Min	p25	p50	p75	Max
CA	15,830	0.481	0.000	0.272	0.468	0.672	1.000
CAPEX	19,895	0.050	0.000	0.002	0.024	0.063	0.482
CASH	19,174	0.129	0.000	0.019	0.063	0.168	0.971
CASHEQ	20,272	0.149	0.000	0.026	0.080	0.203	0.971
CF	19,682	−0.073	−12.193	−0.037	0.030	0.088	0.379
CFO	19,935	−0.022	−5.017	−0.026	0.025	0.090	0.396
DER	18,529	0.509	0.000	0.000	0.149	0.551	11.284
DIV	19,832	0.015	0.000	0.000	0.004	0.023	0.170
DPR	19,203	0.272	−1.146	0.000	0.000	0.436	3.569
DY	16,856	2.010	0.000	0.000	1.081	3.361	15.856
EIR	14,771	11.337	0.000	3.925	6.371	9.464	302.075
EMP	15,845	5.405	0.000	0.048	0.267	1.726	639.904
EQ	19,394	0.482	−4.422	0.321	0.531	0.785	0.991
FINC	6424	21.978	0.000	0.000	0.000	40.213	100.000
fin_cstr	20,232	0.318	0.000	0.000	0.000	1.000	1.000
FTAX	12,583	31.427	−32.012	0.000	0.000	74.830	133.333
INV	19,986	0.072	0.000	0.000	0.006	0.101	0.996
IPO	7243	3.796	0.000	1.000	3.000	6.000	11.000
LEV	20,130	0.168	0.000	0.000	0.103	0.258	1.633
LIQ	14,893	−0.027	−3.432	−0.114	−0.010	0.102	0.720
LTA	20,269	4.305	−4.636	2.792	4.245	5.702	12.441
MTB	16,731	2.587	0.146	0.864	1.326	2.715	50.000
OPM	18,222	−0.305	−15.441	−0.075	0.033	0.118	0.985
PR	19,908	−0.010	−6.088	−0.058	0.036	0.154	5.095
RAT	14,398	1.597	1.000	1.333	1.571	1.875	3.000
RD	6785	0.058	0.000	0.000	0.012	0.061	0.962
REV	19,669	0.776	0.000	0.055	0.602	1.226	3.448
ROA	20,099	−0.101	−12.367	−0.061	0.016	0.058	0.347
SBB	19,498	0.005	0.000	0.000	0.000	0.000	0.146
TAX	19,203	0.178	−1.379	0.000	0.046	0.356	3.066
VOLA	13,611	35.426	2.183	23.771	32.418	45.351	93.358
WK	15,718	0.117	−3.246	−0.030	0.110	0.283	0.905
risk_idio	22,056	0.444	0.006	0.166	0.289	0.548	3.584
risk_industry	29,076	0.482	0.250	0.365	0.394	0.614	0.750
std_CF	28,836	0.215	0.000	0.033	0.070	0.192	5.821
N	29,076						

Table 11 Summary statistics after Modified Baum Correction

	European Union (27)						
	Count	Mean	Min	p25	p50	p75	Max
CA	55, 571	0.506	0.000	0.330	0.510	0.680	1.000
CAPEX	58, 122	0.055	0.000	0.013	0.035	0.070	0.487
CASH	50, 649	0.098	0.000	0.018	0.050	0.121	1.000
CASHEQ	61, 491	0.131	0.000	0.029	0.076	0.174	1.000
CF	54, 491	−0.009	−12.193	0.001	0.053	0.098	0.379
CFO	57, 122	0.016	−5.017	−0.007	0.048	0.102	0.397
DER	57, 896	0.740	0.000	0.048	0.351	0.913	11.384
DIV	57, 441	0.013	0.000	0.000	0.002	0.019	0.170
DPR	56, 185	0.244	−1.146	0.000	0.000	0.378	3.569
DY	48, 704	1.891	0.000	0.000	0.941	3.125	15.915
EIR	50, 255	10.564	0.000	4.158	6.020	8.825	302.075
EMP	52, 602	7.051	0.000	0.115	0.530	2.629	639.904
EQ	60, 340	0.430	−4.422	0.275	0.434	0.632	0.991
FINC	11, 111	23.121	0.000	0.000	1.538	43.872	100.000
fin_cstr	61, 428	0.265	0.000	0.000	0.000	1.000	1.000
FTAX	22, 545	24.121	−32.461	0.000	0.000	44.399	133.395
INV	60, 235	0.110	0.000	0.003	0.061	0.179	0.996
IPO	15, 710	3.648	0.000	1.000	3.000	5.000	11.000
LEV	61, 257	0.208	0.000	0.031	0.170	0.324	1.636
LIQ	45, 084	0.027	−3.432	−0.074	0.030	0.158	0.720
LTA	61, 593	4.732	−4.636	3.188	4.591	6.156	12.586
MTB	50, 049	2.500	0.146	0.884	1.478	2.693	50.438
OPM	58, 349	−0.161	−15.466	−0.024	0.029	0.078	0.998
PR	60, 665	0.004	−6.116	−0.029	0.065	0.159	5.095
RAT	40, 362	1.688	1.000	1.431	1.677	2.000	3.000
RD	21, 663	0.040	0.000	0.000	0.007	0.042	0.962
REV	60, 214	0.911	0.000	0.392	0.833	1.296	3.455
ROA	61, 078	−0.047	−12.367	−0.026	0.023	0.060	0.347
SBB	45, 505	0.004	0.000	0.000	0.000	0.000	0.146
TAX	42, 492	0.246	−1.379	0.000	0.168	0.427	3.144
VOLA	40, 753	34.497	0.000	23.889	32.172	43.348	93.358
WK	55, 347	0.136	−3.311	−0.002	0.129	0.288	0.905
risk_idio	73, 392	0.363	0.000	0.147	0.243	0.422	4.242
risk_industry	84, 504	0.429	0.250	0.319	0.390	0.476	0.750
std_CF	80, 184	0.144	0.000	0.028	0.056	0.123	5.821
N	84, 504						

Table 12 Summary statistics after Modified Baum Correction

	G-7 Countries						
	Count	Mean	Min	p25	p50	p75	Max
CA	169, 518	0.513	0.000	0.302	0.518	0.716	1.000
CAPEX	173, 028	0.055	0.000	0.010	0.030	0.067	0.488
CASH	116, 561	0.159	0.000	0.027	0.083	0.199	1.000
CASHEQ	177, 977	0.206	0.000	0.039	0.116	0.277	1.000
CF	166, 300	−0.245	−12.320	−0.083	0.038	0.084	0.379
CFO	170, 386	−0.092	−5.127	−0.050	0.039	0.094	0.397
DER	154, 541	0.704	0.000	0.005	0.253	0.839	11.397
DIV	172, 454	0.007	0.000	0.000	0.000	0.008	0.170
DPR	171, 276	0.157	−1.146	0.000	0.000	0.196	3.569
DY	148, 178	1.165	0.000	0.000	0.000	1.835	15.906
EIR	134, 187	10.362	0.000	2.137	5.434	9.028	302.489
EMP	143, 592	5.847	0.000	0.101	0.502	2.400	2200.000
EQ	170, 027	0.390	−4.440	0.266	0.473	0.703	0.991
FINC	72, 280	10.117	0.000	0.000	0.000	5.410	100.000
fin_cstr	177, 749	0.389	0.000	0.000	0.000	1.000	1.000
FTAX	72, 415	18.619	−32.440	0.000	0.000	19.847	134.231
INV	173, 952	0.094	0.000	0.000	0.041	0.145	1.000
IPO	35, 918	3.727	0.000	1.000	3.000	6.000	11.000
LEV	171, 967	0.215	0.000	0.005	0.148	0.335	1.637
LIQ	104, 836	−0.009	−3.469	−0.085	0.030	0.178	0.720
LTA	175, 782	4.353	−4.828	2.712	4.605	6.203	13.459
MTB	136, 162	2.865	0.146	0.846	1.499	2.869	50.695
OPM	152, 939	−0.362	−15.500	−0.060	0.017	0.060	1.000
PR	173, 522	−0.007	−6.117	−0.081	0.050	0.146	5.132
RAT	91, 269	1.623	1.000	1.375	1.607	1.895	3.000
RD	108, 117	0.053	0.000	0.000	0.006	0.047	0.975
REV	173, 140	0.881	0.000	0.274	0.790	1.290	3.455
ROA	173, 181	−0.295	−13.028	−0.118	0.010	0.047	0.347
SBB	143, 986	0.005	0.000	0.000	0.000	0.000	0.146
TAX	136, 452	0.254	−1.379	0.000	0.032	0.485	3.147
VOLA	127, 634	41.996	0.000	26.583	37.868	55.283	99.931
WK	161, 456	0.133	−3.320	−0.001	0.159	0.353	0.905
risk_idio	192, 240	0.408	0.000	0.139	0.234	0.461	5.289
risk_industry	247, 692	0.438	0.250	0.358	0.394	0.494	0.750
std_CF	240, 996	0.431	0.000	0.031	0.083	0.300	8.736
N	247, 692						

Table 13 Summary statistics after Modified Baum Correction

	China						
	Count	Mean	Min	p25	p50	p75	Max
CA	22,963	0.531	0.000	0.370	0.542	0.695	1.000
CAPEX	22,851	0.071	0.000	0.020	0.050	0.100	0.486
CASH	12,981	0.190	0.000	0.079	0.147	0.255	0.878
CASHEQ	23,119	0.191	0.000	0.083	0.151	0.256	0.887
CF	21,222	0.045	−11.858	0.034	0.062	0.100	0.379
CFO	22,762	0.054	−4.456	0.009	0.054	0.108	0.395
DER	22,440	0.742	0.000	0.148	0.461	0.975	11.268
DIV	12,414	0.008	0.000	0.000	0.000	0.010	0.170
DPR	12,430	0.164	−1.107	0.000	0.000	0.207	3.484
DY	16,993	1.000	0.000	0.000	0.427	1.383	15.877
EIR	20,620	6.472	0.000	3.463	4.971	6.577	298.788
EMP	19,428	4.658	0.000	0.669	1.597	3.546	552.810
EQ	23,011	0.476	−4.423	0.343	0.486	0.649	0.991
FINC	5375	2.094	0.000	0.000	0.000	0.000	100.000
fin_cstr	23,058	0.104	0.000	0.000	0.000	0.000	1.000
FTAX	8496	4.618	−30.491	0.000	0.000	0.000	128.284
INV	23,032	0.145	0.000	0.058	0.122	0.200	0.946
IPO	6082	2.864	0.000	1.000	2.000	4.000	11.000
LEV	23,034	0.248	0.000	0.091	0.232	0.373	1.605
LIQ	12,795	−0.031	−3.412	−0.146	−0.018	0.113	0.719
LTA	23,114	5.080	−4.791	4.259	5.018	5.866	12.360
MTB	17,750	3.937	0.145	1.738	2.823	4.706	50.612
OPM	22,882	0.027	−15.221	0.023	0.068	0.139	0.998
PR	23,029	0.096	−6.109	0.036	0.087	0.173	5.122
RAT	9116	1.532	1.000	1.250	1.500	1.750	3.000
RD	7247	0.012	0.000	0.000	0.004	0.016	0.852
REV	22,995	0.714	0.000	0.376	0.604	0.918	3.454
ROA	22,891	0.025	−12.967	0.014	0.041	0.080	0.346
SBB	13,246	0.000	0.000	0.000	0.000	0.000	0.144
TAX	10,676	0.240	−1.352	0.075	0.186	0.326	3.127
VOLA	9214	40.294	13.376	34.358	39.330	44.138	97.695
WK	22,811	0.118	−3.299	−0.043	0.118	0.303	0.902
risk_idio	30,048	0.391	0.000	0.201	0.314	0.465	3.923
risk_industry	33,048	0.380	0.250	0.310	0.358	0.390	0.750
std_CF	31,068	0.094	0.000	0.021	0.039	0.071	6.334
N	33,048						

Table 14 Summary statistics after Modified Baum Correction

	India						
	Count	Mean	Min	p25	p50	p75	Max
CA	16,834	0.521	0.000	0.363	0.519	0.682	1.000
CAPEX	16,481	0.076	0.000	0.016	0.049	0.109	0.486
CASH	16,419	0.024	0.000	0.005	0.012	0.029	0.406
CASHEQ	17,038	0.054	0.000	0.012	0.028	0.066	0.435
CF	16,291	0.059	−8.938	0.030	0.066	0.110	0.379
CFO	16,492	0.029	−4.227	−0.020	0.035	0.096	0.396
DER	15,980	1.104	0.000	0.221	0.723	1.471	11.397
DIV	16,279	0.009	0.000	0.000	0.002	0.011	0.168
DPR	16,165	0.141	−1.146	0.000	0.015	0.204	3.502
DY	13,019	1.889	0.000	0.000	1.176	2.829	15.873
EIR	15,101	10.548	0.000	5.544	8.426	11.780	295.963
EMP	5237	3.965	0.000	0.400	1.022	2.797	238.583
EQ	16,899	0.379	−4.345	0.255	0.387	0.557	0.991
FINC	11,050	0.392	0.000	0.000	0.000	0.000	98.965
fin_cstr	17,022	0.153	0.000	0.000	0.000	0.000	1.000
FTAX	11,783	0.404	−32.473	0.000	0.000	0.000	119.108
INV	16,873	0.166	0.000	0.061	0.149	0.238	0.971
IPO	1842	2.618	0.000	1.000	2.000	4.000	11.000
LEV	16,903	0.315	0.000	0.132	0.305	0.456	1.634
LIQ	15,942	0.108	−3.455	0.013	0.119	0.240	0.720
LTA	17,045	3.704	−4.677	2.452	3.676	4.905	10.910
MTB	14,468	1.991	0.146	0.581	1.095	2.179	49.492
OPM	16,662	−0.008	−15.429	0.009	0.043	0.092	0.997
PR	16,901	0.101	−5.865	0.026	0.109	0.202	4.907
RAT	5455	1.435	1.000	1.000	1.368	1.673	3.000
RD	9886	0.004	0.000	0.000	0.000	0.001	0.955
REV	16,805	0.914	0.000	0.507	0.831	1.219	3.445
ROA	16,909	0.029	−6.635	0.006	0.037	0.077	0.347
SBB	14,947	0.001	0.000	0.000	0.000	0.000	0.137
TAX	15,568	0.240	−1.365	0.001	0.182	0.423	3.143
VOLA	13,427	46.809	0.000	40.436	47.024	53.635	81.438
WK	16,713	0.139	−3.313	0.033	0.144	0.274	0.904
risk_idio	10,656	0.273	0.000	0.131	0.205	0.322	3.179
risk_industry	28,704	0.378	0.250	0.313	0.319	0.394	0.750
std_CF	28,476	0.065	0.000	0.022	0.036	0.064	3.575
N	28,704						

References

Acharya, V., Almeida, H., & Campello, M. (2007). Is cash negative debt? *Journal of Financial Intermediation, 16*, 515–554.

Acharya, V., Davydenko, S., & Strebulaev, I. (2012). Cash holdings and credit risk. *The Review of Financial Studies, 25*, 3572–3609.

Almeida, H., Campello, M., & Weisbach, M. S. (2009). The cash flow sensitivity of cash. *The Journal of Finance, 59*, 1777–1804.

Bates, T. W., Kahle, K. M., & Stulz, R. M. (2009). Why do U.S. firms hold so much more cash than they used to? *The Journal of Finance, 64*, 1985–351.

Baum, C. F., Caglayan, M., Ozkan, N., & Talavera, O. (2006). The impact of macroeconomic uncertainty on non-financial firms' demand for liquidity. *Review of Financial Economics, 15*, 289–304.

Baumol, W. J. (1952). The transactions demand for cash: An inventory theoretic approach. *Quarterly Journal of Economics, 66*, 545–556.

Brealey, R. A., & Myers, S. C. (2002). *Principles of corporate finance* (7th ed.). New York, NY: Mcgraw Hill.

Campbell, J. Y., Lettau, M., Malkiel, B. G., & Xu, Y. (2001). Have individual stocks bcome more volatile? An empirical exploration of idiosyncratic risk. *The Journal of Finance, 56*, 1–43.

Cooper, R., & Haltiwanger, J. C. (2006). On the nature of capital adjustment costs. *Review of Economic Studies, 73*, 611–633.

Denis, D. J., & Sibilkov, V. (2010). Financial constraints, investment and the value of cash holdings. *The Review of Financial Studies, 23*, 247–269.

Fazzari, S. M., Hubbard, G., & Petersen, B. (1988). Financing constraints and corporate investment. *Brookings Paper on Economic Activity, 1*, 141–195.

Foley, C. F., Hartzell, J. C., Titman, S., & Twite, G. (2007). Why do firms hold so much cash? A tax-based explanation. *Journal of Financial Economics, 86*, 579–607.

Jensen, M. C. (1986). Agency costs of free cash flow, corporate finance and takeovers. *American Economic Review, 86*, 413–435.

Jensen, M. C., & Meckling, W. H. (1976). Theory of the firm: Managerial behavior, agency costs and ownership structure. *Journal of Financial Economics, 3*, 413–435.

Jung, K., Kim, Y.-C., & Stulz, R. M. (1996). Timing, investment opportunities, managerial discretion, and the security issue decision. *Journal of Financial Economics, 42*, 159–185.

Keynes, J. M. (1936). *The general theory of employment, interest and money*. New York: Harcourt, Brace & World.

Kim, C.-S., Mauer, D. C., & Sherman, A. E. (1998). The determinants of corporate liquidity: Theory and evidence. *Journal of Financial and Quantitative Analysis, 33*, 335–359.

Miller, M. H., & Orr, D. (1966). A model of the demand for money by firms. *Quarterly Journal of Economics, 80*, 413–435.

Myers, S. C. (1977). Determinants of corporate borrowing. *Journal of Financial Economics, 5*, 147–175.

Myers, S. C., & Majluf, N. S. (1984). Corporate financing and investment decisions when firms have information that investors do not have. *Journal of Financial Economics, 13*, 187–221.

Opler, T., Pinkowitz, L., Stulz, R., & Williamson, R. (1999). The determinants and implications of corporate cash holdings. *Journal of Financial Economics, 52*, 3–46.

Opler, T. C., & Titman, S. (1994). Financial distress and corporate performance. *Journal of Finance, 49*, 1015–1040.

Ozkan, A., & Ozkan, N. (2004). Corporate cash holdings: An empirical investigation of UK companies. *Journal of Banking and Finance, 28*, 2103–2134.

Savage, L. (1954). *The foundations of statistics*. New York: Wiley.

Shyam-Sunder, L. & Myers, S. C. (1999). Testing static trade-off against pecking order models of capital structure. *Journal of Financial Economics, 51*, 219–244.

Smith, C. W., & Watts, R. L. (1992). The investment opportunity set and corporate financing, dividend and compensation policies. *Journal of Financial Economics, 51*, 219–244.

Stiglitz, J., & Weiss, A. (1984). Informational imperfections in the capital market and macroeconomic fluctuations. *The American Economic Review, 74*, 194–199.

Williams, J. C. (2012). Cash is dead! Long live cash!, Annual report, Federal Reserve Bank of San Francisco.

Woolridge, J. M. (2010). *Econometric analysis of cross section and panel data* (2nd ed.). Cambridge, MA: The MIT Press.

Might Tobin be Right?

The Role of Market Frictions in Policy-Induced Portfolio Shifts for Growth

Ekkehard Ernst

JEL codes: E44, E52, E63, G11, J63

1 Introduction

Balance-sheet recessions and the effect of private sector develeveraging on aggregate demand have come to play an important part in the discussion of the causes of the Great Recession (Eggertsson and Krugman 2012; Koo 2015). In the face of binding debt sustainability constraints in depressed financial markets, housholds and firms aim at reducing their leverage ratio, thereby withdrawing liquidity from goods markets to increase their savings. Governments—to the extent that they are facing less binding constraints—ought to intervene in such situations to restore a faster growth path through deficit-spending and increasing the stock of public debt. Such macro-economic portfolio shifts can help restore sufficient aggregate demand growth and overcome liquidity traps, in which monetary policy interventions are essentially ineffective. And indeed, most of the recent fiscal policy literature argues that under such conditions, public spending multipliers can be significantly above unity, making such interventions highly desirable to get the economy back on a recovery path (Woodford 2011; Batini et al. 2014).

The underlying stock-perspective that is taken by this literature has, however, wider implications. Indeed, it can be argued that in a dynamically growing economy, the impact of such portfolio shifts on the equilibrium interest rate will affect

E. Ernst (✉)
Research Department, International Labour Organization, Route des Morillons 4, 1211 Geneva, Switzerland
e-mail: ernste@ilo.org

L. Bernard, U. Nyambuu (eds.), *Dynamic Modeling, Empirical Macroeconomics, and Finance*, DOI 10.1007/978-3-319-39887-7_11

the steady state growth rate. In economic models, where growth is exogenously determined, a rising interest rate would only lead to a shift in the capital-labour ratio. However, in economies with endogenously determined growth patterns, a rise in the equilibrium interest rate can attract more savings to boost technological change and hence economic growth. This is essentially the mechanism underlying the current debate on secular stagnation and the flight to safety whereby high economic uncertainty and lack of suitable assets cause investors to oversubscribe to assets with low yields and little long-term economic benefits (Caballero and Farhi 2014).

Portfolio shifts cannot only be induced by public sector debt, however. Money creation also can play an essential role in this respect. For the short run, money has come to play a key role in determining output dynamics in modern macroeconomics. In the presence of product market show rigidities with sluggish price setting, the New Keynesian macroeconomic model how central banks can influence economic activity in the short-run despite the long-run neutrality of money (e.g. Woodford 2003). Even though the resulting short-run trade-off between nominal and real variables is not permanently manipulable, it will cause aggregate demand shocks to have both real and nominal effects, sufficient for monetary policy to be able to reach its inflation target.

The transmission mechanism of these New Keynesian workhorse models runs counter earlier insights in the tradition of cash-in-advance (CIA) models. In particular those versions where both consumer and capital goods where subject to the CIA-constraint were characterized by a negative impact of rising money supply on growth due to portfolio shifts (Stockman 1981). The empirical case for such a negative relationship between inflation and growth is not particularly strong (e.g. Cooley and Hansen 1989) but subsequent developments of this CIA literature have extended the New Keynesian insights, making use of limited participation models (e.g. Lucas 1990; Christiano 1994; Fuerst 1992). These models rely on financial market imperfections more than on price rigidities in imperfectly competitive product markets and seem indeed to better account for certain macroeconomic dynamics than their sticky price counterparts (Christiano et al. 1997), albeit being subject to demanding parameter assumptions.

Money-induced portfolio shifts and monetary non-neutrality even in the longer run, however—the thrust of James Tobin's original insights (see Tobin 1965)—seem to have come out of fashion. In light of the empirical evidence on the positive relationship between inflation and growth, at least at low levels of inflation (see e.g. Ahmed and Rogers 2000) and the possibility of a long-run trade-off between inflation and employment as put forward by Akerlof et al. (2000), Tobin's ideas nevertheless seem to continue to bear relevance. Recently, the possibility of inflation as greasing the wheels of friction-ridden labour markets has finally attracted some attention (Wang and Xie 2012), albeit within a model that relies on a restrictive interpretation of the CIA-constraint with firms being obliged to hold cash in order to pay out workers.

Long-run effects of either variations in public sector debt or monetary balances rely on macroeconomic portfolio shifts, which are relevant also during normal times of economic expansions. This paper demonstrates the working of such mechanism and generating higher economic growth through such macroeconomic portfolio shifts even in the absence of short-term fluctuations. In particular, the paper attempts at taking up the original approach favoured by limited-participation models by using financial market imperfections to motivate both the demand for money and its long-run effects. One simple way of introducing imperfect financial markets in an otherwise standard macro-model is to consider search frictions that extend to financial markets in a similar fashion as the—by now standard—set-up of search frictions on labour markets. This allows a natural role for financial intermediaries and costly adjustment of capacity depending on the liquidity of financial markets. Such a framework would allow to mimic limited participation as part of the funds are not being funnelled to productive investments but stuck in the matching process, while on the other hand providing some appealing characteristics both methodologically and theoretically.

The methodological advantages of a search-theoretic approach lie in the parametric nature of its rigidities that can be modelled by references to the parameters both of the matching function and of the institutional framework in which the economy operates. In this regard, the matching approach has made strong inroads into labour economies precisely because it adopts a black-box approach that avoids introducing real and nominal rigidities in an ad hoc manner. Hence, methodologically such a generalised search approach would be attractive as it would stay away from mixing parametric and ad hoc forms of rigidities such as those in the current New Keynesian macroeconomics literature.

Theoretically, such an approach could prove fruitful due to the specific set-up of search models: the stock-flow link for each individual decision economic agents make in these models. This stock-flow link is a peculiar feature of the search literature—in contrast to the competitive RBC world where households will invest in only one asset—as every decision (to work, to open a vacancy, to supply funds, to search for a new product) is modelled in terms of an asset that is being accumulated (the expected discounted wealth of a wage contract, the expected net present value of a filled job, etc.) through an investment decision. These models allow for far more different types of liquidity effects than in the standard literature creating new interesting transmission mechanisms of shocks. In particular, labour supply and savings decisions no longer depend only on current wages, prices and interest rates but on the future expected liquidity of the labour and financial markets.

Based on such a framework with search and matching on financial markets, money demand arises naturally in the model of this paper without any additional assumptions regarding cash-in-advance or the Sidrauskian "money in the utility function" hypotheses. This allows us to formulate straightforwardly an optimal money supply rule against which different monetary policy regimes must be assessed. In addition, the characteristics and determining factors of such a rule can be analysed, in particular the impact of endogenous (AK-)growth, and possible Mundell-Tobin liquidity effects on growth.

Similar to the role of money in the growth process, fiscal policy—in particular the financing side of fiscal policy—can occupy a central role in models where perfect intermediation between deposits and bonds do not exist. Here, we consider a model where government is perfectly social waste but where government activity in form of the financing through taxes or bonds can modify the optimal growth path. Government spending needs to be backed by taxation (at least in the Ricardian set-up that is considered here) but government bonds will provide liquidity to the financial market that raises incentives for households to save and deposit money rather than to consume their income. Hence, there will be an optimal tax-bond equilibrium with non-trivial public debt such as to maximise the optimal growth path.

The paper starts with a discussion of the search and matching methodology applied to both labour and financial markets, based on earlier work by Amable and Ernst (2005), Becsi et al. (2000), Den Haan et al. (2003), Ernst and Semmler (2010), and Wasmer and Weil (2004). Introducing intertemporally optimising households and firms, the financial and labour market equilibria are determined, first in a stationary model without growth and then in an endogenous growth model based on a simple AK-growth mechanism. Then the paper turns to analyse the implications of portfolio shifts induced by monetary and fiscal policies and discusses optimal inflation rates and public debt ratios. In particular, it is shown that due to the ambiguous role played by the inflation rate, an optimal, positive inflation rate is shown to exist that balances the direct negative impact of inflation on growth with its positive impact on financial market liquidity through the household's portfolio decisions. A similar effect can be shown to exist for tax-financed public debt, albeit being limited to a particular sub-set of financial market equilibria. Finally, a positive interaction between monetary and fiscal policy can be demonstrated. In the appendix, a first attempt to assess the out-of-equilibrium dynamics is undertaken based on a calibrated discrete-time version of the model and its reaction to monetary policy and technology shocks.

2 Equilibrium Money Demand and Unemployment

The models short-term dynamics will be determined both by the nominal rigidities on the product market and by search frictions on financial and labour markets. The long-term steady state behaviour of the model, however, will be exclusively determined by the interaction between search frictions on the latter two markets.

2.1 Financial and Labour Market Search

Three types of agents are considered: entrepreneurs, workers and financiers. Entrepreneurs have ideas but cannot work in production and possess no capital. Worker transform entrepreneurs' ideas into output but have neither entrepreneurial skills nor capital; financiers (or bankers) have access to the financial resources required to implement production but cannot be entrepreneurs nor workers. A productive firm is thus a relationship between an entrepreneur, a financier and a worker. In the following, decentralised economy, however, only the optimal program of firms and households are considered, the activity of financial investors in intermediating deposits into bonds is assumed to be summarised in a matching function of the financial market.

Labor market frictions are present under the form of a matching process à la (Pissarides 2000), with a constant returns matching function $m^L(\mathscr{U}, \mathscr{V})$.[1] Matches between workers and firms depend on job vacancies $\mathscr{V}$ and unemployed workers $\mathscr{U}$. From the point of view of the firms, labor market tightness is measured by $\theta \equiv \frac{\mathscr{V}}{\mathscr{U}}$. Labor market liquidity will be $1/\theta$. The instantaneous probability of finding a worker is thus $m^L(\mathscr{U}, \mathscr{V})/\mathscr{V} = m^L(1/\theta, 1) \equiv q(\theta)$, $q_\theta(\theta) < 0$ while for the worker the instantaneous probability to find a firm writes as $m^L(\mathscr{U}, \mathscr{V})/\mathscr{U} = m^L(\theta) \equiv \theta q(\theta)$, $\frac{\partial(\theta q(\theta))}{\partial \theta} > 0$. Finally, the aggregate matching process yielding N filled jobs, and normalising the labour force to unity, i.e. $L = 1$, we have $\mathscr{U} = L - N \Rightarrow u^l \equiv \frac{L-N}{L} = 1 - n$, with $n \equiv \frac{N}{L}$.

Similarly, in order to increase its capital stock, a firm needs to find external funding on financial markets (no internal funding is supposed to be available), which are subject to search frictions, in parallel to those on labour markets. As with labour market search, matching on financial market is characterised by the constant returns matching function $m^B(\mathscr{D}, \mathscr{B})$ where $\mathscr{D}$ is the deposit amount by households to be intermediated into entreprises' bonds $\mathscr{B}$ to finance an increase of their capital stock. From the point of view of firms, credit market tightness is measured by $\phi \equiv \mathscr{B}/\mathscr{D}$ and $1/\phi$ is an index of credit market liquidity, i.e. the ease with which entrepreneurs can find financing. The instantaneous probability that an entrepreneur will find a banker is $m^B(\mathscr{D}, \mathscr{B})/\mathscr{B} = m^B(1/\phi, 1) \equiv p(\phi)$. This probability is increasing in credit market liquidity, i.e. $p'(\phi) < 0$.

2.2 The Optimal Program of Households

Household hold their wealth in three different assets: firm equity, government bonds or money. The economy's capital account, therefore, writes (in real terms) as:

Assets	Liabilities
K: Firm equity	W: Households' wealth
B: Government bonds	
M: Real money balances	

[1] m^L has positive and decreasing marginal returns on each input.

Households hold part of their financial wealth (b) in government bonds, the rest is invested in equity $(1-b)$.[2] In addition, money demand arises endogenously in this model as search friction on the financial market put a wedge between households' deposits, $\mathscr{D}$, and their transformation into interest bearing titles (either equity, K, or government bonds, B). Hence, the rate of unmatched financial assets can be defined as: $u^f = \frac{W-(K+B)}{W} = \frac{M}{W}$. This can be used to write $K = (1-b)\left(1-u^f\right)W$ and $B = b\left(1-u^f\right)W$.[3]

Households earn different returns on these three assets, denoted as r_K, r_B and r_M. Real money balances do not earn interest but depreciate with the inflation rate, π, hence $r_M = -\pi$, where the inflation rate is supposed to be controlled directly by the monetary authority through the evolution of the money supply, M^S, i.e. $\pi = g^m = \frac{\dot{M}}{M} = \hat{M}$.[4,5] Moreover, households have to pay taxes on capital income, hence the after tax rate of return on capital holdings write as $r_K = (1-\tau_K)\,r_K$, with τ_K: taxes on capital income. Finally, in the absence of stochastic returns, the equilibrium condition for households to hold either bonds or financial assets in non-trivial amounts requires that the return on investment on each of the two assets are equal, i.e. $r_K = r_B$.

Households accumulate wealth through deposits $\mathscr{D} = Y - c$ with income generated from asset returns and gainful employment $Y = r_W W + (1-\tau_w)\,w\cdot N$, with τ_w: taxes on labour income, $r_W = [r_K(1-b) + r_B b](1-\tau_K)\left(1-u^f\right) - \pi u^f$ the average *ex post* return on wealth and N the level of employment:

$$\dot{W} = \mathscr{D} = r_W W + (1-\tau_w)\,w\cdot N - c \tag{1}$$

In addition, households face an employment constraint, which depends on the number of jobs created by firms and determines the evolution of their disposable income:

$$\dot{N} = m^L(\theta) - \sigma\cdot N \tag{2}$$

where σ: the rate of job destruction assumed to be exogenously given. Households, then, maximise their intertemporal utility by determining their optimal path of consumption, c, subject to the job dynamics N according to the optimal household

---

[2]In this deterministic case, the portfolio decisions between firm equity and government bonds are exogenously given. They could be endogenized by moving to a stochastic set-up.

[3]For notational convenience, the time subscript is being dropped here and in the following.

[4]In the spirit of the original (Tobin 1965) model, the analysis here concentrates on changes in the long-run, steady state behaviour, in which any short-term effects of money market interest rates have already played out.

[5]The following notational conventions are being used: $\dot{X} \equiv \frac{\partial X}{\partial t}$, $\hat{X} \equiv \frac{\dot{X}}{X}$.

program:

$$\max_{c} \int_{t=0}^{\infty} u(c, N)\, e^{-\rho t} dt$$

$$\text{s.t. (1) and (2)}$$

with the contemporaneous utility function $u(c) = \frac{(c \cdot N^{-\gamma})^{1-\eta}}{1-\eta}$ where η: the (constant) relative risk aversion, ρ: the intertemporal time preference rate and $-\gamma(1-\eta)$: the disutility from labour.[6]

Using standard dynamic programming techniques, the above optimal program can be solved to yield to the optimal path of consumption summarised in the following proposition:

Proposition 2.1 *In steady state, households optimal consumption evolve according to:*

$$g = \frac{\dot{c}}{c} = \frac{1}{\eta}(r_W - \rho) \tag{3}$$

Proof See Appendix 1, p. 277. ∎

Note that consumption growth accelerates with the rate of return on capital and slows down when capital taxes, inflation or unmatched financial resources increase, in line with earlier findings of the CIA literature.

2.3 *Capital and Labour Input of Intermediate Firms*

2.3.1 The Optimal Program of Firms

Firms post vacancies on the labour market and raise funds for investment on the capital market. Firms transform inputs according to a constant-returns-to-scale production function with labour-saving technological progress, A:

$$F = F(K, AN),\, F_K > 0,\, F_N > 0,\, F_{KK} < 0,\, F_{NN} < 0.$$

[6]In most applications of search-based models to short-term macroeconomics, an additive-separable utility function is assumed to guarantee the characteristics of employment lotteries (see the discussion in Andolfatto 1996; Trigari 2009). Such a utility function is not compatible with long-run steady state (endogenous) growth. Here, we make the assumption that labour supply is exogenously given and that households maximise their intertemporal utility against the expected path of employment opportunities given by the expected employment rate.

Keeping a vacancy open is costly and measured in proportion to the ongoing average wage (opportunity costs), $\zeta \cdot w\mathcal{V}$. Similarly, opportunity costs arise for fund raising and finding a financial investor that is ready to buy the firm's obligations, supposed to be measured by $\kappa \cdot r_K\mathcal{B}$. Once a post filled, the job disappears at an exogenous rate σ. Similarly, installed capital depreciates at rate δ. The firm then maximises profits selecting vacancies, $\mathcal{V} \geq 0$, and issues of bonds, $\mathcal{B} \geq 0$, according to the following optimal program:

$$\max_{N,K,\mathcal{V},\mathcal{B}} \int_0^\infty \nu_1(t)\left(F(K,AN) - wN - r_K K - \zeta \cdot w\mathcal{V} - \kappa \cdot r_K\mathcal{B}\right)e^{-\rho t}dt \quad (4)$$

where ρ stands for the intertemporal preference rate and $\nu_1(t)$ for the household's shadow variable related to its accumulation constraint,[7] which firms maximise against the constraints of employment and capital accumulation:

$$\dot{N} = q(\theta)\mathcal{V} - \sigma \cdot N \quad (5)$$

$$\dot{K} = p(\phi)\mathcal{B} - \delta \cdot K. \quad (6)$$

2.3.2 Determining Wages and Interest Payments Through Nash Bargaining

Wages and interest rates are negotiated on labour and financial markets respectively. Workers enjoy bargaining power β, financial investors pressure for interest rates with bargaining power γ. In order to simplify the bargaining process, we will follow Eriksson (1997), here assuming that the bargaining power describes the bounds between which wages and interest rates have to fall:

- Workers have a fall-back option provided by the social security system, which guarantees a replacement income in form of unemployment benefits, R_0. Hence, wages will be determined as a weighted average of on the one hand marginal productivity, F_N, and the gain of filling a vacancy, $\theta\zeta_0$, and unemployment benefits, R_0, on the other:

$$w = \beta(F_N + \theta\zeta_0) + (1-\beta)R_0$$

- Interest payment will be determined as a part of the marginal productivity of capital and the gain of matching issued debt, assuming that the fall-back option for financial investors is zero (we are assuming a closed economy here, i.e. no investment opportunities on international financial markets). Moreover, interest

[7] This follows the spirit of Merz (1995) determination of the competitive search equilibrium (see Merz 1995, pp. 287–293). It is, indeed, easy to show that this formulation is equivalent to the one where the intertemporal preference rate for firms is replaced by the—commonly used—interest rate for capital. In both cases, it is assumed that capital is ultimately owned by households.

payments will be negotiated before the firm hires its employees, hence the effect of installed capital on raising wage claims are being taken into account:

$$r_K = \gamma\left(F_K - w_K(\theta)N + \phi\kappa_0\right).$$

Unemployment benefits are assumed to be adjusted in line with real wages in order to capture increases in productivity. This will affect the increase in the reservation wage, the vacancy costs and the costs of scheduling new debt over time, we write $R_0 = R \cdot w\ (0 \leq R < 1)$, $\zeta_0 = \zeta \cdot w$ and $\kappa_0 = \kappa \cdot r_K$. We, therefore, can reformulate the two prices as:

$$\begin{aligned} w &= \beta\left(F_N + \theta\zeta \cdot w\right) + (1-\beta) R \cdot w \\ \Leftrightarrow w &= \frac{\beta}{1 - \beta\theta\zeta - (1-\beta) R} F_N \qquad (7) \\ r_K &= \gamma\left(F_K - w_K(\theta) N + \phi\kappa \cdot r_K\right) \\ \Leftrightarrow r_K &= \frac{\gamma}{1 - \gamma\phi\kappa}\left(F_K - w_K(\theta) N\right) \qquad (8) \end{aligned}$$

Hence, wages are only affected by labour market conditions in the current period, while interest rates are influenced by both labour and financial market conditions.[8]

2.3.3 Equilibrium on Capital and Labour Markets

The program set up by (4) subject to (5)–(8) yields an equilibrium on capital and labour markets summarised in the following proposition.

Proposition 2.2 *Firms select the optimal stream of jobs and investment according to:*

$$(1-\beta)(1-R) - \beta\theta\zeta - \frac{\zeta\beta}{q(\theta)}\left(\sigma + \rho - \hat{w} + \eta\frac{\dot{c}}{c}\right) = 0 \qquad (9)$$

$$1 - \gamma(1 + \phi\kappa) - \frac{\kappa\gamma}{p(\phi)} \cdot \left(\delta + \rho + \eta\frac{\dot{c}}{c}\right) = 0 \qquad (10)$$

Proof See Appendix 2, p. 278. ■

[8] In a more elaborate version, one could introduce a fully intertemporal negotiation model where wages and interest rates would also be affected by future developments in their respective markets, as discussed in Layard et al. (1991).

3 Macroeconomic Policies and Long-Run Steady State Shifts

3.1 Steady State with Exogenous Growth

Once the optimal paths for consumption and labour supply determined, the availability of financial funds from which firms can draw their resources derives. This can be used to determine the optimal capital-labour ratio, k, that has been left undetermined in the model thus far. Let multi-factor productivity grow at an exogenously given rate g, i.e. $A(t) = A_0 e^{gt}$. In equilibrium, given a constant population, i.e. $\dot{L} = 0$, all dynamic variables need to grow at the same rate:

$$\frac{\dot{c}}{c} = \frac{\dot{K}}{K} = \frac{\dot{\mathscr{B}}}{\mathscr{B}} = \frac{\dot{A}}{A} = \frac{\dot{w}}{w} = g$$

Note that when growth is exogenously given, the optimal financial and labour market liquidities can be uniquely determined from (9) and (10). This, in turn, allows to derive steady-state employment as:

$$N^* = \frac{m^L(\theta^*)}{\sigma + m^L(\theta^*)}.$$

Moreover, given that $\dot{K}/K = g \Rightarrow p(\phi)\,\mathscr{B}/K - \delta = g$ we have:

$$\frac{\mathscr{B}^*}{K^*} = \frac{g + \delta}{p(\phi^*)}.$$

Now, recall the optimal consumption path $\frac{\dot{c}}{c} = \frac{1}{\eta}(r_W - \rho)$, which leads to the steady state return on equity:

$$r^* = \frac{\eta g + u^f \pi - b\left(1 - u^f\right) r_B + \rho}{(1-b)(1-u^f)}.$$

The exogenously given growth rate of consumption, g, together with the wage- and interest-rate curves (7) and (8), can then be used to determine the steady-state capital intensity as the solution of the set of non-linear equations:

$$r^* = \frac{\eta g + u^f \pi - b\left(1 - u^f\right) r_B + \rho}{(1-b)(1-u^f)} \stackrel{!}{=} \frac{\gamma}{1 - \gamma\phi\kappa}\left[f'(k) - w_K\left(\theta^*\right) N^*\right] \quad (11)$$

$$w^* = \frac{\beta}{1 - \beta\theta^*\zeta - (1-\beta)R}\left(f(k^*) - k^* f'(k^*)\right). \quad (12)$$

The following proposition can be immediately derived from the above equilibrium interest and wage rates:

Proposition 3.1 *When growth is exogenously given, a rise in the inflation rate, π, will increase the equilibrium interest rate, r^*, and lower the steady-state capital-labour ratio, k^*. As a consequence, it will lower the optimal wage level, w^*. Equilibrium liquidity on labour and capital markets, θ^* and ϕ^*, remain unaffected.*

Proof By inspection of Eqs. (11) and (12). ∎

Hence, in line with the CIA literature, the introduction of financial and labour market search frictions does not change the effect of inflation on the macroeconomic equilibrium. With growth given exogenously, inflation is by assumption neutral to the aggregate dynamics. It will, however, have an effect on portfolio decisions through its positive impact on the ex-post warranted real interest rate, r^*. Consequently, the aggregate capital stock will be lower, negatively affecting the base level of negotiated wages. In sum, an increase in inflation will lead to a decrease in economic welfare through a portfolio shift, even so the growth rate remains unaffected.

3.2 Long-Term Shifts of the Steady State

How do these results change when growth is endogenously determined in the model? Existing CIA-models that include assumptions on endogenous growth suggests that superneutrality of money should also hold in these cases (e.g. Grinols and Turnovsky 1998a,b). In addition, public debt does not serve any purpose; instead, the government should be issuing assets so as to help stabilising the economy.

However, these papers do not consider financial market frictions and therefore do not take any portfolio shifts into account that may be induced by monetary and budgetary policies. As seen in the previous section, such portfolios do not have any long-run steady state effects when growth is exogenous. However, they may induce changes in the steady state when growth is endogenous. In particular, changes in the equilibrium interest rate will have immediate consequences for the rate of growth in equilibrium, irrespective of the underlying mechanism that generates endogenous growth (e.g. *AK*, Schumpeterian or multi-sector models with human capital spillovers).

In order to demonstrate the importance of such a mechanism, endogenous growth is introduced in its simplest form in the above model. Here, we follow Romer (1986, 1994) and consider spillovers from installed capital to the overall economy. In order to avoid scale effects (Jones 1995), we consider average capital per worker to be the relevant externality. Hence, the production function for each single firm i writes as:

$$F^i\left(K_i, A_i N_i\right) = \overline{K}^{\alpha}\left(A_i N_i\right)^{\alpha} K_i^{1-\alpha}$$

where $\overline{K} = \frac{\sum_i K_i}{\sum_i N_i} = \frac{K}{N}$ and A_i: a base technology parameter. Hence, in equilibrium where $\overline{A} = A_i = A_j$ and $K_i = K_j$, $N_i = N_j$, $i \neq j$, aggregate production writes as:

$$F(K, A \cdot N) = \sum_i F^i(K_i, A_i N_i) = i \cdot \left(\frac{K}{N}\right)^{\alpha} \cdot (A \cdot N)^{\alpha} K^{1-\alpha} = i \cdot \overline{A} \cdot \overline{K}.$$

The externality now stems from the number of firms, i, rather from the size of the economy. Letting $A = i \cdot \overline{A}$, the equilibrium interest rate (8) now writes as:

$$\begin{aligned} r_K &= \frac{\gamma}{1-\gamma\phi\kappa}\left(F_K - w_K(\theta) N\right) \\ \Leftrightarrow r_K &= \frac{A\gamma}{1-\gamma\phi\kappa}\left(1-\alpha-\frac{\beta\alpha}{1-\beta\theta\zeta-(1-\beta)R}\right) \end{aligned}$$

This can be used to rewrite the optimal growth of consumption (3) as:

$$g = \frac{A\gamma(1-\tau_K)\left(1-u^f\right)}{\eta(1-\gamma\phi\kappa)}\left(1-\alpha-\frac{\beta\alpha}{1-\beta\theta\zeta-(1-\beta)R}\right) - \frac{\rho+\pi u^f}{\eta} \qquad (13)$$

Equation (13) shows that growth depends on both financial and labour market liquidity: Tighter financial markets lead to an increase in interest rates, which helps channeling more resources into savings and hence capital accumulation; this will contribute to a faster increase in technological progress and hence faster growth.[9] On the other hand, the growth rate decreases with labour market tightness due to the strong effects on wage increases.[10]

Plugging (13) into (9) and (10) allows to determine the impact of endogenous growth on the labour and financial market steady states. The long-term equilibrium in endogenous growth can then be derived from:

$$\Theta(\theta, \beta, \zeta, R) - r_K(\phi, \theta)(1-\tau_K)\left(1-u^f\right) = \frac{\eta\sigma+\rho}{\eta-1} - \pi u^f \qquad (14)$$

$$\Phi(\phi, \gamma, \kappa) - r_K(\phi, \theta)(1-\tau_K)\left(1-u^f\right) = \delta - \pi u^f \qquad (15)$$

where $\frac{\partial r_K}{\partial \phi} > 0$, $\frac{\partial r_K}{\partial \theta} < 0$, $\Theta(\theta, \beta, \zeta, R) = \eta q(s, \theta)\frac{(1-\beta)(1-R)-\beta\theta\zeta}{\zeta\beta(\eta-1)}$, $\Theta_\theta < 0$, $\Theta_\beta < 0$, $\Theta_R < 0$, $\Theta_\zeta < 0$ and $\Phi(\phi, \gamma, \kappa)$, $\Phi_\phi < 0$, $\Phi_\gamma < 0$, $\Phi_\kappa < 0$. Given that growth is negatively affected by rising labour market tightness due to an increase in the labour income share ($w(\theta)N$ unambiguously increases with θ), Eq. (14) now allows two

[9]Note that u^f declines with increases in financial market tightness.

[10]Note that in this model, prices are fully flexible and wage increases do not lead to an expansion of aggregate demand.

roots in θ, one stable and one unstable. In the following, we want to concentrate on the stable one.

Proposition 3.2 *Let $\eta > 1$. Then, at the stable root liquidity on the labour market declines but increases on financial markets with an increase in the level of labour productivity, A. Moreover, both labour and financial markets become tighter with increases in the inflation rate, π.*

Proof (Sketch, for Full Proof See Appendix 3, p. 279.) The equilibrium interest rate, r_K, increases linearly with ϕ at slope γ_K. The growth rate, in turn, is positively affected by an increase in the interest rate and a decrease in the amount of unused funds that suffer from the inflation tax:

$$\frac{dg}{d\phi} = \frac{1}{\eta}\left[(1-\tau_K)\left(1-u^f\right)\frac{dr_K}{d\phi} - (r_K(1-\tau_K)+\pi)\frac{du^f}{d\phi}\right] > 0.$$

As regards the productivity level, A, note that it affects both labour market and financial market liquidity in a negative way provided that $\eta > 1$. Given that financial market liquidity impacts positively on labour market liquidity, the sign of $\frac{d\theta}{dA}$ is unambiguously negative, at least for values of η greater than unity. ■

3.3 Monetary Policy

In the long run, the monetary authority is supposed to control the money supply to reach its inflation target. This can be done without further repercussions to the real economy only to the extent that growth is exogenously given. In this case, inflation will have no impact on the rate of accumulation of household wealth. However, both in the case of exogenous and endogenous growth, the depreciation of real money balances will impact on the portfolio balance through the financial market liquidity. In particular, financial market tightness will increase as households prefer to withdraw their funds and to redirect them towards consumption when inflation rises.

In the exogenous growth model, this does not have any further consequences. However, in the endogenous growth model, a reduction in the financial market liquidity will lead to a rise in the equilibrium before-tax interest rate; this causes the steady state growth rate to increase. Hence the following proposition can be proven:

Proposition 3.3 *In the exogenous growth case, money is neutral in the long-run. In the endogenous growth case, however, an optimal inflation rate $\overline{\pi} > 0$ exists such that for $\pi < \overline{\pi}$ the steady state growth rate increases with the inflation rate and decreases for $\pi > \overline{\pi}$, leading to a hump shaped impact of inflation on growth.*

Proof See Appendix 4, p. 283. ■

Similar to the spirit of the original Tobin model, therefore, we get an impact of portfolio choices on growth that is determined by the inflation rate (Walsh 2001, Chap. 2). However, due to the still negative impact of inflation on the overall return of the households' portfolio, there will only be a certain range of inflation rates for which the impact is positive. Hence, to the extent that monetary authorities control the inflation rate directly, there will be a growth maximising money supply rule. This, in turn, is in line with recent research on search models focussing on labour market search only and introducing money demand through a cash-in-advance rule (Wang and Xie 2012) and also matches empirical evidence cited in the introduction.

Moreover, it can be shown that having employment as an additional or alternative objective for the central bank would complicate matters in this model:

Proposition 3.4 *A central bank that maximises a weighted sum of growth and inflation will not opt for the growth-maximising inflation rate but a lower one. Moreover, a central bank that targets both growth and employment (besides inflation) will face a trade-off between employment-maximisation and output growth-maximisation.*

Proof From (14) follows that inflation has a positive (negative) impact on the vacancy rate for $\eta > 1$ ($\eta < 1$). On the other hand, growth will have a negative (positive) impact on employment creation for $\eta > 1$ ($\eta < 1$). Therefore, whatever the value of η, there exists an employment-maximising inflation rate $\pi^E \neq \overline{\pi}$ that is different from the growth-optimizing inflation rate. ∎

3.4 Fiscal Policy

The model only considers Ricardian budgetary policies—i.e. there is no fiscal dominance—and their impact on steady state labour and financial market equilibrium. Under such a fiscal rule, public debt is supposed to remain on a non-explosive path with the government running a balanced budget (most of the time). Unproductive government expenditures and interest payments on government bonds are being financed by tax revenues and changes in government debt:

$$G + r_B B = T + \dot{B}, \quad B < \infty$$

where $T = \tau_K r_K K + \tau_w w N$.

To simplify matters even further, government spending is assumed to be constant and nil ($G = 0$) and the government uses the tax burden, T, to control the size of public debt given that in a Ricardian fiscal regime a rise in public debt cannot be financed through higher deficits. In our set-up, taxes and the tax structure have a particular role to play as taxes on labour and taxes on capital do not work the same way: Indeed, as can be seen from (13), only capital taxes will directly affect the equilibrium growth rates. Labour income taxes, on the other hand, will impact on capital accumulation only indirectly via possible effects of labour market liquidity on the financial market equilibrium. Their main effect, however, will come from the labour supply, thereby principally leading to a level shift instead of a growth effect.

Similarly to the impact of monetary policy, the impact of taxes can be assessed using the full system of financial and labour market equations. While labour market liquidity reacts ambiguously with an increase in τ_K—it increases for $\eta > 1$ and decreases otherwise—financial market tightness, ϕ, increases and so do interest rates as government bonds compete with bonds issued by firms. Hence, *ceteris paribus* an increase in government financing through bonds will lift ϕ. Hence, two competing channels of the impact of taxes on growth arise, opening the possibility for an optimal, growth-maximising stock of public debt. In this regard, a comparison of a no-tax regime ($\tau = 0, \phi^*$) with a regime with positive capital income taxes and positive government debt ($\tau > 0, \phi^{**}$), where $\phi^{**} > \phi^*$ due to government bonds, yields the following proposition. Note that in order to assess the impact of fiscal policy in isolation, we abstract from monetary policy for the moment, setting the inflation rate to zero, i.e. $\pi = 0$.

Proposition 3.5 *For sufficiently low equilibrium financial market tightness $\phi \leq \overline{\phi}$, there exists an interior value for tax-backed bond financing of government spending (i.e. Ricardian fiscal policy). The upper bound of a positive optimal tax policy depends on the share of government revenues financed through capital taxation, $\varpi = \frac{\tau_K}{\tau}$, i.e. $\overline{\phi} = \overline{\phi}(\varpi)$, $\overline{\phi}' < 0$.*

Proof See Appendix 5, p. 284. ■

By introducing a competing asset that raises steady state interest rates, the government can affect the equilibrium growth rate above the rate obtained in a no-government equilibrium. This, however, is only the case for relatively low financial market tightness: the higher it is, the less of an impact will the introduction of government bonds on interest rates have. Unfortunately, there is no guarantee that the optimal tax policy will be positive at the financial market equilibrium and the optimal inflation rate implicitly defined by Proposition 3.3, and no closed-form algebraic solution can be derived. In order to get a sense of the equilibrium values, the following two graphs give an idea on the parameter space (γ, κ) that would allow positive optimal tax policies, one for a fully capital-taxed financed budget (right panel) and one for a budget financed at one quarter by capital taxation, as is common in OECD countries (Fig. 1).

3.5 Interaction Between Monetary and Fiscal Policy

The growth rate of the economy can be maximised through a coordination of monetary and fiscal policies. In this case, we have to introduce seignorage into the government's budget constraint to account for the impact of different levels of inflation (and money supply growth). The new government budget constraint

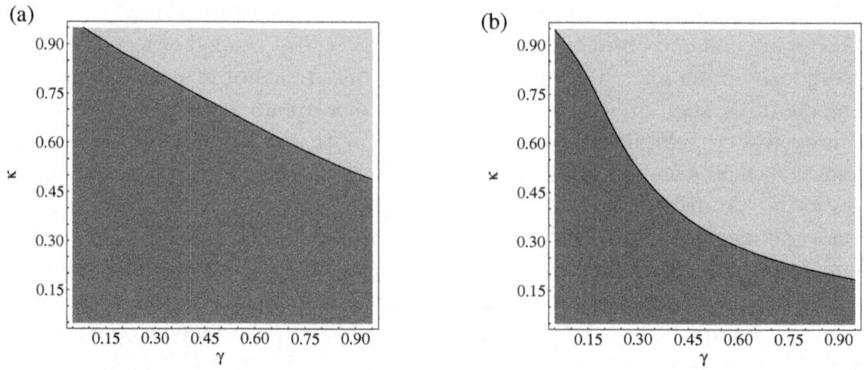

Fig. 1 Allowable parameter space for positive optimal tax rates. (**a**) Full capital taxation financed public deficit (ϖ=1). (**b**) Public deficit financed through capital taxation by 25 %(ϖ=0.25). Note: The *light grey shaded* areas indicate the parameter tupels (κ, γ) for which the financial market equilibrium is consistent with a positive optimal tax rate

including seignorage (s) now writes as:

$$\dot{B} = r_B B - T + G - s$$

Lemma 3.6 *Raising the money supply growth rate increases the optimal rate of taxation* $\left(\tau_K^*, \tau_w^*\right)$ *provided that* $\phi(\overline{\pi}) \leq \overline{\phi}$.

Proposition 3.7 *The growth rate can be maximised when government spending is financed through a combination of seignorage and bonds. The optimal growth rate of money is in this case also higher than in the absence of government spending (and hence the inflation rate is higher).*

Proof See Appendix 6, p. 285. ■

In other words, government inactivity is not optimal and the policy mix can be set in such a way that it improves over the individual instruments.

4 Conclusion

Search and matching frictions on financial and labor markets have been introduced in an otherwise standard dynamic general equilibrium model. In the absence of product market frictions the analysis has emphasised the long-run behaviour of the model at the steady state, both when growth is exogenous and endogenous. In particular, the reaction of the stationary state with respect to changes in monetary and fiscal policies has been analysed.

In this model, households have a demand for money that arises endogenously with the fact that not all deposits will be immediately transformed into profitable savings as financial markets are ridden by search frictions. This creates portfolio effects of inflation as households will face a lower rate of return on their total wealth when inflation increases. Nevertheless, and in the absence of endogenous growth, money is superneutral.

However, when growth is endogenous and depends on the state of the financial market due to the effect the real interest rate has on capital deepening and hence on growth, monetary and fiscal policies are shown to be no longer neutral. As they affect financial market liquidity through portfolio shifts (monetary policy) and the introduction of a competing, risk-free asset (public debt), growth will no longer be unaffected in steady state by macroeconomic policies. In particular, the paper demonstrates the existence of a growth-maximising inflation rate, a (possible) trade-off between growth and employment maximisation, and the possibility for fiscal policies to increase the growth rate in case of very illiquid financial markets (from the point of view of households) by offering tax-backed public debt. Finally, positive interaction between monetary and fiscal policy has been shown to exist due to their complementary effect on the financial market equilibrium, inducing an optimal policy mix that maximises growth over and above the individual contribution of either monetary or fiscal policy.

The paper demonstrates the usefulness of introducing search frictions not only on the labour market but also on the financial market in particular when attempting to generate and analyse portfolio shifts induced by monetary policy. Further developments along these lines would concentrate more on the short-term dynamics, introducing price or information stickiness in order to generate a downward-sloping short-run Phillips curve. Moreover, the endogenous growth mechanism that has been retained for this paper is fairly standard; different mechanisms—for instance Schumpeterian growth—could be introduced instead, in order to assess to what extent the results presented here carry over to these alternative growth mechanisms.

Appendix 1: The Household Program

To solve the problem, the Hamiltonian is set up as follows:

$$\mathcal{H} = \frac{\left(c \cdot N^{-\gamma}\right)^{1-\eta}}{1-\eta} + \nu \left[r_W W - c + (1-\tau_w)\, wN\right].$$

The household's first-order condition can then be determined as:

$$\frac{\partial \mathcal{H}}{\partial c} = 0 \Leftrightarrow c^{-\eta} N^{-\gamma(1-\eta)} = \nu \tag{16}$$

The co-state variable W evolves according to:

$$\dot{\nu}_1 = \rho \nu_1 - \frac{\partial \mathcal{H}}{\partial W} = \rho \nu - \nu r_W \tag{17}$$

Deriving (16) with respect to time, noting that $\dot{N} = 0$ in steady state and substituting $\dot{\nu}_1$ and ν_1 in (17), the household's optimal consumption path can be determined as:

$$-\eta\dot{c}c^{-\eta-1} = c^{-\eta}(\rho - r_W)$$
$$\Leftrightarrow \hat{c} \equiv \frac{\dot{c}}{c} = \frac{1}{\eta}(r_W - \rho) \tag{18}$$

Appendix 2: Equilibrium Investment and Hiring

The firm maximises profits selecting vacancies, $\mathscr{V} \geq 0$, and issues of bonds, $\mathscr{B} \geq 0$, subject to costs of job vacancy creation, $\zeta \cdot w \cdot \mathscr{V}$, and finding a financial investor, $\kappa \cdot r_K \cdot \mathscr{B}$. Accordingly, its optimal program writes as:

$$\max_{N,K,\mathscr{V},\mathscr{B}} \int_0^\infty \nu_1(t)\left(F(K, AN) - wN - r_K K - \zeta \cdot w\mathscr{V} - \kappa \cdot r_K \mathscr{B}\right) e^{-\rho t} dt$$

where ρ stands for the intertemporal preference rate and $\nu_1(t)$ for the household's shadow variable related to its accumulation constraint,[11] which firms maximise against the constraints of employment and capital accumulation, Eqs. (5) and (6).

Capital is predetermined with respect to the wage bargaining process and firms will take the impact of their capital decision on wages into account. The Hamiltonian therefore writes as:

$$\mathscr{H} = \nu_1(t)\left[F(K,N) - w(K)N - r_K K - \zeta \cdot w\mathscr{V} - \kappa \cdot r_K \mathscr{B}\right] + \lambda\left[q(\theta)\mathscr{V} - \sigma N\right]$$
$$+\mu\left[p(\phi)\mathscr{B} - \delta K\right]$$

leading to the first-order conditions:

$$\frac{\partial \mathscr{H}}{\partial \mathscr{V}} = 0 \Leftrightarrow \nu_1(t)\zeta \cdot w = \lambda q(\theta) \Leftrightarrow \lambda = \nu_1(t)\frac{\zeta \cdot w}{q(\theta)}$$
$$\frac{\partial \mathscr{H}}{\partial \mathscr{B}} = 0 \Leftrightarrow \nu_1(t)\kappa \cdot r = \mu p(\phi) \Leftrightarrow \mu = \nu_1(t)\frac{\kappa \cdot r_K}{p(\phi)}$$

[11] This follows the spirit of Merz (1995) determination of the competitive search equilibrium (see Merz 1995, pp. 287–293). It is, indeed, easy to show that this formulation is equivalent to the one where the intertemporal preference rate for firms is replaced by the—commonly used—interest rate for capital. In both cases, it is assumed that capital is ultimately owned by households.

and the evolution of the co-state variables:

$$\dot{\lambda} = \rho\lambda - \frac{\partial \mathcal{H}}{\partial N} = \lambda\left(\rho + \sigma\right) - \nu_1\left(t\right)\left(F_N - w\right)$$

$$\dot{\mu} = \rho\mu - \frac{\partial \mathcal{H}}{\partial K} = \mu\left(\rho + \delta\right) - \nu_1\left(t\right)\left(F_K - N\frac{\partial w^*}{\partial K} - r_K\right)$$

Deriving the first-order conditions with respect to time, one obtains:

$$\dot{\lambda} = \frac{\zeta \cdot \dot{w}}{q\left(\theta\right)} + \dot{\nu}_1\left(t\right)\frac{\zeta \cdot w}{q\left(\theta\right)}$$

$$\dot{\mu} = \frac{\kappa \cdot \dot{r}_K}{p\left(\phi\right)} + \dot{\nu}_1\left(t\right)\frac{\kappa \cdot r_K}{p\left(\phi\right)}$$

and hence:

$$\frac{\zeta \cdot \dot{w}}{q\left(\theta\right)}\nu_1\left(t\right) + \dot{\nu}_1\left(t\right)\frac{\zeta \cdot w}{q\left(\theta\right)} = \nu_1\left(t\right)\frac{\zeta \cdot w}{q\left(\theta\right)}\left(\rho + \sigma\right) - \nu_1\left(t\right)\left(F_N - w\right)$$

$$\frac{\kappa \cdot \dot{r}_K}{p\left(\phi\right)}\nu_1\left(t\right) + \dot{\nu}_1\left(t\right)\frac{\kappa \cdot r_K}{p\left(\phi\right)} = \nu_1\left(t\right)\frac{\kappa \cdot r_K}{p\left(\phi\right)}\left(\rho + \delta\right) - \nu_1\left(t\right)\left(F_K - N\frac{\partial w^*}{\partial K} - r_K\right)$$

which can be rewritten as:

$$F_N - w + \frac{\zeta \cdot w}{q\left(\theta\right)}\left[\hat{w} + \frac{\dot{\nu}_1\left(t\right)}{\nu_1\left(t\right)} - \rho - \sigma\right] = 0$$

$$F_K - N\frac{\partial w^*}{\partial K} - r_K + \frac{\kappa \cdot r_K}{p\left(\phi\right)}\left[\hat{r_K} + \frac{\dot{\nu}_1\left(t\right)}{\nu_1\left(t\right)} - \rho - \delta\right] = 0$$

where $\hat{w} \equiv \frac{\dot{w}}{w}$ and $\hat{r_K} \equiv \frac{\dot{r_K}}{r_K}$.

Using the wage- and interest-rate curves, Eqs. (7) and (8), and noting that in the steady-state equilibrium $\hat{r}_K = 0$ and $\nu_1\left(t\right) = c^{-\eta} \Rightarrow \frac{\dot{\nu}_1(t)}{\nu_1(t)} = -\eta\frac{\dot{c}}{c}$ will give the steady state conditions for labour and financial market liquidity (9) and (10) as given in the proposition.

Appendix 3: Changes in Growth Following Supply and Demand Shocks

Instead of assuming simultaneous negotiations of wages and interest rates, a more realistic assumption would be to consider wages to be set after the capital stock has been built. In this case, interest payment will take into account the effect of the capital stock development on wages, in addition to the marginal productivity of capital and the gain of matching issued debt:

$$r_K = \gamma \left(F_K - N \frac{\partial w^*}{\partial K} + \phi \kappa_0 \right).$$

Hence the equilibrium interest rate writes as:

$$r_K = \gamma \left(F_K - w_K^* (\theta) \cdot N + \phi\kappa \cdot r_K \right) \Leftrightarrow r_K = \frac{\gamma}{1 - \gamma\phi\kappa} \left(F_K - w_K^* (\theta) \cdot N \right)$$

$$r_K = \frac{\gamma}{1 - \gamma\phi\kappa} \left(f'(k) - w_K^* (\theta) \cdot N \right)$$

When endogenous growth is introduced, the rate of return on capital writes as[12]:

$$\begin{aligned} r_K &= \gamma \left(F_K - w_K^* (\theta) \cdot N + \phi\kappa_0 \right) \\ \Leftrightarrow r_K &= \frac{\gamma}{1 - \gamma\phi\kappa} \left(A (1 - \alpha) - w_K^* (\theta) \cdot N \right) \\ \Leftrightarrow r_K &= \frac{A\gamma}{1 - \gamma\phi\kappa} \left(1 - \alpha - \frac{\beta\alpha}{1 - \beta\theta\zeta - (1 - \beta) R} \right) \end{aligned}$$

and we have:

$$\frac{\partial r_K}{\partial \theta} < 0, \frac{\partial r_K}{\partial \phi} > 0$$

---

[12]Note that given the labour productivity externality the cross derivate writes as:

$$\frac{\partial^2 F (K_i, A_i N_i)}{\partial N \partial K} = \frac{\partial}{\partial K} \left(\frac{\partial F}{\partial N} \right) = \frac{\partial}{\partial K} \left(\alpha A_i \cdot \overline{K}^{\alpha} N_i^{\alpha - 1} K_i^{1 - \alpha} \right) = \alpha A_i \cdot \overline{K}^{\alpha} N_i^{\alpha - 1} K_i^{-\alpha}$$

and in equilibrium:

$$\frac{\partial^2 F}{\partial N \partial K} = \alpha A \cdot N^{-1}.$$

Hence, the term containing the optimal wage reaction can be written as:

$$\begin{aligned} w_K^* (\theta) \cdot N &= \frac{\beta}{1 - \beta\theta\zeta - (1 - \beta) R} F_{NK} N = \frac{\beta}{1 - \beta\theta\zeta - (1 - \beta) R} \alpha A \cdot N^{-1} N \\ &= \frac{\beta\alpha A}{1 - \beta\theta\zeta - (1 - \beta) R} \end{aligned}$$

The simultaneous equilibrium on labour and financial markets therefore writes as:

$$(1-\beta)(1-R)-\beta\theta\zeta-\frac{\zeta\beta}{q(\theta)}(\sigma+\rho+(\eta-1)g)=0$$

$$1-\gamma(1+\phi\kappa)-\frac{\kappa\gamma}{p(\phi)}\cdot(\delta+\rho+\eta g)=0$$

Using (3), the equilibrium condition for returns to investment, $r_W = r_B = r_K$ and inputting the equilibrium value for r_K, we obtain:

$$\begin{aligned}&(1-\beta)(1-R)\\&-\beta\theta\zeta-\frac{\zeta\beta}{q(\theta)}\left(\sigma+\frac{\rho}{\eta}+\left(\frac{\eta-1}{\eta}\right)\left[\frac{A\gamma}{1-\gamma\phi\kappa}\left(1-\alpha-\frac{\beta\alpha}{1-\beta\theta\zeta-(1-\beta)R}\right)\right.\right.\\&\qquad\left.\left.(1-\tau_K)\left(1-u^f\right)-\pi u^f\right]\right)=0\\&1-\gamma(1+\phi\kappa)-\frac{\kappa\gamma}{p(\phi)}\cdot\left(\delta+\frac{A\gamma}{1-\gamma\phi\kappa}\left(1-\alpha-\frac{\beta\alpha}{1-\beta\theta\zeta-(1-\beta)R}\right)\right.\\&\qquad\left.(1-\tau_K)\left(1-u^f\right)-\pi u^f\right)=0\end{aligned}$$

and, hence, after rearranging terms:

$$\begin{pmatrix}LL=0\\FF=0\end{pmatrix}$$

$$\Leftrightarrow\begin{pmatrix}\Theta(\theta,\beta,\zeta,R)-\frac{A\gamma}{1-\gamma\phi\kappa}\left(1-\alpha-\frac{\beta\alpha}{1-\beta\theta\zeta-(1-\beta)R}\right)(1-\tau_K)\left(1-u^f\right)=\frac{\eta\sigma+\rho}{\eta-1}-\pi u^f\\\Phi(\phi,\gamma,\kappa)-\frac{A\gamma}{1-\gamma\phi\kappa}\left(1-\alpha-\frac{\beta\alpha}{1-\beta\theta\zeta-(1-\beta)R}\right)(1-\tau_K)\left(1-u^f\right)=\delta-\pi u^f\end{pmatrix}$$

These equilibrium conditions are no longer characterised by a triangular structure. In order to determine the effect of inflation and multi-factor productivity on the labour and financial market liquidity, we will make use of Cramer's rule. Writing $x\in\{A,\pi\}$ and fully differentiating LL and FF we obtain:

$$\begin{pmatrix}\frac{\partial LL}{\partial\theta}&\frac{\partial LL}{\partial\phi}\\\frac{\partial FF}{\partial\theta}&\frac{\partial FF}{\partial\phi}\end{pmatrix}\begin{pmatrix}d\theta\\d\phi\end{pmatrix}=-\begin{pmatrix}\frac{\partial LL}{\partial x}\\\frac{\partial FF}{\partial x}\end{pmatrix}dx$$

Applying Cramer's rule allows to write:

$$\frac{d\theta}{dx}=-\frac{\begin{vmatrix}\frac{\partial LL}{\partial x}&\frac{\partial LL}{\partial\phi}\\\frac{\partial FF}{\partial x}&\frac{\partial FF}{\partial\phi}\end{vmatrix}}{\begin{vmatrix}\frac{\partial LL}{\partial\theta}&\frac{\partial LL}{\partial\phi}\\\frac{\partial FF}{\partial\theta}&\frac{\partial FF}{\partial\phi}\end{vmatrix}},\quad\frac{d\phi}{dx}=-\frac{\begin{vmatrix}\frac{\partial LL}{\partial\theta}&\frac{\partial LL}{\partial x}\\\frac{\partial FF}{\partial\theta}&\frac{\partial FF}{\partial x}\end{vmatrix}}{\begin{vmatrix}\frac{\partial LL}{\partial\theta}&\frac{\partial LL}{\partial\phi}\\\frac{\partial FF}{\partial\theta}&\frac{\partial FF}{\partial\phi}\end{vmatrix}}$$

Independently of the value of η we have:

$$\frac{\partial FF}{\partial \phi} < 0, \frac{\partial FF}{\partial \theta} > 0, \frac{\partial LL}{\partial \phi} < 0, \frac{\partial LL}{\partial A} = \frac{\partial FF}{\partial A} < 0, \frac{\partial LL}{\partial \pi} = \frac{\partial FF}{\partial \pi} > 0$$

and

$$\left|\frac{\partial LL}{\partial \theta}\right| > \left|\frac{\partial FF}{\partial \theta}\right|, \left|\frac{\partial LL}{\partial \phi}\right| < \left|\frac{\partial FF}{\partial \phi}\right|$$

Furthermore:

$$\frac{\partial LL}{\partial \phi} = \begin{cases} > 0 \text{ for } \eta < 1 \\ < 0 \text{ for } \eta > 1 \end{cases}$$

and

$$\frac{\partial LL}{\partial \theta} = \begin{cases} < 0 \text{ for } \eta < 1 \\ < 0 \text{ for } \eta > 1 \text{ at the stable root} \end{cases}$$

Hence, the determinant is positive at the stable root for θ for $\eta > 1$:

$$\begin{vmatrix} \frac{\partial LL}{\partial \theta} & \frac{\partial LL}{\partial \phi} \\ \frac{\partial FF}{\partial \theta} & \frac{\partial FF}{\partial \phi} \end{vmatrix} > 0$$

Moreover, we have:

- For $x = A$:

$$\begin{vmatrix} \frac{\partial LL}{\partial A} & \frac{\partial LL}{\partial \phi} \\ \frac{\partial FF}{\partial A} & \frac{\partial FF}{\partial \phi} \end{vmatrix} = \overset{(-)}{\frac{\partial LL}{\partial A}} \overset{(-)}{\frac{\partial FF}{\partial \phi}} - \overset{(-)}{\frac{\partial FF}{\partial A}} \overset{(-)}{\frac{\partial LL}{\partial \phi}} = \overset{(-)}{\frac{\partial LL}{\partial A}} \left[\overset{(-)}{\frac{\partial FF}{\partial \phi}} - \overset{(-)}{\frac{\partial LL}{\partial \phi}} \right] > 0$$

$$\begin{vmatrix} \frac{\partial LL}{\partial \theta} & \frac{\partial LL}{\partial A} \\ \frac{\partial FF}{\partial \theta} & \frac{\partial FF}{\partial A} \end{vmatrix} = \overset{(+)}{\frac{\partial LL}{\partial \theta}} \overset{(-)}{\frac{\partial FF}{\partial A}} - \overset{(+)}{\frac{\partial FF}{\partial \theta}} \overset{(-)}{\frac{\partial LL}{\partial A}} < 0$$

- For $x = \pi$:

$$\begin{vmatrix} \frac{\partial LL}{\partial \pi} & \frac{\partial LL}{\partial \phi} \\ \frac{\partial FF}{\partial \pi} & \frac{\partial FF}{\partial \phi} \end{vmatrix} = \overset{(+)}{\frac{\partial LL}{\partial \pi}} \overset{(-)}{\frac{\partial FF}{\partial \phi}} - \overset{(+)}{\frac{\partial FF}{\partial \pi}} \overset{(-)}{\frac{\partial LL}{\partial \phi}} < 0$$

$$\begin{vmatrix} \frac{\partial LL}{\partial \theta} & \frac{\partial LL}{\partial \pi} \\ \frac{\partial FF}{\partial \theta} & \frac{\partial FF}{\partial \pi} \end{vmatrix} = \overset{(-)}{\frac{\partial LL}{\partial \theta}} \overset{(+)}{\frac{\partial FF}{\partial \pi}} - \overset{(+)}{\frac{\partial FF}{\partial \theta}} \overset{(+)}{\frac{\partial LL}{\partial \pi}} < 0$$

The above conclusions regarding the effect of monetary and fiscal policies therefore carry over to the Stackelberg case.

Appendix 4: Monetary Policy and the Optimal Inflation Rate

Proof As seen from (13), inflation has a first-order negative impact on the growth rate: $\frac{\partial g}{\partial \pi} = -\frac{u^f}{\eta}$ which is independent from the inflation rate. However, inflation also affects financial market liquidity, as an increase in inflation leads to a portfolio shift by households away from savings towards consumption; totally differencing the financial market equilibrium condition (14) yields (note that in the following equations $u^f \equiv u^f(\phi^*)$ and $r_K \equiv r_K(\phi^*)$):

$$\frac{d\phi^*}{d\pi} = \frac{p(\phi^*)\, u^f}{p(\phi^*)\left(p(\phi^*) + (1-u^f)\, r_K' - (r_K + \pi)\, u^{f\prime}\right) - p'(\phi^*)\,(\delta + r_K(1-u^f) - \pi u^f)} > 0$$

As shown before in Proposition 3.2, both the equilibrium interest rate and the growth rate increase with financial market tightness as less funds are left unused in the household's deposits (u^f decreases with ϕ). Hence, inflation has a positive second-order impact on the growth rate *via* its impact on the financial market liquidity.

Combining the impact of inflation on financial market liquidity and financial market liquidity on growth, the second-order positive effect writes as:

$$\begin{aligned}
&\frac{\partial g}{\partial \phi} \cdot \frac{\partial \phi}{\partial \pi} \\
&= \frac{p(\phi^*)\, u^f\left((1-u^f)\, r_K' - r_K u^{f\prime}\right)}{\eta\left(p(\phi^*)\left(p(\phi^*) + (1-u^f)\, r_K' - (r_K+\pi)\, u^{f\prime}\right) - p'(\phi^*)\left(\delta + r_K\left(1-u^f\right) - \pi u^f\right)\right)} \\
&= \frac{u^f(\phi)}{\eta} \frac{p(\phi^*)\, u^f\left((1-u^f)\, r_K' - r_K u^{f\prime}\right)}{p(\phi^*)\left(p(\phi^*) + (1-u^f)\, r_K' - (r_K+\pi)\, u^{f\prime}\right) - p'(\phi^*)\left(\delta + r_K\left(1-u^f\right) - \pi u^f\right)}
\end{aligned}$$

The growth-maximising inflation rate, therefore, can be determined as:

$$\begin{aligned}
&\overline{\pi} \in \left\{ \pi \,\middle|\, \frac{\partial g}{\partial \phi} \cdot \frac{\partial \phi}{\partial \pi} = \left|\frac{\partial g}{\partial \pi}\right| \right\} \\
\Leftrightarrow\; &\frac{u^f}{\eta} \frac{p(\phi^*)\left((1-u^f)\, r_K' - r_K u^{f\prime}\right)}{p(\phi^*)\left(p(\phi^*) + (1-u^f)\, r_K' - (r_K+\pi)\, u^{f\prime}\right) - p'(\phi^*)\,(\delta + r_K(1-u^f) - \pi u^f)} = \frac{u^f}{\eta} \\
\Leftrightarrow\; &\frac{p(\phi^*)\left((1-u^f)\, r_K' - r_K u^{f\prime}\right)}{p(\phi^*)\left(p(\phi^*) + (1-u^f)\, r_K' - (r_K+\pi)\, u^{f\prime}\right) - p'(\phi^*)\,(\delta + r_K(1-u^f) - \pi u^f)} = 1 \\
\Leftrightarrow\; &p(\phi^*)^2 = p'(\phi^*)\left(\delta + r_K\left(1-u^f\right) - \pi u^f\right) \\
\Leftrightarrow\; &\overline{\pi} = \frac{p(\phi^*)^2 - p'(\phi^*)\left(\delta + r_K\left(1-u^f\right)\right)}{-p'(\phi^*)\, u^f} > 0.
\end{aligned}$$

The impact on employment can be determined in a straightforward way by totally differentiating (15). ■

Appendix 5: Optimal Debt and Growth

In the following we adopt the convention that $\phi(\tau_K) = \phi + \tau$ for convenience only. The financial market equilibrium is unaffected by capital taxes (only indirectly through impact on savings and consumption decision). The negative direct impact of taxes on growth increases with the equilibrium value of ϕ:

$$\frac{\partial g}{\partial \tau_K} = -\frac{1}{\eta} r_K \left(1 - u^f\right) \leq 0, \frac{\partial^2 g}{\partial \tau_K \partial \phi} = -\frac{1}{\eta}\left[\frac{\partial r_K}{\partial \phi}\left(1 - u^f\right) - r_K \frac{\partial u^f}{\partial \phi}\right] < 0$$

This has to be weighted against the positive effect of government bonds on the financial market equilibrium, which increases the growth rate:

$$g = \frac{\gamma\left(1-\tau_K\right)\left(1-u^f\right)}{\eta} r_K(\phi) - \frac{\rho + \pi u^f}{\eta}$$

$$\frac{\partial g}{\partial \phi} = \frac{1-\tau_K}{\eta}\left[\left(1-u^f\right)\frac{\partial r_K}{\partial \phi} - r_K \frac{\partial u^f}{\partial \phi}\right] - \frac{\pi}{\eta}\frac{\partial u^f}{\partial \phi}$$

$$\frac{\partial^2 g}{\partial \phi^2} = \frac{1-\tau_K}{\eta}\left[-2\frac{\partial u^f}{\partial \phi}\frac{\partial r_K}{\partial \phi} - r_K \frac{\partial^2 u^f}{\partial \phi^2}\right] - \frac{\pi}{\eta}\frac{\partial^2 u^f}{\partial \phi^2}$$

given that $\frac{\partial^2 r_K}{\partial \phi^2} = 0$. Under our functional assumption regarding the constant-elasticity-to-scale property of the matching process, $\frac{\partial^2 g}{\partial \phi^2}$ will be negative except for very small values of $\phi < \underline{\phi}$ with $\underline{\phi} > 0$. Moreover, we have that at $\phi = 0$:

$$\frac{\partial g}{\partial \tau_K} = 0, \frac{\partial g}{\partial \phi} \rightarrow \infty$$

Hence, a zero-tax policy cannot be optimal at $\phi = 0$. Given that $\left|\frac{\partial g}{\partial \tau_K}\right|$ is monotonically increasing with ϕ and $\frac{\partial g}{\partial \phi}$ is monotonically decreasing with ϕ, at least from $\phi > \underline{\phi}$ onwards. Hence there will be a $\overline{\phi} > \underline{\phi} > 0$ such that $\frac{\partial g}{\partial \phi} + \frac{\partial g}{\partial \tau_K} = 0$ for $\tau_K > 0$.

When only $\varpi\%$ of public spending is financed through corporate taxation, then the growth rate writes as:

$$g = \frac{\gamma\left(1-\varpi \cdot \tau\right)\left(1-u^f\right)}{\eta} r_K(\phi) - \frac{\rho + \pi u^f}{\eta}$$

and consequently, the direct negative effect of a tax-financed increase in public debt is reduced:

$$\frac{\partial g}{\partial \tau} = -\varpi \frac{1}{\eta} r_K \left(1 - u^f\right) \leq 0, \frac{\partial^2 g}{\partial \tau \partial \varpi} = -\frac{1}{\eta} r_K \left(1 - u^f\right) \leq 0$$

hence $\frac{\partial g}{\partial \phi} + \frac{\partial g}{\partial \tau} (\varpi) = 0$ is satisfied for a financial market equilibrium $\overline{\overline{\phi}} (\varpi) > \overline{\phi} > 0$.

Appendix 6: Interaction Between Monetary and Fiscal Policy

Proof of Lemma The inflation rate does not affect the direct negative impact of corporate taxation on growth. However, it has an impact on the way financial market liquidity affects growth. Recall that the growth rate increases with financial market liquidity at rate:

$$\frac{\partial g}{\partial \phi} = \frac{1 - \tau_K}{\eta} \left[\left(1 - u^f\right) \frac{\partial r_K}{\partial \phi} - r_K \frac{\partial u_f}{\partial \phi} \right] - \frac{\pi}{\eta} \frac{\partial u^f}{\partial \phi}$$

This derivative depends positively on the inflation rate:

$$\begin{aligned} \frac{\partial^2 g}{\partial \phi \partial \pi} &= \frac{\partial^2 g}{\partial \phi^2} \cdot \frac{\partial \phi}{\partial \pi} - \frac{1}{\eta} \frac{\partial u^f}{\partial \phi} \\ &= \frac{(1 - \tau_K) \left(1 - u^f\right) u^f \frac{\partial r_K}{\partial \phi} - ((1 - \tau_K) r_K - \rho) \frac{\partial u^f}{\partial \phi}}{\left(\rho - (1 - \tau_K)(1 - u^f) r_K + \pi \cdot u^f\right)^2} > 0 \end{aligned}$$

which is unambiguously positive for non-negative growth rates (for which $r_K (1 - \tau_K) > r_K (1 - \tau_K) \left(1 - u^f\right) > \rho$ holds). ■

Proof of Proposition This first part follows directly from the lemma above. Moreover, note that

$$\overline{\pi} = \frac{p (\phi^*)^2 - p' (\phi^*) \left(\delta + r_K \left(1 - u^f\right)\right)}{-p' (\phi^*) u^f}$$

depends positively on ϕ^* when the matching function is characterised by constant returns to scale. ■

References

Ahmed, S., & Rogers, J. H. (2000). Inflation and the great ratios: Long term evidence from the U.S. *Journal of Monetary Economics 45*, 3–35.

Akerlof, G. A., Dickens, W. T., & Perry, G. L. (2000). Near-rational wage and price setting and the optimal rates of inflation and unemployment. *Brookings Papers on Economic Activity, 1*, 1–45.

Amable, B., & Ernst, E. (2005). Financial and labour market interactions. Specific investment and market liquidity, CEM Working Paper 93. Bielefeld: University of Bielefeld.

Andolfatto, D. (1996). Business cycles and labor-market search. *American Economic Review, 86*(1), 112–132.

Batini, N., Eyraud, L., Forni, L., & Weber, A. (2014). Fiscal multipliers: Size, determinants, and use in macroeconomic projections. Tech. rep. Washington, DC.: International Monetary Fund (IMF).

Becsi, Z., Li, V., & Wang, P. (2000). Financial matchmakers in credit markets with heterogeneous borrowers, Working Paper 2000-14. Atlanta: Federal Reserve Bank of Atlanta.

Caballero, R., & Farhi, E. (2014). The safety trap, Working Paper 19927. Cambridge, MA.: NBER.

Christiano, L. J. (1994). Modeling the liquidity effect of a money shock. In R. Fiorito (Ed.), *Inventory, business cycles and monetary transmission*. Lecture Notes in Economics and Mathematical Systems (Vol. 413, pp. 61–124). Berlin/Heidelberg: Springer.

Christiano, L. J., Eichenbaum, M., & Evans, C. L. (1997). Sticky price and limited participation models of money: A comparison. *European Economic Review, 41*(6), 1201–1249.

Cooley, T. F., & Hansen, G. D. (1989). The inflation tax in a real business cycle model. *American Economic Review, 79*(4), 733–748.

Den Haan, W. J., Ramey, G., & Watson, J. (2003). Liquidity flows and fragility of business enterprises. *Journal of Monetary Economics, 50*(6), 1215–1241.

Eggertsson, G. B., & Krugman, P. (2012). Debt, deleveraging, and the liquidity trap: A fisher-minsky-koo approach. *Quarterly Journal of Economics, 127*(3), 1469–1513.

Eriksson, C. (1997). Is there a trade-off between employment and growth? *Oxford Economic Papers, 49*, 77–88.

Ernst, E., & Semmler, W. (2010). Global dynamics in a model with search and matching in labor and capital markets. *Journal of Economic Dynamics and Control, 34*(9), 1651–1679.

Fuerst, T. S. (1992). Liquidity, loanable funds, and real activity. *Journal of Monetary Economics, 29*(1), 3–24.

Grinols, E. L., & Turnovsky, S. J. (1998a). Consequences of debt policy in a stochastically growing monetary economy. *International Economic Review, 39*(2), 495–521.

Grinols, E. L., & Turnovsky, S. J. (1998b). Risk, optimal government finance and monetary policies in a growing economy. *Economica, 65*(259), 401–427.

Jones, C. I. (1995). Time series tests of endogenous growth models. *Quarterly Journal of Economics, 110*, 495–525.

Koo, R. (2015). *The escape from balance sheet recession and the QE trap*. Singapore: Wiley.

Layard, R., Nickell, S., & Jackman, R. (1991). *Unemployment. Macroeconomic performance and the labour market*. Oxford: Oxford University Press.

Lucas, R. E. (1990). Liquidity and interest rates. *Journal of Economic Theory, 4*(2), 103–124.

Merz, M. (1995). Search in the labor market and the real business cycle. *Journal of Monetary Economics, 36*, 269–300.

Pissarides, C. (2000). *Equilibrium unemployment theory*. Cambridge, MA: The MIT Press.

Romer, P. M. (1986). Increasing returns and long-run growth. *Journal of Political Economy, 94*(5), 1002–1037.

Romer, P. M. (1994). The origins of endogenous growth. *Journal of Economic Perspectives, 8*(1), 3–22.

Stockman, A. C. (1981). Anticipated inflation and the capital stock in a cash-in-advance economy. *Journal of Monetary Economics, 8*, 387–393.

Tobin, J. (1965). Money and economic growth. *Econometrica, 33*, 671–684.

Trigari, A. (2009). Equilibrium unemployment, job flows and inflation dynamics. *Journal of Money, Credit and Banking, 41*(1), 1–33.

Walsh, C. E. (2001). *Monetary theory and policy*. Cambridge, MA: MIT Press.

Wang, P., & Xie, D. (2012). Real effects of money growth and optimal rate of inflation in a cash-in-advance economy with labor-market frictions, MPRA Paper 42291. Munich: RePEc.

Wasmer, E., & Weil, P. (2004). The macroeconomics of labor and credit market imperfections. *American Economic Review, 94*(4), 944–963.

Woodford, M. (2003). *Interest and prices: Foundations of a theory of monetary policy*. New York: Princeton University Press.

Woodford, M. (2011). Simple analytics of the government expenditure multiplier. *American Economic Journal: Macroeconomics 3*(1), 1–35.

Business Confidence and Macroeconomic Dynamics in a Nonlinear Two-Country Framework with Aggregate Opinion Dynamics

Matthieu Charpe, Carl Chiarella, Peter Flaschel, and Christian R. Proaño

JEL codes: E12, E24, E31, E52

1 Introduction

As it is generally acknowledged, expectations and beliefs about future developments play a crucial role in the current decision making of economic agents. However, as already pointed out by Keynes (1936), such expectations are not only the reflection of economic fundamentals but they are also determined by "animal spirits". In Keynes's own words: "a large proportion of our positive activities depend on spontaneous optimism rather than mathematical expectations, whether moral or hedonistic or economic. Most, probably, of our decisions to do something positive,

M. Charpe
International Labor Organization, Geneva, Switzerland
e-mail: charpe@ilo.org

C. Chiarella
University of Technology Sydney, Ultimo, NSW, Australia
e-mail: carl.chiarella@uts.edu.aus

P. Flaschel
Universität Bielefeld, Bielefeld, Germany
e-mail: pflaschel@wiwi.uni-bielefeld.de

C.R. Proaño (✉)
Otto-Friedrich-Universität Bamberg, Bamberg, Germany
e-mail: christian.proano@uni-bamberg.de

L. Bernard, U. Nyambuu (eds.), *Dynamic Modeling, Empirical Macroeconomics, and Finance*, DOI 10.1007/978-3-319-39887-7_12

the full consequences of which will be drawn out over many days to come, can only be taken as the result of animal spirits – a spontaneous urge to action rather than inaction, and not as the outcome of a weighted average of quantitative benefits multiplied by quantitative probabilities" (Keynes 1936, pp. 161–62).

Despite the potential importance of such "animal spirits" for the economic agents' expectations formation and decision making, they have played only a secondary role in the mainstream macroeconomics literature of the last 40 years given the predominance of the rational expectations paradigm originally proposed by Muth (1961). Nonetheless, since the outbreak of the 2007–2008 global financial crisis, alternative approaches to the rational expectations paradigm based on bounded rationality and social norms have become increasingly interesting to the great majority of the economics profession [see, for example, the recent edited volume by Frydman and Phelps (2013)].

In this context, a particularly interesting field of research is the modelling of bounded-rationality based-aggregate opinion dynamics. Indeed, given the social context in which economic agents interact, the beliefs and opinions by one agent are quite likely to affect other agent's decision making, contributing thus to the formation of an aggregate opinion which is in turn likely to affect the agents' decisions through a process of "social learning" (cf. Acemoglu and Ozdaglar 2011). Lux (2009) provides for example empirical evidence which suggests that survey data on business confidence may be the result of a social process of opinion formation among the respondents, instead of being the representation of rational forecasts of future economic developments.

From a theoretical perspective, a parsimonious and elegant modeling of such opinion dynamics was proposed by Weidlich and Haag (1983). In this framework—which has been widely applied in the behavioral finance following the work by Lux (1995)—agents' beliefs are modelled through a binary choice problem where the stochastic transitions of agents between two types of beliefs alternatives due to exogenous factors and group pressure. The great majority of studies in the behavioral finance literature along these lines, however, restrict their analysis to the financial markets, and do not consider in an explicit manner how heterogenous behavioral expectations at the microeconomic level may interact with macroeconomic fundamentals in an international setup.[1] However, given the quite advanced integration of international goods and financial markets, the analysis of international macroeconomic interactions has become increasingly relevant both from the academic as well as from the policy perspective.

[1]Indeed, the analysis of such international interactions has been based over the last decades—and hand in hand with increasing predominance of the neoclassical rational expectations approach following the work by Lucas (1976)—on the presumption of rational agents, see e.g. Frenkel and Razin (1985) and Corden (1985), with the outcome that the resulting theoretical frameworks—being the New Open Economy Macroeconomics (NOEM) approach developed by Obstfeld and Rogoff (1995) the most recent incarnation—suggest an inherent and often misleading stability of such international interactions.

In this light, the main objective of the present paper is to fill in this gap in the behavioral macro-finance literature by explicitly investigating the role of the state of confidence for the macroeconomic dynamics of two interacting economies using the opinion dynamics approach by Weidlich and Haag (1983) and Lux (1995).[2] Particularly, the overall state of confidence in the world (two-country) economy is assumed to influence not only the dynamics of the nominal exchange rate, but also the dynamics of the real economy through the determination of aggregate investment. This novel feature allows us to consider far richer international macroeconomic interactions than most standard models. Further, it features wage-price dynamics that interact with output and employment fluctuations—leading to a Goodwin (1967)-type of distributive cycle—, as well as debt dynamics due to a credit-financed investment behavior. The resulting framework is both advanced as well as flexible enough to generate various types of persistent fluctuations, and also complex dynamics. This model also provides a platform for the analysis of economic policy as a necessary stabilization mechanism. The fluctuations shown in the paper preserve the insight of seminal partial contributions of the literature such as the Goodwin distributive cycle, the related inflation–stagflation cycle, debt cycles, and more.

The remainder of the paper is organized as follows. In Sect. 2 we present the theoretical framework representing two large open economies interacting with each other through international trade of goods as well as through financial linkages. In Sect. 3 we describe the model's intensive form and its steady state. Various numerical simulations of the model focusing among other things on the role of the state of confidence for the macrodynamics of the two economies are discussed in Sect. 4. Finally, we draw in Sect. 5 some concluding remarks from this study.

2 The Model

In the following we consider two countries which interact with each other through the international trade of goods and services as well as financial assets. For the sake of simplicity, we assume however that only consumption goods are internationally traded, while investment and government consumption goods are country-specific, and that there is no labor mobility across countries. Further, we assume that

[2] See Taylor and O'Connell (1985) and Franke and Semmler (1989), for two early contributions thematizing the role of the state of confidence for the economy's dynamics and stability, as well as Franke (2012) and Flaschel et al. (2015) for other macroeconomic models along the lines of the present paper.

while the behavioral equations and macroeconomic adjustments mechanisms of both economies are characterized by the same structure, they may differ in some specific numerical parameter values in order to allow for different types of economic asymmetries.[3]

2.1 The Private Sector

In a standard manner, both economies consist of a household sector, an entrepreneurial sector and a government sector. The household sector consists of asset holders—who own the means of production and obtain an income out of the firms' profits—and workers who provide their skills and labor for a wage income on a continuous basis. Asset holders and workers are assumed to have different marginal consumption propensities out of their respective income streams: While workers are assumed to consume the totality of their labor income, the asset holders' marginal consumption propensity out of their capital income c_c is assumed to be significantly smaller than one.

Let p denote the aggregate price level, Y the aggregate output, w the nominal wage, L^d the firms' labor demand, r the nominal interest rate, Λ the firms' debt level, K the domestic capital stock, δ the capital depreciation rate and ρ^g be the firms' gross profit rate, defined as

$$\rho^g = \frac{pY - wL^d - r\Lambda - \delta pK}{pK}. \tag{1}$$

with analogous expressions for the foreign economy being denoted by the superscript * in the following.

Aggregate domestic private consumption is then described by

$$pC = wL^d - pT_w + c_c(\rho^g pK + r\Lambda - pT_c), \quad c_c \leq 1 \tag{2}$$

where C represents aggregate private consumption in real terms, T_w the taxes paid by the workers out of their labor income, T_c the taxes paid by the asset holders out of their gross profits, and $\rho^g pK + r\Lambda - pT_c$ the asset holders' capital income after taxes. The distribution of income between workers and asset holders is thus non-trivial for aggregate demand, and by extension for the overall macroeconomic development in both domestic and foreign countries as aggregate private consumption is assumed to be a function of the workers' and asset holders' current income and asset holders feature a lower marginal propensity to consume than the workers.

---

[3]For a more elaborate framework along the same modeling lines where the interconnectedness of a high-income and a low-income labor markets is explicitly modeled, see Charpe et al. (2015).

Concerning the activities of the entrepreneurial sector in both countries (which is own by the respective asset holders, as previously mentioned), we assume as in previous related work a production technology with fixed coefficients and that labor demand by firms is determined by

$$L^d = Y/x, \tag{3}$$

with Y denoting aggregate output, K the capital stock and x the average labor productivity.

The growth rate of capital stock ι (i.e. aggregate investment by the entrepreneurial sector I in terms of domestic capital units K) is in turn assumed to depend on the firms' expected gross profit rate ρ^{ge}, on the real interest rate $r-\hat{p}$ (where $\hat{p}$ represents the price inflation rate), on the utilization rate of the workforce of firms $u = Y/Y^p$, where Y^p represents the firms' potential output which in turn is determined by the given output-capital ratio $y^p = const.$, i.e. $Y^p = y^pK$),[4] on the firms' debt to capital ratio λ, (all these variables as deviations from their corresponding steady state levels), and in addition on a given trend term γ which determines all other trend terms considered in this model, i.e.

$$\iota = I/K = i_\rho(\rho^{ge} - \rho^g_o) - i_r(r - \hat{p} - \bar{r}) + i_o(u - u_o) - i_\lambda(\lambda - \lambda_o) + \gamma + \delta \tag{4}$$

where $\dot{K} = I - \delta$, where I represents net aggregate investment, $\dot{K}$ the change in the firms' capital stock, and $\rho^{ge} = \rho^g + \psi$, where ψ is the state of confidence in the economy. Accordingly, the growth rate of capital is assumed to be driven by expectations of profitability, financing costs, the rate of capacity utilization relative to its normal level, the firms' indebtedness (also relative to its "normal" level), and by an exogenous term γ and the rate of depreciation of capital δ. The state of confidence in the economy is thus assumed to affect not only the dynamics of the exchange rate, as it will be discussed below, but also the real economy through its impact on the expected profit rate and by extension on aggregate investment.

The state of confidence of the economy—which determines the expected profit rate, since $\rho^{ge} = \rho^g + \psi$—is determined by

$$\psi = \psi_a(u - u_o) + \psi_b(u^* - u^*_o) - \psi_\eta(\eta - 1)^2 + \psi_z z. \tag{5}$$

The state of confidence ψ is assumed to depend positively on the current state of the business cycle in both countries represented by the respective capacity utilization rates u and u^* and negatively on a largely over- or undervalued real exchange rate $\eta = p/(sp^*)$ (cf. Charpe et al. 2015), with s being the nominal exchange rate. The economic rationale for this specification is straightforward: large deviations from the PPP-consistent level of one are considered as reflection of structural

[4]Note that by denoting with y the actual output-capital ratio, it holds for the utilization rate of capital that $u = Y/Y^p = y/y^p$.

macroeconomic imbalances between the two economies, and therefore as a threat for global macroeconomic stability. Lastly, the state of confidence is also augmented with a self-reference component as confidence is related to the share of chartist agents z (to be described in detail below).

We assume that firms distribute all profits to asset owners and that firms are not confronted with any type of credit rationing. It follows that new credit is given by the level of aggregate investment according to the following rule:

$$\dot{\Lambda} = (\alpha_f^o + \alpha_f^1(\lambda_o - \lambda))pI \tag{6}$$

The wage-price inflation dynamics is of the kind considered e.g. by Chiarella and Flaschel (2000), Flaschel and Krolzig (2006) and are described by

$$\hat{w} = \beta_{we}(e - e_o) + \kappa_w(\sigma\hat{p} + (1 - \sigma)(\hat{p}^* + \hat{s})) + (1 - \kappa_w)\pi^c + \hat{x} \tag{7}$$

$$\hat{p} = \beta_{pu}(u - u_o) + \kappa_p(\hat{w} - \hat{x}) + (1 - \kappa_p)\pi^c \tag{8}$$

where $\hat{w}$ and $\hat{p}$ represent the wage- and price inflation rates, $\hat{x} = const.$ the growth rate of labor productivity, $v = \omega/x$ the wage share and π^c the economy's long-run inflation rate, which for the sake of simplicity will be assumed to be equal to the central bank's target inflation rate in the following. Accordingly labor- and goods market disequilibria have a positive direct impact on wage and price inflation, respectively, as well as an indirect impact through the cross-over inflationary expectational terms, with the respective weights κ_{wp} and κ_{pw}. In this context, following the assumption that only consumption goods are traded internationally, the firms's expectations are given by a weighted average of the growth rate of domestic unit labor costs $\hat{w} - \hat{x}$, and long-term inflation π^c. By contrast, nominal wage inflation takes into account not only the domestic price inflation, but also the imported inflation given by $\hat{p}^* + \hat{s}$ by the extent $(1 - \sigma)$, the share of foreign goods in the domestic households's aggregate consumption C.

Through straightforward algebra it can be easily confirmed that the structural wage–price inflation equations given by Eqs. (7) and (8) can be reduced to one equation which describes the law of motion of the labor share v (under the assumption that $\hat{x} = 0$)

$$\hat{v} = \kappa[(1-\kappa_{pw})\beta_{we}(e-e_o)-(1-\kappa_{wp})\beta_{pu}(u-u_o)+\kappa_{pw}(1-\kappa_{pw})(1-\sigma)(-\hat{\eta})] \tag{9}$$

where $\kappa = 1/(1 - \kappa_w\kappa_p)$ and $\hat{\eta} = \hat{p} - \hat{s} - \hat{p}^*$.

Finally, aggregate output is assumed to be determined in a delayed manner in reaction to the aggregate demand

$$Y^d = C + I + G + \delta K + NX \tag{10}$$

where G represents the government expenditures (to be described below), and NX the net exports, given by

$$NX = (1 - \sigma^*)C^*/\eta - (1 - \sigma)C. \tag{11}$$

Finally, aggregate output is assumed to be determined by

$$\dot{Y} = \beta_y(Y^d - Y) + nY \tag{12}$$

with ρ^g given by Eq. (1), $I = \iota K$, with ι given by Eq. (4), NX by Eq. (11) and G by Eq. (20).

2.2 FX Market Dynamics

Concerning the determination of the nominal exchange rate, we assume in the following that its growth rate is determined by two sources: real trade-linked demand and supply of foreign currency, as well as by financial trading.

The expectations driving financial trading are modeled through an aggregate opinion dynamics approach along the lines of Lux (1995), and more recently, Franke (2012). Accordingly, financial trading in the FX market can be assumed to be characterized by two main forecasting strategies: *fundamentalism* (denoted by a superscript f), and *chartism* (denoted by a superscript c). Fundamentalist traders expect the nominal exchange rate to stabilize in the long run, or equivalently, that the growth rate of nominal exchange rate becomes zero—its steady state position in the present model—while chartists traders use a simple adaptive rule to forecast the depreciation rate of the nominal exchange rate $\hat{s}$.

Given that agents have heterogeneous expectations, it is not obvious a priori what *market* expectations should be. Indeed, while in standard equilibrium models with efficient markets heterogeneous information and beliefs are spontaneously aggregated and made uniform under the pressure of market forces, this is not the case in our framework given the implicit bounded rationality underlying our modeling approach.

For the sake of simplicity, we assume as is Flaschel et al. (2015), the following simple law of motion for the expected depreciation rate[5]:

$$\dot{\pi}_s^e = \beta_{\pi_s^e}\left[\left(1-\frac{1+z}{2}\right)(0-\pi_s^e)+\frac{1+z}{2}(\hat{s}-\pi_s^e)\right]$$
$$= \beta_{\pi_s^e}\left[\frac{1+z}{2}\hat{s}-\pi_s^e\right], \tag{13}$$

where $\beta_{\pi_s^e} > 0$ represents an adjustment speed parameter, π_s^e the market expectation of growth rate of the nominal exchange rate, and z is the relative share of chartists in the whole population of chartists and fundamentalists, defined as

$$z = \frac{n}{N} \in [-1, 1], \quad 1 - z = \frac{N_f}{N}, \quad 1 + z = \frac{N_c}{N}, \tag{14}$$

where, N_c is the number of speculative traders using chartism and N_f those with fundamentalist expectations, with $N_c + N_f = 2N$ as in Franke (2012), and $n = \frac{N_c - N_f}{2}$. The distribution of chartists and fundamentalists in the population is thus described by the difference in the size of the two groups normalised by N.[6]

We denote by $p^{f\to c}$ the transition probability of a fundamentalist becoming a chartist, and viceversa for $p^{c\to f}$, and we assume that these transition probabilities are determined by

$$p^{f\to c} = \exp(\psi), \tag{15}$$
$$p^{c\to f} = \exp(-\psi). \tag{16}$$

where ψ represents the state of confidence as described by Eq. (5). Accordingly, an increase in ψ is assumed to increase the probability that a fundamentalist becomes a chartist, and to decrease the probability that a fundamentalist becomes a chartist. Further, the evolution of z is assumed to be determined by

$$\dot{z} = \beta_z((1-z)p^{f\to c} - (1+z)p^{c\to f}) \tag{17}$$

[5]Obviously, this is only one possible formalization of the dynamics of aggregate expectations in markets with heterogeneous agents, and alternative approaches can be proposed (see, for example, the approach adopted by De Grauwe and Grimaldi (2006) in their analysis of the behaviour of agents on foreign exchange markets). Yet, we regard Eq. (13) as a very parsimonious way of capturing *both* the influence of aggregate observed variables *and* the role of heterogeneity and self-driving forces in expectation formation.

[6]As in Franke (2012), we assume that N is large enough so that the intrinsic noise from different realisations when individual agents apply their random mechanism can be neglected.

Inserting this expressions in Eq. (17) delivers[7]

$$\dot{z} = \beta_z((1-z)\exp(\psi) - (1+z)\exp(-\psi)). \tag{18}$$

We thus assume that exchange rate expectations are driven in a complex manner through the relative importance of chartists and fundamentalists in the FX market.[8]

Given this nominal exchange rate expectations formation mechanism, the dynamics of the nominal exchange rate is assumed to be driven by:

$$\hat{s} = \beta_s\left[\beta_i(r^* - r + \pi_s^e) - nx\right] + \bar{\pi} - \bar{\pi}^* \tag{19}$$

where the first term represents arbitrage possibilities results from deviations of the expected rates of return of foreign and domestic bonds from the Uncovered Interest Rate Parity (UIP) to an extent β_s, and by nx, the net exports to domestic capital ratio, which represent the excess supply for foreign currency and its appreciative effect on the domestic currency.[9,10]

Given the strong nonlinearities in the opinion dynamics module and also in the rate of return function of the 4D dynamics (despite the simple linear behavioral rules we have adopted), we shall address these questions by means of numerical simulations in the next section. They will show that interesting irregular and persistent fluctuations in the real and financial variables of the model can be generated, quite in contrast to what is possible in such a model type under the assumption of the homogeneous rational expectations of the mainstream literature.

2.3 The Government Sector

The government's behavior is described by

$$G = T_w + T_c - g_c(u - u_o)K \tag{20}$$

[7]This law of motion is derived in detail in Franke (2012) and its use of the exponential terms guarantees that neither the chartist nor the fundamentalist group can be totally eliminated in the FX-market.

[8]As fundamentalist form their expectations in a simple regressive way, these expectations do not really show up in the formation of averages and the average expectation of their formation.

[9]The economic rationale for this exchange rate specification can be related with the economy's balance of payments—which comprises the trade as well as the capital accounts—, namely

$$NX + (r^* - \hat{\eta})B_b - (r - \hat{p})B_a^* = \eta\dot{B}_b - \dot{B}_a^*,$$

where B_b and B_a the domestic economy's holdings of foreign and domestic bonds, respectively. As it should be clear, trading recorded in the trade and the capital accounts does not have to follow the same rationale or rules, and may therefore be modelled independently.

[10]Note that capital gain expectations are zero in the inflation-free steady state of the whole model.

Government expenditures follow an anti-cyclical fiscal policy, which is financed through the issuance of new government bonds. This anti-cyclical behavior fluctuates around a balanced-budget level determined by the tax revenues $T_w + T_c$. With respect to monetary policy, the following Taylor rule is assumed

$$r = \bar{r} + r_p(u - u_o) + r_p(\hat{p} - \bar{\pi}). \tag{21}$$

Accordingly, the monetary authorities simply raise the interest rate above a given level $\bar{r}$ if domestic prices are higher than their steady state value and vice versa. This fixing of the rate of interest r at each moment of time only requires that money supply M (not explicitly modeled here) is endogenous and adjusted to the money demand of asset holders via open market operations in the form of short-term government bonds B.

3 The Model's Intensive Form and Steady State

We now reformulate the two-country framework just outline in terms of domestic capital units, so that $y = Y/K$, $y^d = Y^d/K$, $t_w = T_w/K$, $t_c = T_c/K$, $\iota = I/K$, $g = G/K$, and $k = K^*/K$. The dynamic laws of motion which describe the evolution of the two-country system over time read on the intensive form level:

$$\begin{aligned}
\dot{y} &= \beta_y(y^d - y) - y(\iota - \delta - \gamma) \\
\hat{\omega} &= \kappa\left[\beta_w(e - \bar{e})(1 - \kappa_p) - \beta_p(u - \bar{u})(1 - \kappa_w\gamma_w)\right. \\
&\quad \left. + \kappa_w(1 - \gamma_w)(\hat{p}^* + \hat{s})(1 - \kappa_p) - \pi(1 - \gamma_w)\kappa_w(1 - \kappa_p)\right] \\
\dot{\lambda} &= (\alpha_f^o + \alpha_f^1(\lambda_o - \lambda))\iota - [\iota + \hat{p}]\lambda \\
\dot{y}^* &= \beta_y(y^{d,*} - y^*) - y^*(\iota^* - \delta^* - \gamma^*) \\
\hat{\omega}^* &= \kappa^*\left[\beta_w^*(e^* - \bar{e}^*)(1 - \kappa_p^*) - \beta_p^*(u^* - \bar{u}^*)(1 - \kappa_w^*\gamma_w^*)\right. \\
&\quad \left. + \kappa_w^*(1 - \gamma_w^*)(\hat{p} - \hat{s})(1 - \kappa_p^*) - \pi^*(1 - \gamma_w^*)\kappa_w^*(1 - \kappa_p^*)\right] \\
\dot{\lambda}^* &= (\alpha_f^o + \alpha_f^1(\lambda_o - \lambda^*))\iota^* - [\iota^* + \hat{p}^*]\lambda^* \\
\hat{s} &= \beta_s\left[\beta_i(r^* - r + \pi_s^e) - nx\right] + \bar{\pi} - \bar{\pi}^* \\
\dot{z} &= \beta_z((1 - z)\exp(\psi) - (1 + z)\exp(-\psi)) \\
\dot{\pi}_s^e &= \beta_{\pi_s^e}\left[\frac{1 + z}{2}\hat{s} - \pi_s^e\right] \\
\hat{k} &= \iota^* - \delta^* - (\iota - \delta),
\end{aligned}$$

with

$$
\begin{aligned}
y^d &= \sigma(vy - t_w + c_c(\rho^g + r\lambda - t_c)) + \iota + g + \delta \\
&\quad + k(1-\sigma^*)(v^*y^* - t_w^* + c_c(\rho^{g*} + r^*\lambda^* - t_c^*))/\eta \\
\rho^g &= y - vy - r\lambda - \delta \\
\iota &= i_\rho(\rho^g + \psi - \rho_o^g) - i_r(r - \hat{p} - \bar{r}) + i_u(y/y^p - \bar{u}) - i_\lambda(\lambda - \lambda_o) + \gamma + \delta \\
g &= t_w + t_c - g_c(y/y^p - \bar{u}) \\
r &= \bar{r} + r_y(y/y^p - \bar{u}) + r_p\hat{p} \\
nx &= (1-\sigma^*)c^*k/\eta - (1-\sigma)c \\
\eta &= \frac{p}{s + p^*} \\
u &= \frac{y}{y^p} \\
e &= \frac{y}{x_0} \\
c &= vy - t_w + c_c(\rho^g + r\lambda - t_c) \\
\psi &= \psi_a(y/y^p - \bar{u}) + \psi_b(y^*/y^{p*} - \bar{u}^*) - \psi_\eta(\eta - \eta_0)^2 + \psi_z z \\
y^{d*} &= \sigma(v^*y^* - t_w^* + c_c(\rho^{g*} + r^*\lambda^* - t_c^*)) + \iota^* + g^* + \delta \\
&\quad + (1-\sigma)\eta(vy - t_w + c_c(\rho^g + r\lambda - t_c))/k, \\
\rho^{g*} &= y^* - v^*y^* - r^*\lambda^* - \delta \\
\iota^* &= i_\rho(\rho^{g*} + \psi - \rho_o^{g,*}) - i_r(r^* - \hat{p}^* - \bar{r}^*) + i_u(u^* - \bar{u}^*) - i_\lambda(\lambda^* - \lambda_o^*) + \gamma^* + \delta^* \\
g^* &= t_w^* + t_c^* - g_c(y^*/y^{p*} - \bar{u}^*) \\
r^* &= \bar{r}^* + r_y(u^* - \bar{u}^*) + r_p\hat{p}^* \\
c^* &= v^*y^* - t_w^* + c_c^*(\rho^{g*} + r^*\lambda^* - t_c^*)
\end{aligned}
$$

and

$$
\begin{aligned}
\hat{p} &= \kappa\left[\beta_p(u - \bar{u}) + \kappa_p\beta_w(e - \bar{e}) + \kappa_p\kappa_w(1-\gamma_w)(\hat{p}^* + \hat{s}) + \pi\kappa_p\kappa_w(\gamma_w - 1)\right] + \pi \\
\hat{p}^* &= \kappa^*\left[\beta_p^*(u^* - \bar{u}^*) + \kappa_p^*\beta_w^*(e^* - \bar{e}^*) + \kappa_p^*\kappa_w^*(1-\gamma_w^*)(\hat{p} - \hat{s})\right. \\
&\quad \left. + \pi^*\kappa_p^*\kappa_w^*(\gamma_w^* - 1)\right] + \pi^*.
\end{aligned}
$$

The model's steady state expressions representing the balanced growth path of the two economies can be calculated in a recursive manner using this intensive form representation. Setting $\hat{v} = 0$, we get the steady state for output $y = \bar{u}y^p$, and the steady state for the employment rate $e = y/x$. Using the equation for the debt dynamics and setting $\iota = \gamma$ we get the steady state for the debt to capital ratio $\lambda = \alpha_f^o$. Setting $\dot{y} = 0$, we get the steady state for aggregate demand $y^d = y$. The steady state for public spending is equal to the sum of the tax revenues: $g = \tau_w + \tau_c$. Setting $\bar{r}$ exogenously and $\eta = 1$, we get the steady state for the labour share ω and the steady state for the rate of profits using the following two equations:

$$y^d = vy - t_w + c_c(\rho^g + r\lambda - t_c) + \iota + g + \delta$$

$$\rho^g = y - vy - r\lambda - \delta.$$

Solving for the labour share, we get:

$$v = \frac{x}{y(1-c_c)}\left(y^d - c_c(y - \delta - t_c) + t_w - \iota - g - \delta\right).$$

Assuming that the interest rates are equal across the two countries at the steady state, setting $nx = 0$, gives the steady state of the relative size between the two countries $k = \frac{\eta(1-\sigma)c}{(1-\sigma^*)c^*}$. The definition for k implies that the two countries have the same size when the parameters are identical.

As most of the equilibrium conditions determining the economy's balanced growth path are self-explanatory and have also been studied in detail in related work such as Chiarella et al. (2005), there is no need to discuss them explicitly here. It is however noteworthy that while both higher tax rates on asset-holders and on workers reduce v_o, the effect of an increase in the propensity to consume of asset-holders is ambiguous, because it depends ultimately on a complex combination of various parameters.[11]

Further, it should be pointed out that we have neglected in the presentation of the model all stock-flow relationships or budget equations which do not have an impact on its dynamics, most importantly the government budget constraint which is still set aside by way of the assumed taxation rules. Also, the stocks of assets—up to the capital stock and the debt to capital ratio of firms do not yet influence the laws of motion of the model, since investment only depends on the loan rate and not on Tobin's q.

[11]This is an important feature of the present model as in purely demand-driven models the spending multiplier is typically an increasing function of the marginal propensities to consume. However, the present framework does not only consider aggregate demand, but also income distribution.

4 Numerical Simulations

It should be clear that due to the high dimensional nature of the model, a huge range of outcomes can be expected in theory, ranging in particular from simple Hopf-bifurcation-generated limit cycles or persistent business fluctuations to truly complex attractors and trajectories converging towards them. Therefore, we restrict the following analysis to three scenarios: (1) a baseline scenario where the influence of the state of confidence and the opinion dynamics is absent, (2) a scenario where the state of confidence reacts to business cycle developments and in turn affects aggregate investment, (3) a scenario where the exchange rate depreciation expectations have a higher speed of adjustment and (4) opinion dynamics come into play.[12]

When not stated elsewise we assume symmetric economies using the parameter values summarized in Table 1. Our choice of parameters, while not be based entirely on econometric estimations, reflects nonetheless the modeling philosophy we pursue here. So for example is c_c, the asset holders' marginal propensity to consume out of their current income, lower than the workers' marginal propensity as usually assumed in Kaleckian models thematizing the income distribution conflict, see e.g. Chiarella et al. (2005). Also, we assume that workers are confronted with a higher relative tax burden than capitalists ($t_w > t_c$), as it is the case in the majority of industrialized economies. Further, we assume that the responsiveness of aggregate investment with respect to expected profits is larger than to the capacity utilization, to the real interest rate and to deviations from a given debt target ($i_\rho > i_u$, $i_\rho > i_r$ and $i_\rho > i_\lambda$). In contrast, the parameters concerning the wage and price Phillips curves are based on the econometric results for the U.S. economy by Franke et al. (2006) and Flaschel and Krolzig (2006), and reflect in particular the empirical finding that wage inflation is more reactive to labor market disequilibria than price inflation with respect to goods market disequilibria. Further, concerning the nominal interest rate set by the monetary authorities, we assume a standard value for $r_p = 1.5$, as well as $r_y = 0.4$. Finally, we assume both in the nominal exchange rate and the aggregate opinion dynamics equations parameters which do not generate an explosive behavior of the whole system, and leave a proper estimation of such an issue for further research. Comparative statics are performed below with respect to the set of parameters driving the opinion dynamic and the state of confidence.

In the following we discuss by means of alternative numerical simulations some of the properties of this two-country framework.

[12]All simulations were computed using the SND software, see Chiarella et al. (2002). The programming code is available upon request from the authors.

Table 1 Calibration parameters

Asset holders' marginal propensity of consumption	$c_c = 0.1$
Workers' marginal propensity of consumption	$c_w = 1$
Home bias	$\sigma = 0.75$
Trend gross investment	$\gamma = 0.125$
Depreciation rate of capital	$\delta = 0.075$
Workers' paid taxes per domestic capital unit	$t_w = 0.2$
Asset holders' paid taxes per domestic capital unit	$t_c = 0.1$
Output-capital ratio	$y_p = 1$
Labor productivity	$x_a = 1.1$
Steady state capacity utilization rate	$u_o = 1$
Parameter in the new credit equation	$\alpha_f^o = 0.3$
Parameter in the new credit equation	$\alpha_f^1 = 0.1$
Investment responsiveness to expected profit rate	$i_\rho = 2$
Investment responsiveness to interest rate	$i_r = 0.5$
Investment responsiveness to capacity utilization	$i_u = 0.1$
Investment responsiveness to debt-capital ratio	$i_\lambda = 0.0$
Goods markets disequilibrium adjustment parameter	$\beta_y = 6$
Parameter in the wage Phillips curve	$\beta_{we} = 0.9$
Parameter in the wage Phillips curve	$\kappa_w = 0.5$
Parameter in the price Phillips curve	$\kappa_p = 0.5$
Parameter in the price Phillips curve	$\beta_{pu} = 0.2$
Parameter in the nominal exchange rate equation	$\beta_s = 0.5$
Parameter in the nominal exchange rate equation	$\beta_i = 2.5$
Steady state nominal interest rate	$\bar{r} = 0.03$
Steady state inflation rate	$\bar{\pi} = 0.0$
Countercyclical fiscal policy parameter	$g_c = 1.1$
Interest responsiveness to capacity utilization	$r_y = 0.4$
Interest responsiveness to price inflation	$r_p = 1.5$
State of confidence responsiveness to capacity utilization	$\psi_a = \psi_b = 0$
State of confidence responsiveness to ex rate volatility	$\psi_n = 0$
State of confidence responsiveness to chartist population	$\psi_z = 1$
Adjustment parameter in the opinion dynamics	$\beta_{\pi_s^e} = 0$
Adjustment parameter in the opinion dynamics	$\beta_z = 0$

4.1 Baseline Simulation

Figure 1 gives an overview of the transmission channels of the model. The calibration corresponds to the list of parameters presented in the previous section and the shock is a decline in the interest rate from 3 to 2.7 % in the home country. As expected, the decline in the interest rate in the home country affects economic activity in an expansionary manner via its effect on the domestic aggregate investment. Further, the interest rate decline also affects the profitability of firms

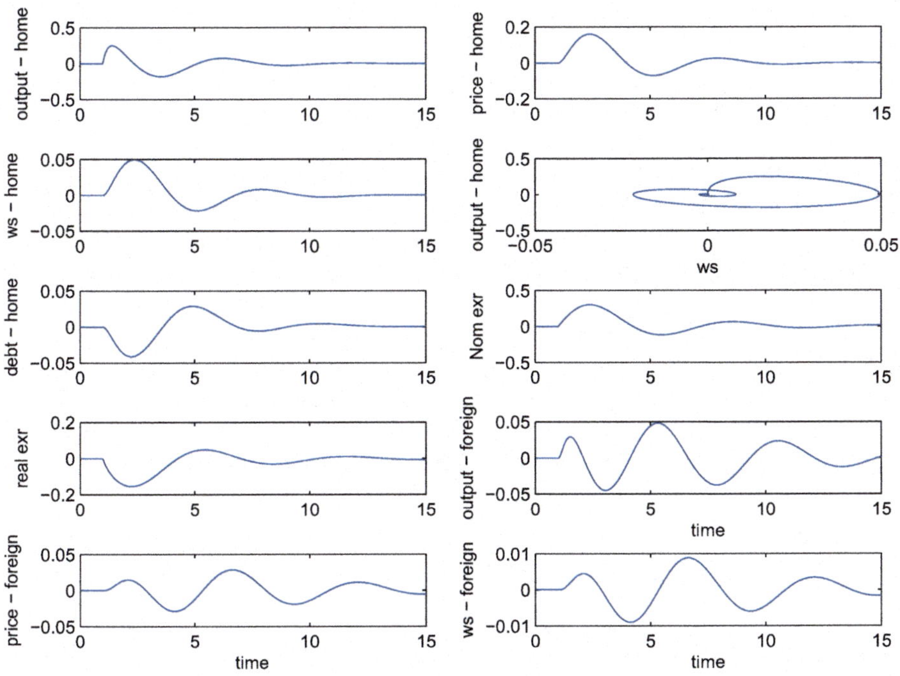

Fig. 1 Dynamic adjustments to an exogenous decline in the nominal interest rate in the home country

in an indirect manner through its effect on the expected profit rate via the interest payment channel. In our calibration, this second effect is non-trivial as the parameter i_ρ is set equal to 2.

Due to the increased level of economic activity nominal wage and price inflation rise, as well as the labour share as nominal wage inflation is assumed to be more reactive than price inflation to its corresponding demand pressure (the flexibility of nominal wage inflation is governed by the parameter $\beta_w = 0.9$, while price inflation reacts to economic activity at a speed $\beta_p = 0.2$). The interaction between labour costs and aggregate demand generates a cycle along the lines of Goodwin (1967). In the upper part of the cycle, the fast increase in the labour costs tend to reduce profitability and, by extension, aggregate investment, putting an end to the expansionary phase of the cycle. The Goodwin cycle is illustrated by the clockwise dependance between the labour share (on the horizontal axis) and economic activity (on the vertical axis) in the graph of the third row, first column.

Following the negative shock on the domestic interest rate the nominal exchange rate depreciates as it is mainly driven by the interest rate differential with $\beta_i = 2.5$ in the baseline calibration. The depreciation of the nominal exchange rate tends to increase aggregate demand further via an improvement in the trade balance. Further, as a result of the fast rise in prices in the expansionary phase the competitiveness of the home country (measured by the real exchange rate) declines. This competitiveness effect together with the locomotive effect related to the fast growth in the home country generates positive spill-over effects on the foreign country where output also expands. This spill-over effects are however delayed, with the biggest increase taking place over the second business cycle.

Finally, the inflationary pressure in the expansionary phase of the cycle leads to a quick increase in the interest rate since $r_p = 1.5$, a dampening effect which is supported by the counter-cyclical fiscal policy behavior according to which fiscal authorities reduces public spending in the boom stabilizing the level of aggregate demand.

The resulting expansion in economic activity in the domestic economy affects neither the state of confidence $\psi_a = \psi_b = 0$, nor the population of chartist relative to fundamentalist agents as the parameters for driving the opinion dynamic β_z is set at zero in the baseline calibration nor the role of exchange rate expectations $\beta_{\pi_s^e} = 0$. The role of these different feedback channels are studied below.

4.2 The Role of the State of Confidence

Figure 2 illustrates the transmission channels associated with the state of confidence in the economy ψ. Figure 2 shows the reaction of the main macroeconomic variables following a decline in the interest rate from 3 to 2.85 %. The difference with the previous simulation is an increase in the parameter for the state of confidence from $\psi_a = \psi_b = 0$ to $\psi_a = \psi_b = 0.1$.

As Fig. 2 clearly illustrates, when the state of confidence reacts to the level of economic activity in both economies the dynamic adjustment of the two economies is more volatile than in the baseline simulation. This is due to the reinforcing effect of higher economic activity on the state of confidence, and subsequently, on the expected profit rate and on aggregate investment. Further, as in the previous case, the foreign economy profits also from the reduction in the nominal interest rate in the domestic economy, as the foreign country benefits from both higher demand in the home country which translates into higher imports. This positive effect takes place despite the fact that the reduction of the interest rate at home leads to a depreciation of the nominal exchange rate, which is detrimental to the foreign country. There are signs of non-linearities in the model. In the home country, oscillations are dampening over time. Contrastingly, the amplitude of oscillations increase between the first and the second cycle in the foreign country before losing magnitude.

It should be noted that since expectations do not adjust $\beta_{\pi_s^e} = 0.0$, the opinion dynamics part of the model does not affect the dynamics of the economy. The importance of exchange rate expectations is illustrated in the next exercise.

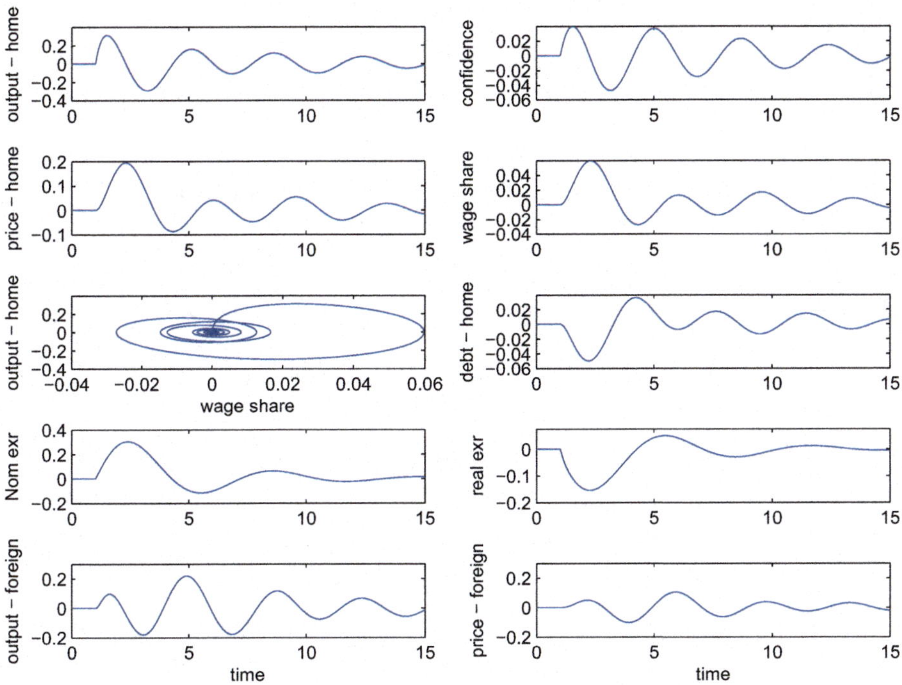

Fig. 2 Dynamic adjustments to an exogenous decline in the nominal interest rate in the home country for $\psi_a = \psi_b = 0.1$

4.3 Exchange Rate Expectations and FX Market Population Dynamics

Figure 3 illustrates the importance of expectations for the dynamic of the exchange rate. In the present figure, the speed of expectations adjustment $\beta_{\pi_s^e}$ is increased from 0 to 5.5, keeping the other parameters as described in Table 1.

As it can be clearly observed in Fig. 3, while exchange rate depreciation expectations did not play a role in the previous simulations, when $\beta_{\pi_s^e} = 5.5$ these depreciation expectations feature persistent oscillations with a slowly decreasing magnitude. The lower left hand figure shows that nominal exchange rate fluctuations are positively related to the exchange rate expectations. Following the decline in the interest rate, the initial depreciations of the nominal exchange rate feeds exchange rate expectations, which then further depreciate the nominal exchange rate as described in Eq. (19).[13]

[13]See Proaño (2009, 2011) for related studies of two-country dynamics using different behaviorally-founded exchange rate specifications.

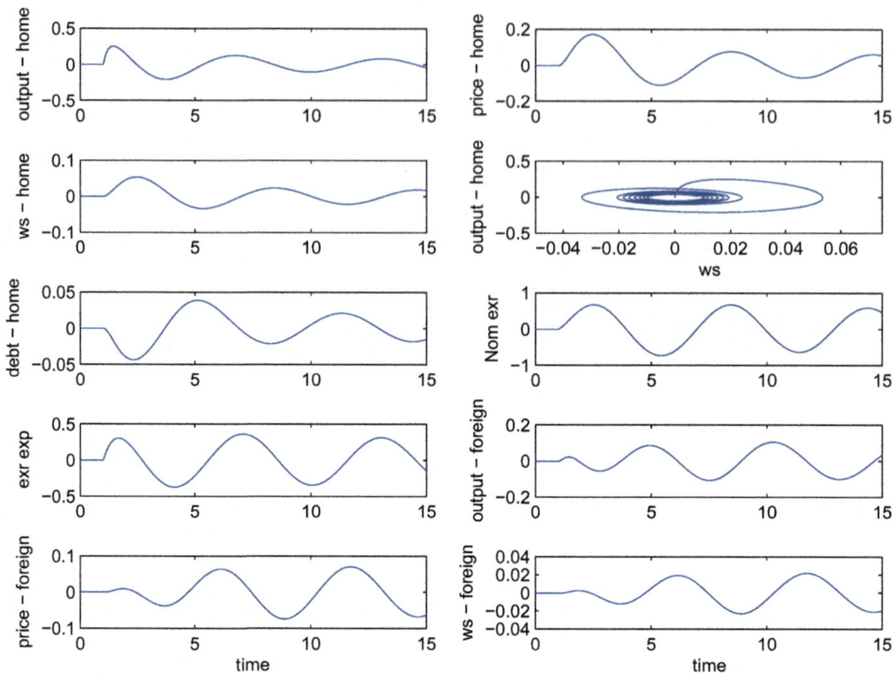

Fig. 3 Dynamic adjustments to an exogenous decline in the nominal interest rate in the home country for $\beta_{\pi_s^e} = 5.5$

Obviously, larger nominal exchange rate fluctuations affect the volatility of output in the domestic and the foreign economy as the graphs in Fig. 3 clearly illustrate. Indeed, as it can be clearly observed, when $\beta_{\pi_s^e} = 5.5$ the fluctuations of domestic and foreign output are larger than in the alternative case. As the expectations add to the depreciation of the nominal exchange rate, the economy further expands through an improvement of the external sector. Further, the domestic economy plays the role of a locomotive for the foreign economy. However, the transmission of shocks takes time and oscillations in the foreign economy are small at first and larger over time.

Finally, it should be pointed out that despite the sizeable speed of adjustment of expectations ($\beta_{\pi_s^e} = 5.5$), the two-country framework remains remarkably stable. This is due to the specification of the wage-price dynamic and the monetary policy, which by construction induce a strong stabilizing forces in the dynamic behavior of the system, at least for a certain range of parameters.

4.4 Are Aggregate Opinion Dynamics Stabilizing?

Finally, we investigate to what extent the endogenous opinion dynamics mechanism contributes to the dynamic stability of the two-country model when the latter is subject to a negative shock. This exercise is particularly interesting due to the nonlinear structure of the opinion dynamics module. Specifically, the blue line in Fig. 4 illustrates the dynamics resulting from a negative shock to the domestic interest rate, with exchange rate expectations $\beta_{\pi_s^e} = 5.5$, opinion dynamics $\beta_z = 25$ and state of confidence $\psi_a = 10$. The green dashed line corresponds to the previous case $\beta_{\pi_s^e} = 5.5$, opinion dynamics $\beta_z = 0$ and state of confidence $\psi_a = 0$. In order to study the impact of the state of confidence on exchange rate expectations, we switch-off the direct effect of the state of confidence on the investment function by replacing ρ^{ge} by ρ^g in the investment equation [Eq. (4)].

The expansionary effect of the cut in the interest rate leads to an increase in the chartist population via the improved state of confidence in the economy. At the peak, the size of the chartist population reaches 0.5. Note that the rise in the chartist population is limited by the non-linear specification of the opinion dynamics, since the state of confidence ψ in the opinion dynamics equation in Eq. (18) is weighted

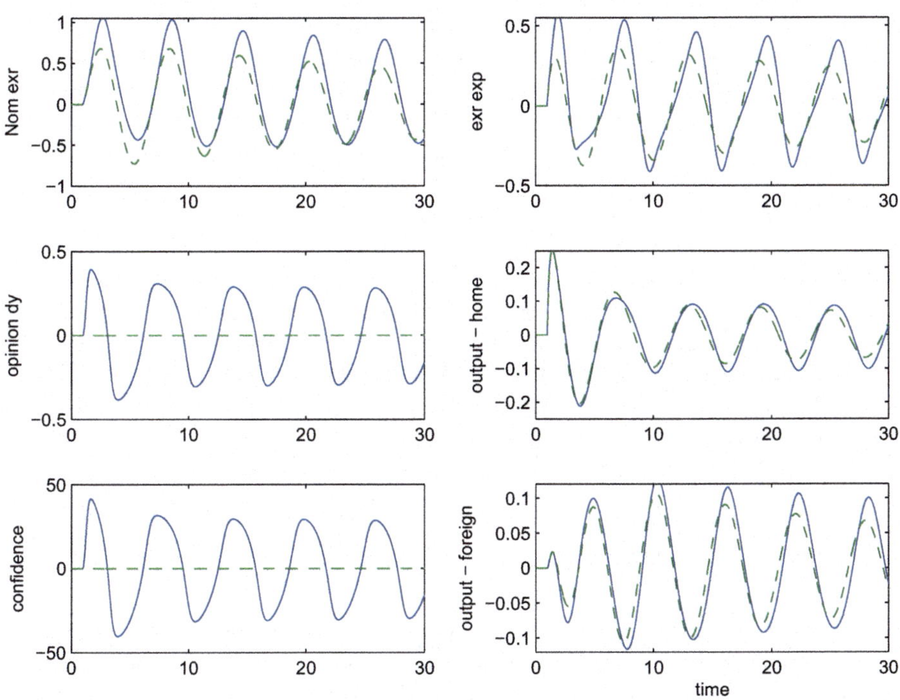

Fig. 4 Dynamic adjustments to a negative shock: the role of aggregate opinion dynamics for macroeconomic stability

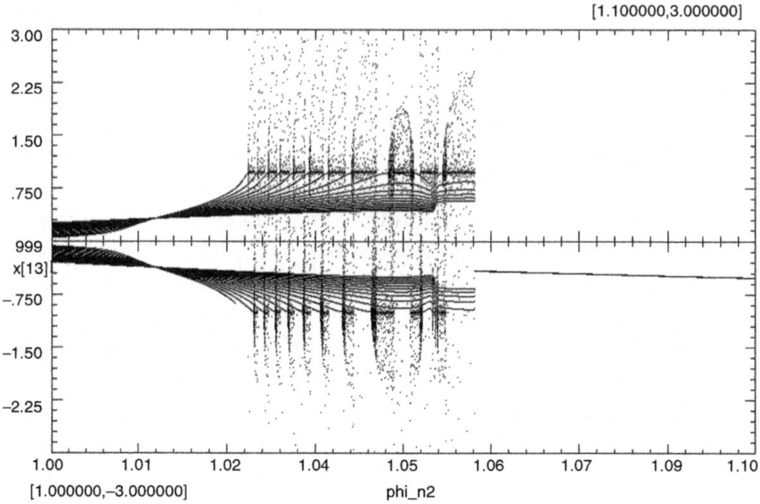

Fig. 5 Bifurcation diagram: x-axis: $1 < \psi_z < 1.1$ y-axis z

by $(1 - z)$, which implies that as the size of the chartist population increases the speed at which opinion shifts from fundamentalist to chartist declines.

The increase in the chartist population directly feeds back into the exchange rate expectations leading to a further depreciation of the nominal exchange rate. In Fig. 4, exchange rate expectations and nominal exchange rate display greater volatility in the presence of active opinion dynamics. It follows that output fluctuations are magnified in both the domestic and foreign countries.

An important transmission channel of the opinion dynamics into the real economy is the parameter ψ_z governing the self-reflection element in the equation for the state of confidence. For value of $\psi_z > 0$, changes in the opinion dynamics and in the state of confidence are reinforcing each others. Figure 5 shows the bifurcation diagram for ψ_z comprised between 1 and 1.1. The bifurcation diagram plots the local maximum and minimum of the opinion dynamics z. For values of ψ_z larger than 1.02, the system displays complex dynamic with z exciting the economically meaningful maximum $(+1)$ and minimum (-1).

5 Concluding Remarks

In this paper we investigated the role of the state of confidence for the macroeconomic dynamics of two large open economies interacting with each other through real and financial channels from a behavioral macrofinance perspective based on the work on aggregate opinion dynamics by Weidlich and Haag (1983) and Lux (1995).

Our analysis highlighted the international dimension of bounded rationality through the modeling of a state of confidence which depended on the business cycles of the domestic and foreign economy. Accordingly, our framework allowed for a detailed analysis of an international transmission channel seldom investigated in the literature, namely how foreign economy developments may affect not the *expectations*, but the *state of confidence* at home. As our numerical simulations showed, this channel does not only have sizable effects for the dynamics of the two economies, but may even bring about persistent cycles or unstable dynamics if the state of confidence becomes too sensitive to business cycle developments at home and abroad. Further, our framework was flexible enough to generate various types of persistent fluctuations which preserve the insight of basic seminal contributions of the literature such as the Goodwin distributive cycle, the related inflation–stagflation cycle, and debt cycles in particular.

As the importance of bounded rationality, social norms and other "non-rational" behavior by the economic agents for the dynamics of the economy at the macroeconomic level are becoming increasingly accepted by the economics profession, this and other related approaches seem a promising field of research. In particular, the empirical quantification of these theoretical framework seems an interesting and worthwhile task to pursue in the future.

References

Acemoglu, D., & Ozdaglar, A. (2011). Opinion dynamics and learning in social networks. *Dynamic Games and Applications, 1*(1), 3–49.

Charpe, M., Flaschel, P., Krolzig, H. M., Proaño, C. R., Semmler, W., & Tavani, D. (2015). Credit driven investment, heterogenous labor markets and macroeconomic dynamics. *Journal of Economic Interaction and Coordination, 10*(1), 163–181.

Chiarella, C., & Flaschel, P. (2000). *The dynamics of Keynesian monetary growth: Macrofoundations*. Cambridge: Cambridge University Press.

Chiarella, C., Flaschel, P., & Franke, R. (2005). *Foundations for a disequilibrium theory of the business cycle. Quantitative analysis and qualitative assessment*. Cambridge: Cambridge University Press.

Chiarella, C., Flaschel, P., Khomin, A., & Zhu, P. (2002). *The SND package. Applications to Keynesian monetary growth dynamics. Dynamic economic theory series* (Vol. 22). Frankfurt am Main: Peter Lang.

Corden, W. (1985). Macroeconomic policy interaction under flexible exchange rates: A two-country model. *Economica, 52*, 9–23.

De Grauwe, P., & Grimaldi, M. (2006). *The exchange rate in a behavioral framework*. Cambridge: MIT Press.

Flaschel, P., Hartmann, F., Malikane, C., & Proaño, C. R. (2015). A behavioral macroeconomic model of exchange rate fluctuations with complex market expectations formation. *Computational Economics, 45*(4), 669–691.

Flaschel, P., & Krolzig, H. M. (2006). Wage–price Phillips curves and macroeconomic stability: Basic structural form, estimation and analysis. In: C. Chiarella , P. Flaschel , R. Franke, & W. Semmler (Eds.), Quantitative and empirical analysis of nonlinear dynamic macromodels. Amsterdam: Elsevier.

Franke, R., Flaschel, P., & Proaño, C. R. (2006). Wage–price dynamics and income distribution in a semi-structural Keynes–Goodwin model. *Structural Change and Economic Dynamics, 17*(4), 452–465.

Franke, R. (2012). Microfounded animal spirits in the new macroeconomic consensus. *Studies in Nonlinear Dynamics and Econometrics, 16*(4), 1–39.

Franke, R., & Semmler, W. (1989). Debt financing of firms, stability and cycles in a dynamical macroeconomic growth model. In W. Semmler (Ed.), *Financial dynamics and business cycles: New perspectives* (pp. 38–64). New York: M.E. Sharpe.

Frenkel, J. A., & Razin, A. (1985). A two-country model of macroeconomic policy interaction. In W. Buiter, R. Marston (Eds.), *International economic policy coordination* (pp. 37–83). Cambridge: Cambridge University Press.

Frydman, R., & Phelps, E. S. (2013). Rethiking expectations. The way forward for macroeconomics. Princeton: Princeton University Press.

Goodwin, R. (1967). A growth cycle. In C. H. Feinstein (Ed.), *Socialism, capitalism and economic growth* (pp. 54–58). Cambridge: Cambridge University Press.

Keynes, J. M. (1936). *The general theory of employment, interest and money*. London: Macmillan.

Lucas, R. E. J. (1976). Econometric policy evaluation: A critique. In *Carnegie-Rochester Conference Series on Public Policy* (Vol. 1, pp. 19–46).

Lux, T. (1995). Herd behaviour, bubbles and crashes. *Economic Journal, 105*, 881–889.

Lux, T. (2009). Rational forecasts or social opinion dynamics? Identification of interaction effects in a business climate survey. *Journal of Economic Behavior & Organization, 72*(2), 638–655.

Muth, J. F. (1961). Rational expectations and the theory of price movements. *Econometrica, 29*, 315–335.

Obstfeld, M., & Rogoff, K. (1995). Exchange rate dynamics redux. *The Journal of Political Economy, 103*(3), 624–660.

Proaño, C. R. (2009). (De-)stabilizing international macroeconomic interactions in an estimated model of the U.S. and the Euro area. *International Economics and Economic Policy, 6*(4), 421–443.

Proaño, C. R. (2011). Exchange rate determination, macroeconomic dynamics and stability under heterogeneous behavioral FX expectations. *Journal of Economic Behavior and Organization, 77*(2), 177–188.

Taylor, L., & O'Connell, S. (1985). A Minsky crisis. *Quarterly Journal of Economics, 100*, 871–886.

Weidlich, W., & Haag, G. (1983). *Concepts and models of a quantitative sociology. The dynamics of interacting populations.* Berlin: Springer.

Self-Falsifying Prophecies

Rajiv Sethi

1 Introduction

One of the ironies of the Great Recession is that it occurred at a time when contractions of such severity were widely believed to be a thing of the past. In a 2004 speech on the Great Moderation, Ben Bernanke attributed the "remarkable decline in the volatility of both output and inflation" in large measure to an improved monetary policy framework, and concluded that the decline was therefore likely to persist. Along similar lines, Robert Lucas argued in his 2003 Presidential Address to the American Economic Association that macroeconomics as originally conceived had succeeded because the "central problem of depression prevention has been solved."

In retrospect, these beliefs appear to be incorrect, or at best premature. But these were beliefs about economic performance, and hence about the aggregate effects of human actions. Since economic behavior is clearly contingent on beliefs, the possibility arises that the resulting outcomes were not just inconsistent with the expressed beliefs, but were in fact a direct consequence of them. That is, high levels of output volatility were a result of behaviors induced by a widespread belief that we had entered a low volatility environment. The underlying beliefs in this case turned out to be *self-falsifying*.

More generally, beliefs about the stability of the macroeconomic environment—*regime perceptions*—affect economic behavior, and hence determine actual levels of stability or *regime realizations*. A fixed point of this mapping from perceptions to

R. Sethi (✉)
Department of Economics, Barnard College, Columbia University, 3009 Broadway, New York, NY 10027, USA
e-mail: rs328@columbia.edu

L. Bernard, U. Nyambuu (eds.), *Dynamic Modeling, Empirical Macroeconomics, and Finance*, DOI 10.1007/978-3-319-39887-7_13

realizations is a level of perceived volatility that is self-fulfilling. Using a nonlinear business cycle model to illustrate, I argue that there may exist no such fixed point, in which case all regime perceptions are self-falsifying. This occurs because volatility need not be continuous in the underlying parameters of the model.

The argument depends crucially on nonlinearity. The steady states of linear models can lose stability at critical parameter values, but since local and global properties coincide in such models, the unstable regime involves explosive trajectories of little economic interest. Accordingly, linear models of business cycles typically assume a parameter range consistent with global stability, and rely on repeated shocks to generate persistent fluctuations. In contrast, nonlinear models can generate trajectories that are locally unstable at a steady state but remain bounded, so both stable and locally unstable regimes are of economic interest.

Deterministic nonlinear models can generate a variety of dynamic behavior, ranging from global stability to highly erratic motion. Different parameter values correspond to different regimes, and the slightest movement of a parameter close to a bifurcation value can dramatically alter the qualitative nature of the trajectories generated by the model. Which of the various regimes is likely to prevail will then depend on the choice of underlying parameters.

This raises the question of what determines the parameters. In many models, parameters include reaction coefficients, representing speeds of adjustment to observed disequilibrium in markets. These coefficients may therefore be expected to evolve as the economy moves forward and new information comes to light. In particular, the parameters will be sensitive to regime perceptions.

As the economy persists within a given regime, the belief that such a regime constitutes a condition of normality will gather currency, and this will tend to shift the model parameters. Whether the economy will settle indefinitely into a specific regime depends, then, on whether or not there exists a set of parameters which are based on perceptions of the very same regime that they give rise to. It may well be the case that no regime is permanent in this sense once the laws governing the evolution of parameters has been specified. This possibility arises because the shift from a given regime to an adjacent one can involve a discontinuous change in the volatility of the resulting time series. In a sense to be made precise below, it is possible that one regime might be "too unstable", while another is "too stable" to persist indefinitely. When there is no intermediate possibility, one would expect the economy to switch interminably between regimes.

2 Literature

Zarnowitz (1985) and Gabisch and Lorenz (1989) have stressed the importance of distinguishing between exogenous (or shock-dependent) and endogenous (or shock-independent) theories of economic fluctuations. The shock-dependent approach, which necessarily includes all linear models, dates back to the work of Frisch (1933) and Slutzky (1937), drawing on ideas in Wicksell.[1] This approach remains dominant in contemporary business cycle theory.

Shock-independent models also have a long history, dating back to contributions by Samuelson (1939), Kaldor (1940), Hicks (1950), Goodwin (1951), and Rose (1967); see Semmler (1986), Foley (1987), and Flaschel et al. (1997) for more recent work in this tradition. Such models can give rise to trajectories with characteristics that linear models cannot possibly generate.[2] For instance, Leijonhufvud's (1973) corridor hypothesis states that an economy may be stable in the face of small shocks but unstable in the face of large disturbances, a feature that cannot arise in linear models but does in the nonlinear models of Tobin (1975) and Semmler and Sieveking (1993).[3]

The idea that perceptions of stability in the macroeconomic environment can be self-falsifying appears repeatedly in the work of Hyman Minsky (1975, 1982, 1986), especially in his seemingly paradoxical claim that stability is itself destabilizing: "Stability—even of an expansion—is destabilizing in that more adventuresome financing of investment pays off to the leaders, and others follow" (Minsky 1975, pp. 126–7). Along similar lines, Paul Volcker (1978) has argued that "a long period of prosperity breeds confidence, and confidence breeds new standards of what is prudent and what is risky" (p. 29), and the "aggressive risk-taker profits handsomely; the rewards of caution seem less evident as memories of hard times recede... Other things equal, a reduction in a firm's *perception of the riskiness of the economic environment* will lead it to prefer to finance by debt issuance" (pp. 45–47, emphasis added).

The model presented here attempts to capture such effects in a simple, stylized manner, by exploring the consequences of introducing regime perception as a parameter determinant in a variant of Kaldor's (1940) trade cycle model. The baseline model used, following Dana and Malgrange (1984), is a two-dimensional discrete time difference equation system with output and the capital stock as state variables.[4] Endogenizing a parameter to allow for the effects of regime perceptions results in a system of much higher dimension. Since the model deals only with the goods market, it is ill-equipped to capture the financial mechanisms explored

[1]Wicksell observed that idiosyncratic shocks could result in damped, periodic oscillations using a metaphor: "If you hit a wooden rocking horse with a club, the movement of the horse will be very different to that of the club" (translated and quoted in Frisch (1933)).

[2]Goodwin (1951) was emphatic on this point, claiming that "economists will be led, as natural scientists have been led, to seek in nonlinearities an explanation of the maintenance of oscillation."

[3]While the earliest nonlinear models were not based on explicit agent optimization, shock-independent models have also been developed in the overlapping generations framework (Benhabib and Day 1982, Grandmont 1985) and in the context of optimal growth (Boldrin and Montrucchio 1986). See Boldrin (1988) for a survey and further discussion.

[4]For other interesting formalizations and extensions of the Kaldor model, see Chang and Smyth (1971), Varian (1979), and Skott (1989).

informally by Minsky and Volcker.[5] Nevertheless, the model's simplicity and the rich dynamics it is capable of generating makes it an ideal vehicle for making the methodological point that self-fulfilling regime perceptions may fail to exist, resulting in alternation between stable and unstable regimes.[6]

If all prophecies are self-falsifying, then convergence to a path on which expectations are fulfilled will not generally be possible, even with sophisticated learning rules, and systematic forecasting errors cannot be attributed to failures of rationality in any meaningful sense. This idea has previously been explored in the context of learning models by Marcet and Sargent (1989), Howitt (1992) and Evans and Honkapohja (1995) among others.[7] Such models are capable of generating a failure to converge to rational expectations equilibrium, but do not typically generate the alternation between periods of high and low volatility that arise in the model explored here.

3 A Kaldorian Model of the Business Cycle

The discrete-time version of Kaldor's model explored by Dana and Malgrange (1984) consists of the following two equations:

$$y_{t+1} = \frac{1}{1+g}(y_t + \alpha(I(y_t, k_t) - S(y_t, k_t) + G_0)) \tag{1}$$

$$k_{t+1} = \frac{1}{1+g}(k_t(1-\delta) + I(y_t, k_t)) \tag{2}$$

where the state variables, y_t and k_t represent detrended levels of output and the capital stock respectively, g is the trend growth rate, I and S are linear homogeneous investment and savings functions, δ is the rate of depreciation, and G_0 is the initial level of public expenditure, which grows at the constant rate g. The parameter α represents the speed with which output responds to excess demand in the product market.

Specifying a linear consumption function and an S-shaped (Kaldorian) investment function, which are then calibrated using French quarterly data, Dana and Malgrange show that the system has a unique fixed point in the positive quadrant.

[5]For models that do capture these financial mechanisms to a degree, see Taylor and O'Connell (1985), Sethi (1992), Skott (1995), Keen (1995), and the papers collected in Semmler (1991).

[6]Such alternation arises also in Sethi (1996), in a model of speculation in an asset market. But the mechanism there is quite different, involving transfers of wealth from stabilizing fundamental strategies (which thrive in unstable markets) and destabilizing momentum-based strategies (which prosper when markets are stable and price movements track fundamentals).

[7]See Evans and Honkapohja (2001) for an overview and García-Schmidt and Woodford (2015) for an important recent contribution.

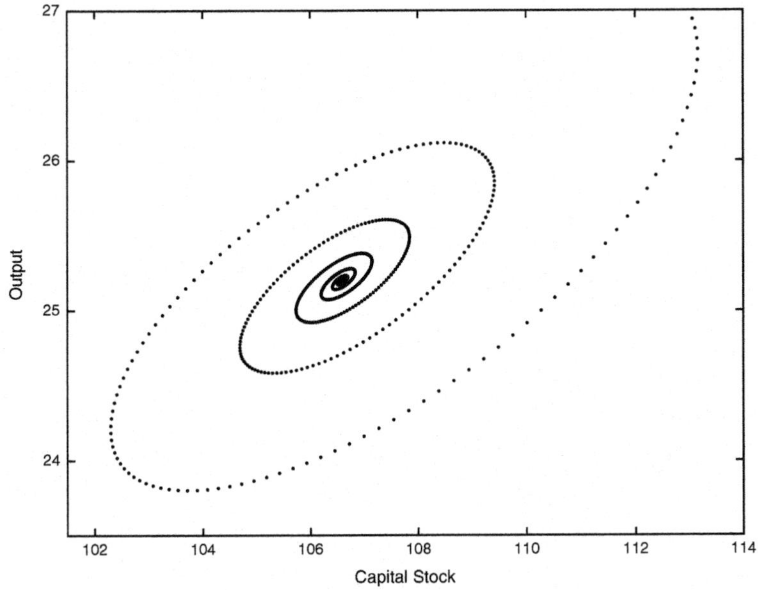

Fig. 1 Convergence to the unique steady state ($\alpha = 0.6$)

They then proceed to examine by means of computer simulations the trajectories generated by the model for various values of the parameter α. Their parameters and function specifications are as follows:

$$g = 0.016,\ \delta = 0.007,\ G_0 = 2,\ S = 0.15y,\ I = kf(\rho),$$

where $\rho = y/k$ is the output-capital ratio and

$$f(\rho) = \left(0.01 + \frac{0.026}{1 + \exp(-9(4.23\rho - 1))}\right). \tag{3}$$

Five distinct regimes are identified as α varies: (i) convergence to the steady state, (ii) periodic (limit cycle) motion, (iii) 'choppy' periodic motion, (iv) 'intermittent chaos', and (v) divergence.

Although the more erratic regimes are of considerable interest in their own right, attention shall be restricted here only to the first two possibilities. The transition from the convergent to the limit cycle regime occurs as α rises through a bifurcation value $\alpha_0 = 0.73$.[8] The convergent regime is illustrated in Fig. 1, and occurs for all $\alpha < \alpha_0$.

[8]Dana and Malgrange report a slightly different bifurcation value $\alpha_0 = 0.76$.

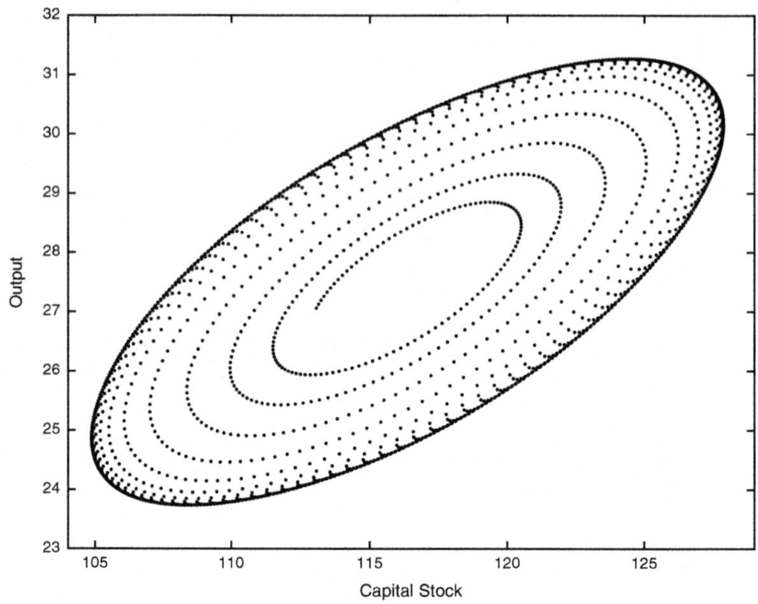

Fig. 2 Convergence to a limit cycle ($\alpha = 0.8$)

As the parameter α rises through α_0, the steady state loses stability, and instead of spiraling inwards to a steady state, the trajectory converges to a stable limit cycle. This regime is illustrated in Fig. 2.

Which (if any) of these regimes will prevail once the remainder of the model has been specified depends on the parameter α. Since this is a reaction coefficient, one cannot expect it to remain invariant with respect to the regime which is widely believed to be prevailing. If variability in the reaction coefficient is to be allowed for, then it is necessary to formulate a hypothesis concerning the manner in which agents process the information that becomes available to them as the economy evolves, and the way in which they respond to this information.

4 Perceived and Realized Volatility

One rather rudimentary way of endogenizing the reaction coefficient α is the following. Suppose that agents measure the volatility of observed output over some historical range, and react to changes in this magnitude by altering their adjustment speeds. More specifically, suppose that if the observed fluctuations in output are large, the reaction to disequilibrium is less rapid. This is a fairly sensible hypothesis: if firms face excess demand in an economy with a history of large fluctuations, they will not expect the excess demand to be sustained over a long horizon, and consequently will react with less enthusiasm to favorable conditions. Similarly, if the historical record indicates a large measure of stability, reactions will be stronger and more immediate. These considerations may be made more precise by specifying the following relationship between the adjustment parameter α and perceived volatility:

$$\alpha_{t+1} = h(s_t), \tag{4}$$

where $h' < 0$ and s_t is the perceived volatility of (detrended) output at time t, after y_t has been observed. The precise manner in which s_t is computed is discussed below. A variety of measures are conceivable, ranging from the standard division of output to the amplitude of fluctuations within the specified historical window. The only requirement is that in the stable regime, perceived volatility—however defined—tends asymptotically to zero, whereas in the periodic regime it tends to some strictly positive constant which depends on α.

The realized (as opposed to perceived) level of volatility, for any given value of α, will tend asymptotically to some magnitude $\sigma(\alpha)$, which depends on properties of the limiting trajectory associated with α. For values of α below the bifurcation point, it is obvious that $\sigma(\alpha) = 0$, since all trajectories converge to the unique steady state. As α rises through the bifurcation value the limiting trajectory becomes a periodic orbit, which has strictly positive volatility. Furthermore, in the neighborhood of the bifurcation value, limit cycles corresponding to higher values of α have greater amplitude, and as α tends to α_0 from above, the amplitude of the limit cycles tends to zero. This follows from the fact that the bifurcation occurring at α_0 is a *local* bifurcation.[9] Consequently, the function $\sigma(\alpha)$ is continuous and locally monotonic in the neighborhood of the bifurcation value.

Since the function $h(s)$ defined in (4) is strictly decreasing and $\sigma(\alpha)$ is increasing in the limit cycle regime, there will be at most one value of α that is self-fulfilling, in the sense that

$$\alpha = h(\sigma(\alpha)). \tag{5}$$

Assume the following:

Assumption 1 *There exists $\bar{s} > 0$ such that $h(s) > \alpha_0$ for all $s < \bar{s}$*

This states simply that if perceived output volatility is sufficiently low, then reaction to disequilibrium will be rapid enough for the coefficient α to exceed the bifurcation value α_0. Bearing in mind that α_0 is given independently of the expectation formation rule (and is approximately equal to 0.73 in the model), this appears to be quite unrestrictive and in accordance with intuition.

[9]Specifically, a Hopf bifurcation. See Dana and Malgrange (1984) for details.

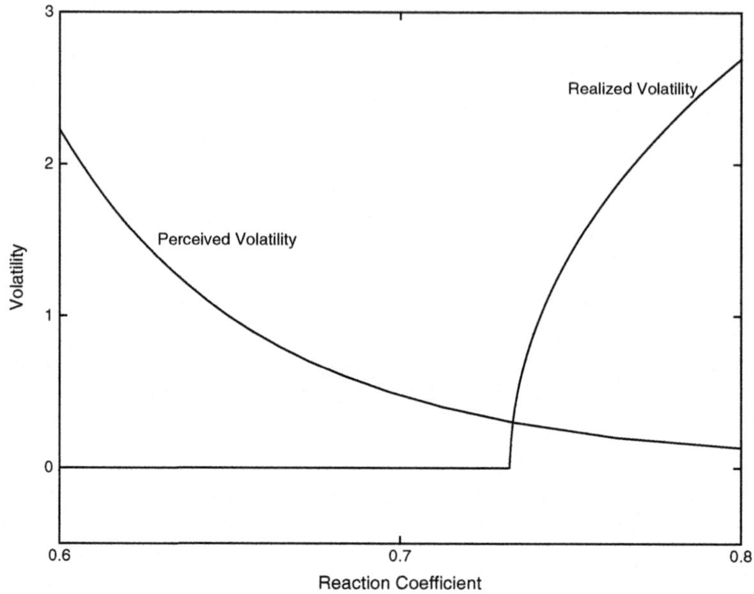

Fig. 3 Determination of the viable trajectory

Assumption 1 implies that convergence of the system to a fixed point cannot occur. To see why, note that convergence requires $\alpha < \alpha_0$, and realized volatility in this case is $\sigma(\alpha) = 0$. Hence $h(\sigma(\alpha)) > \alpha_0$ by the assumption, so (5) cannot possibly hold. Intuitively, if convergence did occur, then sooner or later perceived volatility would tend to zero and the reaction coefficient α would rise through the bifurcation value, switching the economy into the periodic regime. In other words, as a consequence of the fact that convergence itself sets in motion forces leading to a switch in regime, one may say the convergent regime is not viable in the long-run—it cannot persist indefinitely. There may however, be a viable limit cycle, which is based on self-fulfilling perceptions of volatility. Its determination is illustrated in Fig. 3, for the special case in which the standard deviation of output over a finite historical range is used as a measure of volatility.

The economic reasoning behind the non-viability of the convergent regime is the following. The degree of perceived volatility, through its impact on the reaction coefficient, influences the nature of the resulting trajectory, and hence the degree of actual volatility. If there is a sharp discrepancy between perceived and actual volatility, then perceived volatility will eventually shift, thus causing a shift in the trajectory and hence in actual volatility. A regime can only be viable in the long-run if, when it is perceived to occur, the parameters of the model take values which actually causes it to occur. This is why the convergent regime is not viable in the long-run: convergence implies diminishing actual volatility, while diminishing perceived volatility is not ultimately consistent with convergence. This is also why the periodic regime may be viable: roughly speaking, there may exist a degree of volatility which, when perceived, gives rise to the same degree of volatility in terms of the actual trajectories.

In the model as currently specified, the amplitude of the periodic trajectories tend to zero as the reaction coefficient approaches the bifurcation value from above. This is a direct consequence of the fact that a local bifurcation occurs at α_0, implying that the function $\sigma(\alpha)$, which relates realized volatility to the parameter α, is continuous in a neighborhood of α_0. This continuity makes it possible for a self-fulfilling trajectory to exist.

There may however, arise situations in which neither regime is viable. In cases where there is a discontinuous jump in trajectory volatility at the bifurcation point, it is quite possible that no self-fulfilling level of volatility can be attained. This will occur if the volatility of the "smallest" limit cycle is sufficiently large.

To illustrate, consider a slightly different specification for the investment function in order to make the model linear in the vicinity of the steady state. Specifically, when the output-capital ratio is within some interval (ρ_1, ρ_2), a linear specification is adopted in such a way as to maintain continuity of the investment function. Suppose that

$$I = \begin{cases} k(a\rho - b) & \text{if } \rho \in (\rho_1, \rho_2) \\ kf(\rho) & \text{otherwise} \end{cases} \tag{6}$$

where $f(\rho)$ is as defined in (3). In order to ensure continuity of I, we must have

$$a\rho_i - b = f(\rho_i)$$

for $i = 1, 2$, which uniquely determines the coefficients a and b.

With this modification, the steady state is shifted slightly, as is the bifurcation value which takes the economy from a convergent to a periodic regime. More importantly, as long as the model is linear in a neighborhood of the steady state, the transition from a stable to a cyclic regime corresponds to a global bifurcation. That is, the function $\sigma(\alpha)$ is now characterized by a jump at the bifurcation value. The increase in volatility that occurs once the bifurcation value is crossed is not gradual, as in the baseline model, but sudden and discontinuous. For precisely this reason, the periodic regime may itself be non-viable. This will happen, roughly speaking, if the "smallest" limit cycle has sufficiently great amplitude, so that its persistence leads to a downward shift in the reaction coefficient, shifting the economy into the convergent regime. This possibility is illustrated in Fig. 4 for the special case of $(\rho_1, \rho_2) = (0.23, 0.24)$, again using the standard deviation of output as a measure of volatility.

It should be emphasized that local linearity is not a necessary condition for a model to display such discontinuities, although it is certainly sufficient, and has the virtue of making the phenomenon appear self-evident.

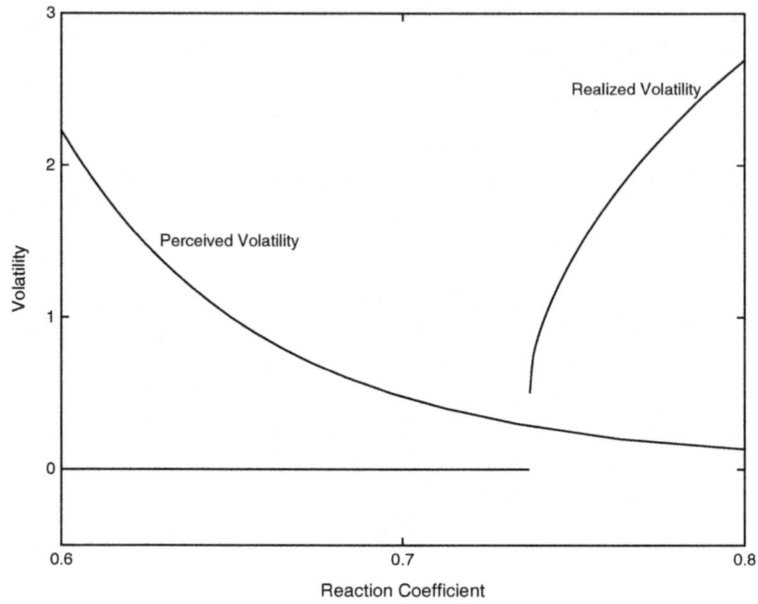

Fig. 4 Nonexistence of a self-fulfilling prophecy

5 Alternating Regimes

If neither regime is viable in the long-run, then what does a typical trajectory look like? This cannot be determined unless some expectation formation procedure by means of which agents attempt to determine the nature of the prevailing regime is made explicit. An individual firm may choose its own behavior but the regime within which it operates is determined by the simultaneous choices of a large number of agents whose own choices are, from the point of view of a single agent, effectively impossible to monitor or control. Information about the "average" choice can, however, be obtained by observing the time series generated as the economy evolves. Let us suppose that, having observed y_t, individuals compute some measure of volatility based on this and the previous n output observations:

$$s_t = \tau(y_t, \ldots, y_{t-n}) \tag{7}$$

For instance, s_t could be the standard deviation of output computed based on the $n+1$ most recent observations. This is clearly a crude specification, but nevertheless one which allows regime perception to evolve over time and to have an impact on regime realization.

The law of motion characterizing the parameter α has already been specified in (4). Updating (1–2) to take account of the endogenity of the reaction coefficient

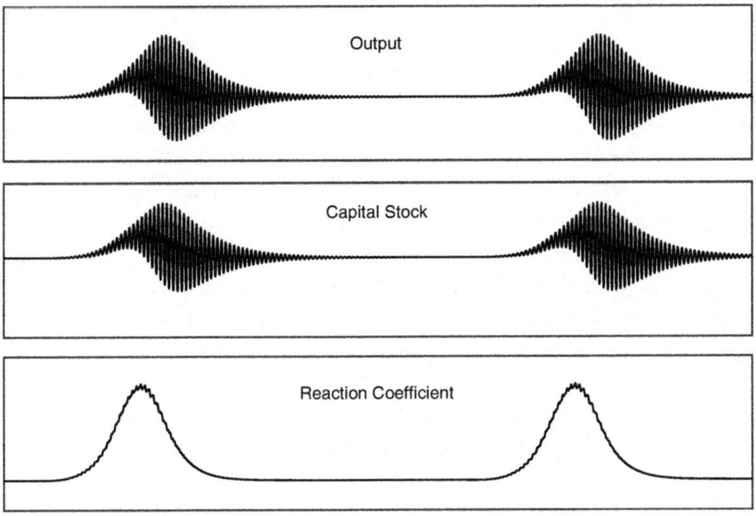

Fig. 5 Alternating stable and unstable regimes

α_t, we obtain the following $n+2$ dimensional nonlinear difference equation system:

$$y_{t+1} = \frac{1}{1+g}(y_t + \alpha_t(I(y_t, k_t) - S(y_t, k_t) + G_0)) \tag{8}$$

$$k_{t+1} = \frac{1}{1+g}(k_t(1-\delta) + I(y_t, k_t)) \tag{9}$$

$$\alpha_{t+1} = h(\tau(y_t, \ldots, y_{t-n})), \tag{10}$$

with the investment function being given by (6). This specification allows us to describe the dynamics using numerical methods. In the case where no self-fulfilling volatility measure exists, the dynamics are characterized by perpetual regime switching, with the parameter α crossing the bifurcation value repeatedly.[10]

This is illustrated in Fig. 5, which shows time series for detrended output, capital, and the reaction coefficient, and the phenomenon of perpetual regime switching is evident. Persistence of the convergent regime cannot be sustained indefinitely, since the damped oscillations cause perceived volatility to decline and eventually raise the reaction coefficient through the bifurcation value. Similarly, as the periodic regime persists, the perceived amplitude of oscillations eventually comes to exceed the minimum value consistent with a value of α above α_0. The movement depicted in the figure arises from a fully deterministic dynamical system, and appears highly irregular as a consequence of the fact that the system is of a very high order.

[10]The system does have a steady state in which perceived volatility is zero, α lies above the bifurcation value α_0, and output and the capital stock grow at the constant rate g. But this steady state is locally unstable and cannot be reached from any other point under the dynamics.

6 Conclusion

The goal of this paper is to make a simple methodological point. When the stability properties of a system change as a parameter crosses a threshold value, there can be a discontinuous change in the level of resulting volatility. If the parameter itself depends on the level of volatility, there may be no belief that is self-fulfilling. As a consequence, the system necessarily alternates between stable and unstable regimes. A period of stability results in destabilizing behavioral adjustments, while the experience of instability induces stabilizing behavioral changes.

This kind of pattern has been described informally by Minsky and Volcker, although the mechanisms they had in mind involved balance sheet effects rather than the investment accelerator that is at the heart of Kaldor's model. Such endogenous regime switching, with stability giving way to instability and vice versa, also seems broadly consistent with recent historical experience.

Acknowledgements This paper is dedicated to Willi Semmler, who introduced me to nonlinear dynamics, and who alerted me to the important distinction between shock-dependent and shock-independent theories of economic fluctuations. For comments on an earlier version of this paper I thank Peter Skott, Gil Skillman, Roberto Veneziani, and other participants at the 2016 Analytical Political Economy Workshop at Queen Mary University of London.

References

Benhabib, J., & Day, R. (1982). A characterization of erratic dynamics in the overlapping generations model. *Journal of Economic Dynamics and Control, 4*, 37–55.

Bernanke, B. (2004). The great moderation. Remarks by Governor Ben S. Bernanke At the meetings of the Eastern Economic Association, Washington, DC, February 20.

Boldrin, M. (1988). Persistent oscillations and chaos in economic models: notes for a survey. In P. W. Anderson, K. J. Arrow, & D. Pines (Eds.), *The economy as an evolving complex system*. Redwood, CA: Addison-Wesley.

Boldrin, M., & Montrucchio, L. (1986). On the indeterminacy of capital accumulation paths. *Journal of Economic Theory, 40*, 26–39.

Chang, W. W., & Smyth, D. J. (1971). The existence and persistence of cycles in a non-linear model: Kaldor's 1940 model ee-examined. *Review of Economic Studies, 38*, 37–44.

Dana, R. -A., & Malgrange, P. (1984). The dynamics of a discrete version of a growth cycle model. In P. Ancot (Ed.) *Analyzing the structure of econometric models*. Amsterdam: M. Nijhoff.

Evans, G. W., & Honkapohja, S. (1995). Local convergence of recursive learning to steady states and cycles in stochastic nonlinear models. *Econometrica, 63*, 195–206.

Evans, G. W., Honkapohja, S. (2001). *Learning and expectations in macroeconomics*. Princeton, NJ: Princeton University Press.

Flaschel, P., Franke, R., & Semmler, W. (1997). *Dynamic macroeconomics: Instability, fluctuation, and growth in monetary economies*. Cambridge, MA: MIT Press.

Frisch, R. (1933). Propagation problems and impulse problems in dynamic economics. In *Economic essays in honor of Gustav Cassel*. London: Allen and Unwin.

Foley, D. K. (1987). Liquidity-profit rate cycles in a capitalist economy. *Journal of Economic Behavior & Organization, 8.3*, 363–376.

Gabisch, G., & Lorenz, H. -W. (1989). *Business cycle theory: A survey of methods and concepts* (2nd edn.). Berlin: Springer.

García-Schmidt, M., & Woodford, M. (2015). Are Low Interest Rates Deflationary? A Paradox of Perfect-Foresight Analysis. Working Paper 21614, National Bureau of Economic Research.

Goodwin, R. M. (1951). The nonlinear accelerator and the persistence of business cycles. *Econometrica, 19*, 1–17.

Grandmont, J. M. (1985). On endogenous competitive business cycles. *Econometrica, 53*, 995–1046.

Hicks, J. R. (1950). *A contribution to the theory of the trade cycle*. Oxford: Clarendon.

Howitt, P. (1992). Interest rate control and nonconvergence to rational expectations. *Journal of Political Economy, 100*, 776–800.

Kaldor, N. (1940). A model of the trade cycle. *Economic Journal, 50*, 78–92.

Keen, S. (1995). Finance and economic breakdown: modeling Minsky's "financial instability hypothesis." *Journal of Post Keynesian Economics, 17*, 607–635.

Leijonhufvud, A. (1973). Effective demand failures. *Swedish Journal of Economics, 75*, 27–48.

Lucas, Jr, R. E. (2003). Macroeconomic priorities. *American Economic Review, 93*, 1–14.

Marcet, A., & Sargent, T. J. (1989). Convergence of least squares learning mechanisms in self-referential linear stochastic models. *Journal of Economic Theory, 48*, 337–368.

Minsky, H. P. (1975). *John Maynard Keynes*. New York: Columbia University Press.

Minsky, H. P. (1982). *Can "it" happen again: Essays on instability and finance*. Armonk, New York: M.E. Sharpe.

Minsky, H. P. (1986). *Stabilizing an unstable economy*. New Haven: Yale University Press.

Rose, H. (1967) On the non-linear theory of the employment cycle. *Review of Economic Studies, 34*, 153–173.

Samuelson, P. A. (1939). A synthesis of the principle of acceleration and the multiplier. *Journal of Political Economy, 47*, 786–797.

Semmler, W. (1986). On nonlinear theories of economic cycles and the persistence of business cycles. *Mathematical Social Sciences, 12*, 47–76.

Semmler, W. (Ed.) (1991). *Financial dynamics and business cycles: New perspectives*. New York: ME Sharpe.

Semmler, W., & Sieveking, M. (1993). Nonlinear liquidity-growth dynamics with corridor-stability. *Journal of Economic Behavior & Organization, 22*, 189–208.

Sethi, R. (1992). Dynamics of learning and the financial instability hypothesis. *Journal of Economics, 56*, 39–70.

Sethi, R. (1996). Endogenous regime switching in speculative markets. *Structural Change and Economic Dynamics, 7*, 99–118.

Skott, P. (1989). Effective demand, class struggle and cyclical growth. *International Economic Review, 30*, 231–247.

Skott, P. (1995). Financial innovation, deregulation and Minsky cycles. In G. A. Epstein, & H. M. Gintis (Eds.), *Macroeconomic policy after the conservative era: studies in investment, saving and finance*. Cambridge: Cambridge University Press.

Slutzky, E. (1937). The summation of random causes as the source of cyclic processes. *Econometrica, 5*, 105–146.

Taylor, L., & O'Connell, S. A. (1985). A minsky crisis. *Quarterly Journal of Economics, 100*, 871–886.

Tobin, J. (1975). Keynesian models of recession and depression. *American Economic Review, 65*, 195–202.

Varian, H. R. (1979). Catastrophe theory and the business cycle. *Economic Inquiry, 17*, 14–28.
Volcker, P. A. (1978). *The rediscovery of the business cycle*. New York: Free Press.
Zarnowitz, V. (1985). Recent work on business cycles in historical perspective. *Journal of Economic Literature, 23*, 523–580.

About the Editors

Lucas Bernard, Ph.D., is a financial economist and chairman of the Department of Business at the New York City College of Technology, the City University of New York. In addition to having studied at the Swiss Finance Institute, he also holds graduate degrees in mathematics from the City University of New York and in computer science from New York University's Courant Institute of Mathematical Sciences. His doctoral dissertation, concerning endogenous models of credit default, was written at the New School for Social Research in New York. Prof. Bernard is interested in the ways in which economics blends together human values, politics, rationality, and irrationality and, more specifically, how this interplay reveals itself in society at large.

Unurjargal Nyambuu, Ph.D., is an economist and professor in the Department of Social Science at the New York City College of Technology, the City University of New York. Previously, she was a research fellow in the Finance and Risk Engineering Department of New York University. Dr. Nyambuu has also served as an economist with the Central Bank of Mongolia, where she worked with International Monetary Fund (IMF) and World Bank (WB) colleagues on the effects of international macroeconomic policy on local economies. Additionally, she is a consultant for the International Labour Organization, a United Nations affiliate based in Genève, Switzerland, where she studies the economic impact of "green" policy.

L. Bernard, U. Nyambuu (eds.), *Dynamic Modeling, Empirical Macroeconomics, and Finance*, DOI 10.1007/978-3-319-39887-7

Printed by Printforce, the Netherlands